The Early Modern Hispanic World

Iberia stands at the center of key trends in Atlantic and world histories, largely because Portugal and Spain were the first European kingdoms to "go global." *The Early Modern Hispanic World* engages with new ways of thinking about the early modern Hispanic past, as a field of study that has grown exponentially in recent years. It focuses predominantly on questions of how people understood the rapidly changing world in which they lived – how they defined, visualized, and constructed communities from family and city to kingdom and empire. To do so, it incorporates voices from across the Hispanic World and across disciplines. The volume considers the dynamic relationships between circulation and fixedness, space and place, and how new methodologies are reshaping global history, and Spain's place in it.

Kimberly Lynn is Associate Professor of Early Modern Europe at the Department of Liberal Studies, Western Washington University.

Erin Kathleen Rowe is Assistant Professor of Early Modern Spain at the Department of History, Johns Hopkins University.

The Early Modern Hispanic World

Transnational and Interdisciplinary Approaches

Edited by

KIMBERLY LYNN

Western Washington University

AND

ERIN KATHLEEN ROWE

Johns Hopkins University

CAMBRIDGE
UNIVERSITY PRESS

CAMBRIDGE
UNIVERSITY PRESS

One Liberty Plaza, 20th Floor, New York, NY 10006, USA
University Printing House, Cambridge CB2 8BS, United Kingdom

Cambridge University Press is part of the University of Cambridge.

It furthers the University's mission by disseminating knowledge in the pursuit of education, learning and research at the highest international levels of excellence.

www.cambridge.org
Information on this title: www.cambridge.org/9781107109285

First published 2017

Printed in the United States of America by Sheridan Books, Inc.

A catalogue record for this publication is available from the British Library

Library of Congress Cataloguing in Publication data
Names: Lynn, Kimberly, editor of compilation. | Rowe, Erin Kathleen, 1974–
editor of compilation. Title: The early modern Hispanic world :
transnational and interdisciplinary approaches / edited by Kimberly Lynn
(Western Washington University) and Erin Kathleen Rowe (Johns Hopkins University).
Description: Cambridge, United Kingdom ; New York, NY :
Cambridge University Press, 2016. | Includes bibliographical references and index.
Identifiers: LCCN 2016025943 | ISBN 9781107109285 (hardback : alkaline paper)
Subjects: LCSH: Spain – History – House of Austria, 1516–1700 –
Historiography. | Spain – History – 18th century – Historiography. |
Portugal – History – Period of discoveries, 1385–1580 – Historiography. |
Portugal – History – Modern, 1580– – Historiography. | Spain – Colonies –
Historiography. | Portugal – Colonies – Historiography. | Spain – Social conditions –
Historiography. | Portugal – Social conditions – Historiography. | Social change –
Spain – Historiography. | Social change – Portugal – Historiography. |
BISAC: HISTORY / Europe / General.
Classification: LCC DP171 .E36 2016 | DDC 909/.097124604 – dc23
LC record available at https://lccn.loc.gov/2016025943

ISBN 978-1-107-10928-5 Hardback

For Richard L. Kagan,
Scholar, Mentor, Friend

FIG. 0.1. Richard L. Kagan, Casa Corberó, Barcelona, 2009, photograph by Joaquín Bérchez

Contents

Figures

Maps

Tables

Notes on Contributors

Ida Altman is Professor of History at the University of Florida. She is the author of *Emigrants and Society: Extremadura and Spanish America in the Sixteenth Century* (1989), *Transatlantic Ties in the Spanish Empire: Brihuega, Spain and Puebla, Mexico, 1560–1620* (2000), *The Early History of Greater Mexico*, co-authored with Sarah Cline and Juan Javier Pescador (2004), and *The War for Mexico's West: Indians and Spaniards in New Galicia, 1524–1550* (2010).

James S. Amelang teaches Early Modern History at the Universidad Autónoma of Madrid. The author of *Honored Citizens of Barcelona* (1986) and *The Flight of Icarus* (1998), he has published widely on the social, urban, and cultural history of early modern Spain and Italy. His current project is to complete the *Oxford History of Early Modern Spain.*

Benjamin Ehlers is Associate Professor of History at the University of Georgia. He is the author of *Between Christians and Moriscos: Juan de Ribera and Religious Reform in Valencia, 1568–1614* (2006). His current research addresses the Spanish encounter with Islam in the Habsburg–Ottoman era, comparing the community of Moriscos in Valencia with the diaspora of Spanish soldiers, merchants, clerics, and captives in North Africa.

Sir John Elliott was Regius Professor of Modern History at Oxford University from 1990 to 1997, and is an Honorary Fellow of Oriel College, Oxford. His books include *Imperial Spain* (1963), *The Revolt of the Catalans* (1963), *The Old World and the New, 1492–1650* (1970), *The Count-Duke of Olivares* (1986), *Empires of the Atlantic World: Britain* *and Spain in America, 1492 to 1830* (2006), and, most recently, *History in the Making* (2012).

Mercedes García-Arenal is Research Professor at the Centro de Ciencias Humanas y Sociales, CSIC, Madrid (Consejo Superior de Investigaciones Científicas, Spanish Council for Scientific Research). Her collaboration with Gerard Wiegers, *A Man of Three Worlds. Samuel Pallache, a Moroccan Jew between Catholic and Protestant Europe* (2003; Spanish edition 1999), is especially well known. Her most recent books (in English) are *Messianism and Puritanical Reform. Mahdis of the Muslim West* (2006), with Fernando Rodríguez Mediano, *The Orient in Spain. Converted Muslims, the Forged Lead Books of Granada and the Rise of Orientalism* (2013), and as editor, with Gerard Wiegers, *The Expulsion of the Moriscos of Spain. A Mediterranean Diaspora* (2014). Her edited book *After Conversion: Iberia and the Rise of Modernity* is forthcoming.

Xavier Gil is Professor of Early Modern History at the University of Barcelona and a member of the Real Academia de la Historia. His field of interest is political thought and culture in the Iberian world in the sixteenth and seventeenth centuries. His books include *Claves del absolutismo y del parlamentarismo, 1603–1715* (1991), *Tiempo de política. Perspectivas historiográficas sobre la Europa moderna* (2006, reprint, 2010), and, as editor with Jon Arrieta and Jesús Morales, *La diadema de la Corona. Vizcaya, Navarra y la Corona de Aragón en la Monarquía española, siglos XVI–XVIII* (2016).

A. Katie Harris is Associate Professor of History at the University of California, Davis. She is the author of *From Muslim to Christian Granada: Inventing a City's Past in Early Modern Spain* (2007).

Kimberly Lynn is Associate Professor of Humanities at Western Washington University. She is the author of *Between Court and Confessional: The Politics of Spanish Inquisitors* (2013).

Fernando Marías is Professor of Art History at the Universidad Autónoma of Madrid and a member of the Real Academia de la Historia. His many publications include *La arquitectura del Renacimiento en Toledo* (1983–6), *El largo siglo XVI* (1989), *El Hospital Tavera de Toledo* (2007), and as editor, with Felipe Pereda, *El Atlas del Rey Planeta. La 'Descripción de España y de las costas y puertos de sus reinos' de Pedro Texeira (1634)* (2002). His work on El Greco includes *Las ideas artísticas de El*

Greco (1981), *El Greco y el arte de su tiempo. Las notas de El Greco a Vasari* (1992), *El Greco, biografía de un pintor extravagante* (1997), and *El Greco, Life and Work. A New History* (2013); while his work on Velázquez includes *Otras Meninas* (1995), *Las Meninas* (1999), and *Velázquez, pintor y criado del rey* (1999).

Sara Tilghman Nalle is Professor Emerita of History at William Paterson University. She is the author of *God in La Mancha: Religious Reform and the People of Cuenca, 1500–1650* (1992) and *Mad for God: Bartolomé Sánchez, the Secret Messiah of Cardenete* (2001), translated as *Loco por dios: Bartolomé Sánchez, el mesías secreto de Cardenete* (2009).

Geoffrey Parker is Distinguished University Professor and Andreas Dorpalen Professor of European History at the Ohio State University. His many books include *The Army of Flanders and the Spanish Road. The Logistics of Spanish Victory and Defeat in the Low Countries Wars, 1567–1659* (1972; revised edition 2004), *Philip II* (1978; third edition 2002), *The Military Revolution. Military Innovation in the Rise of the West, 1500–1800* (1988; expanded edition 2002), *The Grand Strategy of Philip II* (1998), and, as editor with Richard L. Kagan, *Spain, Europe, and the Atlantic World. Essays in Honour of John H. Elliott* (1995). His most recent work includes *Global Crisis: War, Climate Change and Catastrophe in the Seventeenth Century* (2013) and *Imprudent King: A New Life of Philip II* (2014).

Felipe Pereda is Fernando Zóbel de Ayala Professor of Spanish Art at Harvard University. His books include *La arquitectura eloquente* (1999), *El Atlas del Rey Planeta* (2002), with Fernando Marías, and *Images of Discord: Poetics and Politics of the Sacred Image in Fifteenth-Century Spain* (2016; Spanish edition 2007).

María M. Portuondo is Associate Professor in the Department of the History of Science and Technology at the Johns Hopkins University. She is the author of *Secret Science: Spanish Cosmography and the New World* (2009), and is currently writing a book on the natural philosophy of Benito Arias Montano.

Allyson M. Poska is Professor of History at the University of Mary Washington. She is the author of *Regulating the People: The Catholic Reformation in Seventeenth-Century Spain* (1998), *Women and Authority in Early Modern Spain: The Peasants of Galicia* (2005), and *Gendered Crossings: Women and Migration in the Spanish Empire* (2016).

Erin Kathleen Rowe is Assistant Professor of History at the Johns Hopkins University. She is the author of *Saint and Nation: Santiago, Teresa of Avila, and Plural Identities in Early Modern Spain* (2011), and is currently working on a project exploring the devotion to black saints in the early modern Hispanic world.

Marta V. Vicente is Associate Professor in History and Women, Gender and Sexuality Studies at the University of Kansas. She is editor, with Luis Corteguera, of *Women, Texts and Authority in the Early Modern Spanish World* (2003), and the author of two monographs, *Clothing the Spanish Empire: Families and the Calico Trade in the Early Modern Atlantic World* (2006) and *Debating Sex and Gender in Eighteenth-Century Spain* (forthcoming).

Elizabeth R. Wright is Professor of Spanish at the University of Georgia. She has published a bilingual, annotated edition of neo-Latin poetry, *The Battle of Lepanto*, in collaboration with Sarah Spence and Andrew Lemons (2014). Other publications include: *The Epic of Juan Latino: Dilemmas of Race and Religion in Renaissance Spain* (2016); *Spanish Golden Age Theater in Mexican Translation* (co-editor with Louise Burkhart and Barry Sell, 2008); a critical edition of Lope de Vega's *Los ramilletes de Madrid* (2012); and *Pilgrimage to Patronage: Lope de Vega and the Court of Philip III* (2001).

Acknowledgments

The editors would like to thank all those who helped this book become a reality. We are grateful to Shreve Simpson, Geoffrey Parker, and John Elliott, who offered sage advice at several important junctures. We are particularly appreciative of Debbie Gershenowitz's support for this project at Cambridge University Press, and we thank the anonymous readers for their useful advice. The expert translation work of Consuelo López-Morillas was invaluable. Western Washington University provided generous support for the volume, and funding from Johns Hopkins University likewise facilitated the manuscript's preparation. Finally, we thank Richard L. Kagan for his encouragement in this, as in so many other endeavors, and all those friends, colleagues, and students of Richard's who have shaped this book and our careers, far too numerous to be encased in just one volume.

Preface

Richard L. Kagan, an Appreciation[1]

Geoffrey Parker

Richard Kagan has published twelve scholarly monographs and almost one hundred articles and book chapters; he has co-edited a Festschrift for his doctoral advisor; he has been the dedicatee of at least one book by a colleague; he has also served as doctoral advisor to more than twenty-one graduates who have so far published sixteen scholarly monographs of their own. He has also appeared (thus far) in two published memoirs of other scholars – one of which even the most persistent librarian and bibliographer might otherwise miss.[2] In assessing Richard Kagan's contribution to the world of scholarship, I have drawn upon these memoirs, and upon the recollections of his doctoral advisees, as well as a correspondence with him that spans fifty years. My only regret is that I lack space to document here at least some of the "Richard" stories that only the ill-informed would regard as exaggerated.[3]

[1] This appreciation incorporates my address, "Kagan: the Missing Years," delivered at "Kagan's Kaleidoscope," a surprise sixty-fifth birthday symposium organized for Richard by his advisees at Johns Hopkins University in 2008, as well as other tributes from the three-day bacchanalia that followed. I thank A. Katie Harris, Kimberly Lynn, Erin Rowe, Shreve Simpson, and other survivors of the 2008 conference for further information and comments. I also thank Richard for fifty years of friendship and a lot of fun.

[2] Pierre Goubert, *Un parcours d'historien. Souvenirs 1915–1995* (Paris: Fayard, 1996), 200–1, a passage about Goubert's stay at Princeton in 1969–70, where he met "Richard et Bob," whom he described as two "anormaux" (an intriguing term that one might translate as "freaks"). "Bob" was of course Robert Darnton; but although Goubert never provided a surname for "Richard," instead using only the nickname he devised for this "anormal" ("le petit voisin"), he noted that Richard's father was "a major manufacturer of barbed wire," and that his wife was "an erudite (and very beautiful) specialist in Persian art." Richard also appeared in Dauril Alden's biography of Charles Boxer: see below.

[3] Some of the best "Richard" stories feature in our correspondence, which will be deposited with the rest of my papers in the Library of St Andrews University, Scotland.

George Kagan, an American of Ukrainian origin, owned a factory in New Jersey that made barbed wire, and he had two sons: Robert, the older, and Richard, born in 1943. Like many entrepreneurs, George intended his two sons to succeed him in control of the business: Bob would be the CEO and Richard the sales manager, ceaselessly traveling the world in search of new clients. In 1956, just as Richard prepared to enter high school, for which he needed to choose a foreign language to study, George returned from a business trip to Latin America and announced that Spanish "was the language of the future, especially for a prospective businessman," and his younger son obediently began to study Spanish and Spanish culture (as luck would have it, with a teacher trained in Spain rather than Spanish America). Little did George (or anyone else) realize that this choice would play a crucial role in seducing his son "away from the world of business and into the groves of academe."[4]

After high school, Richard went to Columbia University, still intending to enter the family business. Thanks to the broad training in both the sciences and the liberal arts on which Columbia then insisted, he took a wide range of courses in history, the history of art and architecture, literature, and the social sciences. He also regularly visited the museums of New York, especially the Frick, the Cloisters, the newly opened Guggenheim, and the Metropolitan Museum of Art. He later wrote, "I am not sure where as a student I learned the most – in Columbia's lecture halls, or in my museum visits and walks through New York's streets."

Two events in 1964 changed Kagan's ambitions. First, he persuaded his father to underwrite a two-month summer trip to Europe to study Renaissance art and architecture in Florence, an opportunity that he exploited to visit other Italian cities as well as England, the Netherlands, France, Germany, and Greece (though, curiously, not Spain). Second, when he returned to Columbia for his senior year he encountered a new faculty advisor: Orest Ranum, a young, charismatic history professor specializing in early modern France, and "probably the first Columbia professor who took a personal interest in my undergraduate education." Inspired by Ranum, Kagan chose history as his "major," and enrolled

[4] Unless otherwise attributed, this and all subsequent quotations come from Kagan's autobiographical memoir "Por Qué España," in *La historia de España en Primera Persona*, ed. Jaume Aurell (Madrid: Editorial Base, 2012), 149–70 (I quote from Richard's original English text of this memoir); and from his valedictory speech at The Johns Hopkins University, "Why Spain? Reflections on a Historian's Career at Hopkins and Beyond," delivered on 1 May 2013 (both texts graciously supplied by the author).

in an advanced undergraduate seminar on the history of seventeenth-century Europe. For Kagan,

> That seminar constituted another turning point in my life. To begin with, it was an intense learning experience, and far superior to any other course in which I had enrolled. Together with the other students, about a dozen in total, I was encouraged to work independently, to express my thoughts and opinions about the readings, and to write research papers on topics of my own choosing.

Kagan thus learned as an undergraduate some basic skills that would serve him well throughout his academic career. Moreover, he later realized:

> What was really important about that seminar is that I was the only student in it able to read Spanish, then a language, especially at Columbia and most other elite Ivy League universities, that lacked the social and intellectual cachet of French, not to mention German and Italian ... What at first seemed a liability became an opportunity in as much as I was able to write an initial paper on the Count-Duke of Olivares, and then another on El Greco and his times.

Under Ranum's tutelage, Kagan became fascinated with the history of early modern Europe, and especially Spain, and contemplated for the first time the possibility of a career that did not involve selling barbed wire. Ranum, for his part, realized that Kagan possessed the potential to become a professional historian, and sent a copy of his paper on Olivares to John H. Elliott, a lecturer in early modern history at Cambridge University and author of two recent path-breaking books, *The Revolt of the Catalans* and *Imperial Spain*, together with a request to accept its author as a research student. Elliott agreed, and George Kagan (though disappointed by the prospect that his son might not after all enter the family business), provided both his blessing and his generous support.

In October 1965, aged twenty-two, Richard Kagan arrived in Cambridge to begin his graduate studies. He joined five others whose theses Elliott had also agreed to direct: James Casey, from Queen's University Belfast (the only other "outsider"); Peter Bakewell, who had just graduated in Spanish language and literature at Cambridge; and Robert Evans, Albert Lovett, and me, who had graduated in History the previous summer. Ranum's choice of Cambridge for his star pupil's graduate studies made perfect academic sense, because in the mid-1960s few universities in the United States offered advanced training in early modern Spanish history; but Richard faced many disadvantages in migrating directly from an American undergraduate education, with its emphasis on breadth, to graduate studies in England, with its emphasis on depth. Apart from Bakewell, the rest of us had studied nothing but History since we left high school, and three of us had even attended Elliott's dazzling undergraduate lecture course on early modern Europe. Our collective historical knowledge must have intimidated Richard almost as much as his expertise in other fields intimidated us.[5]

For Richard had been busy since learning that he would come to Cambridge. Elliott had already suggested a possible dissertation topic – the education of the aristocracy in early modern Spain – while Kagan's uncle Herbert McClosky, a distinguished political scientist at the University of California at Berkeley, provided a methodology. McClosky had devised elaborate survey instruments to study political attitudes, and he impressed upon his nephew that History had to forge an alliance with the Social Sciences, otherwise its importance as a discipline was likely to fade. He also encouraged the young Kagan to study the quantitative methods that he had applied to the study of politics in the United States, and to read Max Weber, the German sociologist widely admired by social scientists in the United States, but not in England (Cambridge University did not create a Department of Sociology until 2004).[6] Kagan dutifully familiarized himself with Weber's broad ideas about the evolution of patrimonial governance towards bureaucracy, and the emergence of "specialists" skilled both in administration and the law, and he sought ways to apply them to early modern Spain. His preliminary research on the history of Spain's universities suggested that the creation of a royal bureaucracy composed of specialists (*letrados* as they were known in sixteenth-century Spain) facilitated the rise of the modern state, just as postulated by Weber, and in July 1966 he set off to test his theory in the archives.

Richard displayed an amazing capacity for hard work. For just over a year he:

> Lived in the equivalent of an archival bubble, doing my own thing, but that was not all easy. Detailed catalogues were few; lighting poor; xerox machines virtually unknown; and the cold – were it not for the occasional shot of some cheap Spanish brandy, then about a nickel a glass, I don't think that I could have been able to tolerate one unheated archive in Valladolid whose broken window panes allowed snowflakes to land on the sixteenth-century documents I was trying to read.

Richard's dedication earned the respect of the archive staff, and brought some unexpected benefits. One day at the Archivo Histórico Nacional in

[5] For more on Kagan's traumatic first encounter with Cambridge, see his essay "John Elliott y la historiografía española," in *John Elliott: El oficio de historiador*, eds. Antonio Passola, María José Vilalta, and Roberto Fernández (Lleida: Editorial Milenio, 2001), 33–42.

[6] On Herbert McClosky (1916–2006), see the obituary posted at http://berkeley.edu/news/media/releases/2006/03/16_mcclosky.shtml (accessed April 2, 2014).

Madrid, an elderly *portero* appeared at his desk, carrying an enormous but uncataloged bundle of documents, and said: "Request this one: it's important for your topic." He was right.

Richard returned to Cambridge in summer 1967 and started writing his thesis, and once again his remarkable dedication brought benefits: he completed his thesis within one year and secured an academic post at Indiana University's Department of History, already very strong in Latin American history, which wished to hire a scholar who specialized in Spanish history (at a time when very few departments of history in the United States produced them). A few days after his thesis "defense" (a *viva* as it was called) in August 1968, Kagan loaded his research notes and other possessions onto the "North German Lloyd line's luxury flagship, T. S. BREMEN" which carried him to New York. Thence he made his way to "Bloomington, the capital of the mid-West, bastion of Republicanism, and standard-bearer of all that is crass in American Civilization."

Almost immediately, I received a stream of letters from Richard, railing about the miseries of academic life in the Mid West:

> The town itself blinks and winks with pink and yellow neon signs advertising 10 cent hamburgers and pizza joints. The movie houses traffick in the banal while the drive-in everythings choke the roads with cars and exhaust fumes. The university campus, although well-landscaped and covered with wonderful trees, still manages to overwhelm one with its large, limestone, Fascist-like architecture. My own office is perched precariously on the seventh floor of an over-heated, overcrowded, under-planned building of classrooms and offices.[7]

He derived some solace in the presence in Bloomington of Charles R. Boxer, who longed to flee the town on "Home Game" Saturdays, but lacked a car. The biography of Boxer by Dauril L. Alden includes two paragraphs on how Richard and his car saved Boxer from the hell of Home Games, leading the distinguished historian of the Dutch and Portuguese overseas empires to dub his savior "King Richard."[8] Indiana University also gave Kagan his first experience of lecturing to large student audiences and directing graduate research, two skills at which he soon excelled, and he still found time to write book reviews and articles, as well as to present papers at conferences. Then Lawrence Stone, founding director of Princeton University's Shelby Cullom Davis Center for Historical Studies, and one of Kagan's thesis examiners, invited Richard

[7] Kagan to Parker, November 9, 1968.

[8] Dauril Alden, *Charles R. Boxer, An Uncommon Life* (Lisbon: Ramer, 2001), 444–5, confirmed by Kagan.

to join him for a year as a Visiting Fellow at the Center. Kagan would later recall that Stone:

took me under his wing as a disciple, and although he knew relatively little about Spain, he taught me about the value of comparative history, the necessity for methodological rigor and critical thinking, together with an insistence on asking big questions and addressing broad themes. Equally important was his enthusiasm, indeed passion for history ... Lawrence also proved instrumental in advancing my professional career.

Then, in 1972, The Johns Hopkins University invited Kagan to teach in its Program in Atlantic History and Culture (the first of many such programs in the United States). Should he accept? Baltimore, he knew, "was no Garden of Eden":

It was [then] a poor city, still suffering from the effects of the riots sparked by [Martin Luther] King's assassination in 1968, and rapidly losing population. It had certain advantages over Bloomington, however: fresh fish and seafood for one thing; proximity to the riches of Library of Congress; Amtrak; and most importantly, a stellar history department and the possibility of participating regularly in its famous Seminar, an institution through which I have encountered historiographies, bibliographies, and methodologies from which I have never ceased to learn.

He therefore accepted, and has never looked back.

Kagan spent his first years at Hopkins preparing an article for publication in *Past and Present*, and turning his thesis into his first book: *Students and Society in Early Modern Spain*.[9] He also embarked, together with a team led by Lawrence Stone, on a study (generously funded by the US Department of Health, Education and Welfare) of the history of universities in comparative perspective, measuring change over time. Kagan's assignment was to chart matriculation trends in southern Europe (France, Italy, Portugal, and Spain) to match the data gathered by Stone and others for northern Europe and the United States, and to this end he worked in the archives of every university town in southern Europe. The project led him both to computer analysis, "coding" tens of thousands of "punch cards" (cutting-edge technology in the 1970s), and also to his second research topic.[10]

[9] Richard L. Kagan, "Universities in Castile 1500–1700," *Past & Present* 49 (1970): 44–71; and *Students and Society in Early Modern Spain* (Baltimore, MD: Johns Hopkins University Press, 1974).

[10] For an impressive example of Kagan's research on this project, see his article, "Le università in Italia 1500–1700," *Società e storia* 28 (1985): 275–317, published in English as "Universities in Italy, 1500–1700," in *Les universités dans l'Europe moderne*, eds. R. Chartier and J. Revel (Paris, 1986), I, 153–86.

Kagan's data revealed a rapid increase in litigation, lawyers, law faculties, and legal education throughout southern Europe: what, he wondered, lay behind these deveopments? Could it all stem merely from the rise of a bureaucracy, as suggested in *Students in Society*, or might it be something broader, such as an institutionalization of conflict-resolution and a concomitant demand for various sorts of legal specialists, lawyers above all? Kagan published some initial findings in another article in *Past and Present*, and in his second book, *Lawsuits and Litigants in Castile, 1500–1700* (1981), which focused on the legal profession, the royal judges, and the manner in which thousands of ordinary Castilians resorted to litigation in order to resolve their disputes. It proved another stunning milestone of quantitative history.[11]

Then, in 1979 Kagan experienced his "cultural turn." Inspired by a reproduction of *The Burial of the Count of Orgaz*, by El Greco of Toledo, Kagan's undergraduate thesis had addressed the topic of "Dreams and Reality in Sixteenth-Century Spain," using a combination of literary, artistic, and historical sources. Subsequently, while researching *Lawsuits and Litigants*, he found rich material in the archives of Toledo and began to contemplate a study, provisionally entitled "Coping with Crisis," of how the city's magistrates had managed (or mis-mismanaged) their community during the calamitous seventeenth century. Then, one evening over dinner, Jonathan Brown of New York University's Institute of Fine Arts described his labor with a team of experts to prepare a major international exhibition centered on El Greco. The team lacked one important member, Brown sighed: a historian familiar with sixteenth-century Toledo who could describe the environment in which the artist had lived and worked. Could Kagan fill that gap? It took only a few hours of reflection before Richard decided to abandon "Coping with Crisis" and sign on to *El Greco of Toledo*. This involved a crash course in Spanish art and art history, during which, "I learned to present lectures with slides; and in doing so, added an all-important visual component to my undergraduate courses"; as well as immersing himself in documents in and

[11] Kagan, "Law Students and Legal Careers in Eighteenth-Century France," *Past & Present* 68 (1975): 38–72; and *Lawsuits and Litigants in Castile, 1500–1700* (Chapel Hill: University of North Carolina Press, 1981). He also published several other studies on the subject: see, for example, Kagan, "Pleitos y poder real: La Chancillería de Valladolid, 1500–1700," *Cuadernos de Estudios Históricos* 2 (1978): 291–316; and "A Golden Age of Litigation: Castile 1500–1700," in *Disputes and Settlements: Law and Human Relations in the West*, ed. J. Bossy (Cambridge: Cambridge University Press, 1984), 145–66.

about Toledo, notably the archives of the Toledo tribunal of the Spanish Inquisition (for which El Greco served as interpreter, when the accused claimed to be a native Greek speaker). Apart from a splendid chapter in the exhibition catalog, published in 1982, the project forged friendships with two eminent art historians: Fernando Marías, with whom Kagan would undertake other important projects, and Marianna Shreve Simpson, whom he married the following year.[12]

By then, Kagan had already embarked upon a new project that combined art history and history in an effort to study the meaning and importance of urban values. It began with El Greco's painting *View of Toledo*, a haunting image of the Imperial City in which the artist altered the urban landscape for pictorial effect. Kagan's essay in *El Greco of Toledo* unraveled some of the secrets of that picture, but it also awakened his curiosity about the way Spanish cities used texts and images for promotional, even propagandistic purposes. This involved not only city views but also a sub-genre of geography known as "chorography," the study of a place and its representation, whether in images or in texts, and an examination of the competing definitions of what constituted a city: was it the structure (*urbs*) or its inhabitants (*civitas*)? In 1989 Kagan published *Cities of the Golden Age: The Views of Anton Van den Wyngaerde*, reproducing and analyzing the series of stunning city views commissioned by Philip II in the 1560s and originally placed on display in Madrid's Alcázar and other royal palaces, explicitly for propagandistic aims. A decade later, with the collaboration of Fernando Marías, he published *Urban Images of the Hispanic World, 1493–1793*, a volume that considered the dichotomy of *urbs* versus *civitas* in representing the city and the importance of topographical accuracy in city views in Spain, Spanish America, and the Philippines.[13]

In between these two ventures, Kagan completed yet another project. In researching *El Greco of Toledo*, he found references to a young street-prophet named Lucrecia de León whose dreams criticizing both the person and the politics of Philip II in the 1580s attracted the attention first of the citizens of Madrid and then of the Inquisition. Kagan thereupon located her trial records, which included an indexed transcript of four

[12] Kagan, "The Toledo of El Greco," in *El Greco of Toledo* (Boston, MA: Little, Brown, 1982), 35–73.

[13] *Ciudades del Siglo de Oro. Las vistas de Anton van den Wyngaerde* (Madrid: El Viso, 1986; English edition published by the University of California Press in 1989); *Imágenes urbanas del mundo hispánico, 1493–1780* (Madrid: El Viso, 1999; English edition published by Yale University Press in 2000).

hundred of her dreams, some predicting that the Armada of 1588 would meet with disaster as a punishment for Philip's murder of his son Don Carlos and his wife Isabel de Valois twenty years before, and transcripts of Lucrecia's conversations with others. Thanks to the popularity of books like Carlo Ginzburg's *The Cheese and the Worms*, micro-histories had begun to enjoy great success; and when Ginzburg visited Johns Hopkins, Kagan sought his advice on whether or not to undertake not just another subject but another genre. Ginzburg approved, and *Lucrecia's Dreams* appeared in 1990, garnering excellent reviews and much scholarly attention in the United States and beyond (including Spanish and Japanese translations). Kagan's only regret was that, after enduring the savage punishment decreed by the inquisitors (one hundred lashes and two years of confinement in a convent), Lucrecia's fate remains unknown.[14]

In 2004, together with Abigail Dyer, Kagan published another work based on Inquisition records. *Inquisitorial Inquiries. Brief Lives of Secret Jews and Other Heretics* presented a collection of what he has termed "involuntary autobiographies" designed for classroom use in colleges and universities in the United States. It is "to date, by far my best-selling work";[15] but by then, his research focus had shifted again: to the local histories written in the Spanish Golden Age. As with "city views," he began with Toledo, and specifically with Pedro Salazar de Mendoza and a handful of other scholars whose local pride led them to write histories that celebrated the grandeur and importance of the Imperial City; but he soon discovered similar histories written about other Spanish cities on both sides of the Atlantic. He set out to relate these "particular" histories to the "general" or "national" histories of the kind written by Juan de Mariana and by historians who received royal patronage or occupied the office of *cronista del rey*. So began a decade-long project that culminated in 2009 with the publication of *Clio and the Crown: The Politics of History in Medieval and Early Modern Spain*, a work that explored the meaning and importance of "official" (that is, sponsored) history.[16]

[14] *Lucrecia's Dreams: Politics and Prophecy in Sixteenth-Century Spain* (Berkeley, CA: University of California Press, 1990, revised edition, 2004); *Los sueños de Lucrecia – Política y profecía en la España del siglo XVI* (Madrid: Nerea, 1991).

[15] Kagan and Dyer, *Inquisitorial Inquiries: The Brief Lives of Secret Jews and Other Heretics* (Baltimore, MD: The Johns Hopkins University Press, 2004); revised Spanish edition, *Vidas infames. Criptojudios y otros herejes ante la Inquisición* (Madrid: Editorial Nerea, 2010); expanded and revised English edition (Baltimore, MD: The Johns Hopkins University Press, 2011).

[16] *Clio and the Crown. The Politics of History in Medieval and Early Modern Spain* (Baltimore, MD: Johns Hopkins University Press, 2009); revised Spanish edition,

Simultaneously, Richard became interested in yet another forgotten genre: how Spanish history has been studied in the United States, and, more broadly, the changing image of Spain in the United States. Starting in the early 1990s, he read the works of earlier generations of American Hispanists, William Hickling Prescott and Washington Irving, George Ticknor and Henry Charles Lea, followed by the works of lesser-known historians and a seemingly infinite number of journalists and travel writers about Spain. The first fruit of this interest was Kagan's best-known article to date, "Prescott's Paradigm: American Historical Scholarship and the Decline of Spain," published in the *American Historical Review* in 1996 – an essay that (Kagan believes) convinced readers in Spain's Ministry of Foreign Affairs to recommend the award of an *encomienda* in the Order of Isabel la Católica, an honor of which Kagan is justly proud.[17] His article also aroused great interest among his colleagues. The most memorable panel at the annual meeting of the Society for Spanish and Portuguese Historical Studies in 1999 was "The Prescott Paradigm: a Debate," in which Kagan presented a sequel to his article, in which he surveyed the state of the field and its future, followed by comments from James Amelang, Isabel Burdiel, Sabine MacCormick, and David Nirenberg. Then, at the society's annual meeting fourteen years later, Richard's students organized a panel entitled "In honor of Richard L. Kagan: rethinking 'Prescott's Paradigm': a new paradigm for a new millennium?" Almost a hundred Hispanists, including Sir John Elliott, attended.[18]

Kagan also edited a volume entitled *Spain in America: The Origins of Hispanism in the United States* (2002) and since then he has published several essays about the teaching of Spanish history, the collecting of Spanish art in the United States, and the entangled themes of Hispanophilia and

Los cronistas y la corona (Madrid: Marcial Pons, 2010); awarded the "Premio de Investigación Humanística Real Sociedad Menéndez Pelayo" for 2011. This was not Kagan's first engagement with the topic – in 2004 he published *El Rey recatado. Felipe II, la historia y los cronistas del rey* – nor will it be his last: he plans to write a micro-history based on an early seventeenth-century lawsuit between the count of Puñonrostro and the royal chronicler, Antonio de Herrera y Tordesillas, which will address questions relating to the establishment of historical truth.

17 Richard L. Kagan, "Prescott's Paradigm. American Historical Scholarship and the Decline of Spain," *American Historical Review* 101 (1996): 423–46 (Castilian translation, "El paradigma de Prescott:la historiografía norteamericana y la decadencia de España," *Manuscrits* 16 (1998): 229–53.

18 *The Bulletin of the Society for Spanish and Portuguese Historical Studies* for 1999, pp. 22–9, published Kagan's presentation in full, together with summaries of the four comments and of the ensuing debate.

Hispanophobia in the era of the Spanish-American War of 1898. His further research and reflection on this theme will figure in his next book, provisionally entitled *The Spanish Craze: The Discovery of the Arts and Cultures of the Hispanic World in the United States, 1890–1930*.[19]

Kagan developed this historiographical interest "partly because of the failure of the *American Historical Review*, on the occasion of that journal's 100th anniversary, to include a historiographical essay devoted to Spain and the Spanish world alongside the others it had commissioned for France, Italy, etc," but "largely to help my graduate students gain a perspective on the place of Spanish history within the broader constellation of historical practice in the United States." Note that phrase, "Largely to help my graduate students": how many historians can say the same about one of their research fields? In Kagan's case, moreover, this is an understatement, because throughout his career he has advanced knowledge of Spanish history through mentorship as well as authorship.

Hugh Trevor-Roper once wrote to a colleague, "A Professor reveals his quality not merely in his own writing but in the work of his pupils, in the 'school' he creates," and he went on to lament that, all too often, "students wishing to do historical research are not directed to problems: they are told to find some neglected pool and paddle in it."[20] Now, mentoring graduate students is hard work. It should not be confused with "advising," which focuses upon getting a student through a program so that s/he can receive a degree: "A mentor shares his knowledge of a field; she offers advice, encouragement, criticism, and feedback; he helps students become contributing members of a field or discipline."[21] The "school of Kagan" now includes dozens of scholars whose research he has guided, directly or

[19] For a foretaste, see Kagan, "The Spanish Craze: The Discovery of Spanish Art and Culture in the United States," in *When Spain Fascinated America*, ed. Ignacio Suárez-Zuloaga (Madrid: Fundación Zuloaga, 2010), 25–46; "The *Spanish Craze* in the United States: Cultural Entitlement and the Appropriation of Spain's Cultural Patrimony, ca. 1890-ca. 1930," *Revista Complutense de Historia de America* 36 (2010): 37–58; "Blame it on Washington Irving: New York's Discovery of the Art and Architecture of Spain," in *Nueva York: 1613–1945*, ed. Edward Sullivan (New York: New York Historical Society and Scala, 2010), 155–71; and "The Spanish Turn: The Discovery of Spanish Art in the United States, 1887–1920," in *Collecting Spanish Art: Spain's Golden Age and America's Gilded Age*, eds. Inge Reist and José Luis Colomer (New York: The Frick Collection Studies in the History of Art Collecting in America; Penn State University Press, 2012), 3–23.

[20] *One Hundred Letters from Hugh Trevor-Roper*, eds. R. Davenport-Hines and A. Sisman (Oxford: Oxford University Press, 2014), 59, letter to J. C. Masterman, December 13, 1956, on the qualities required of the next Regius Professor of Modern History at Oxford.

[21] www.grad.washington.edu/mentoring/gradfacultymentor.pdf.

indirectly, starting with seminars and meetings in which he helped them grasp the content and contours of the field. They read the key works – what Richard calls "the old chestnuts" – and explored the latest approaches, learning to see beyond the content, the dates, and names, to an understanding of the field as a whole, its development and its relationship to other areas of inquiry, and eventually, to his students' place within it.

Richard always regarded his students as "junior colleagues-in-training," and all of them have benefitted from his generosity, which starts long before they set forth for the archives. One advisee recorded that, "One of the things I liked best about Richard is that anytime a professor or important scholar came to the university, he would always introduce all the students present, talk about our research and treat us with respect."[22] Through seminar readings and discussions, his students also collaborated on many of his projects (and the "acknowledgments" section of his books usually thanks his students, alongside established scholars, for their input). As part of helping them become members of the larger scholarly community, Richard always cultivated a spirit of collaboration and community among his students. Even when they were chasing after the same grants, they cooperated rather than competed, reading draft after draft of each other's Fulbright application essays in Richard's grant-writing workshops.

Richard's remarkable generosity stretched far beyond his own advisees. Chuck Jago, another of John Elliott's advisees, recalls Richard as "the sleuth who discovered new archives and new caches of documents" and "the one we all depended on for our thesis work and our subsequent work as historians." I can say the same. Richard could be relied on to supply almost anything – a reference, a photocopy, an elusive item from Marcial Pons – and it duly appeared, either directly or indirectly through his impressive network of friends, advisees, and colleagues. Thus in 1966, just after he and I started our research in Spanish archives, Richard sent me the call number and a partial transcript of a document he had just read in the Biblioteca Nacional that mentioned the transport of silver bullion from Milan to the Netherlands, the subject on which I was then working. He added, "Hope it proves of import." (It did.)[23]

[22] This and subsequent quotations, unless otherwise stated, come from the moving address delivered by Professor A. Katie Harris, one of Richard's advisees, at "Kagan's Kaleidoscope" in 2008. Names have been redacted to protect the guilty.

[23] Video tribute recorded for "Kagan's Kaleidoscope" by Charles J. Jago, September 12, 2008, and sent to Geoffrey Parker; Kagan to Parker, October 7, 1966.

I had always suspected that Richard showed the same generosity towards his other colleagues, but lacked proof until 2008, when his advisees organized "Kagan's Kaleidoscope," a symposium in honor of his sixty-fifth birthday. Not only did more than a score of his students come to Baltimore to celebrate his achievements and present papers (many of them included in this volume), but almost every senior Hispanist either attended in person (James Amelang, Jonathan Brown, John Elliott, Xavier Gil, Lu Ann Homza, David Nirenberg, Benjamin Schmidt, and Stuart Schwartz) or delivered a recorded tribute via videolink (Jim Casey, Chuck Jago, David Lagomarsino, Óscar Mazín, and Teo Ruiz): the list reads like a *Who's Who* of Spanish, American, British, and Canadian scholars on the world of the Spanish Habsburgs.

Perhaps Kenneth Andrien, an expert in Colonial Latin American history, provided the best testimony to Richard's remarkable gifts as a friend, scholar, mentor, and networker. When informed of Richard's plan to retire, Andrien offered the following prediction:

> I cannot believe that Richard is retiring. Historians of his stature are like Andean Gods, they do not die, they just become part of the natural landscape, what the Andeans call a *huaca*. Richard will become a great huaca with strong spiritual power for his people to come and worship.[24]

[24] Kenneth Andrien to Geoffrey Parker, email, February 16, 2013, on hearing that Richard's advisees had organized a panel at ASPHS in honor of his retirement.

Introduction: Mapping the Early Modern Hispanic World

Kimberly Lynn and Erin Kathleen Rowe

In 1996, an essay appeared in the *American Historical Review* by Richard L. Kagan that presented the history of the study of Spain in the United States, beginning with the seminal nineteenth-century figure, William H. Prescott. Kagan argued that Prescott created a paradigm for understanding Spanish history largely based upon the Black Legend – that is, the belief that Spain had sunk into decline due to religious fanaticism and corruption – and that this paradigm shaped understandings of Spain's history in the United States for a century.[1] It was only beginning in the mid-twentieth century that scholars began chipping away at Prescott's paradigm. The mid-1990s marked a particularly propitious moment to examine the history of Spain, as the paradigm Kagan described was in the process of unraveling, to be replaced with new paradigms. Kagan likened the up-and-coming generation of Hispanists to "iconoclasts, bent on replacing the old images that had formed the Spain envisioned by their forbearers and, in some cases, their own teachers as well."[2] In addition to Kagan's essay, more work appeared that highlighted the crucial work being done to reshape the scholarly image of early modern Spain, including an important Festschrift volume in honor of Sir John Elliott.[3] From this period on, the development of the study of Spanish history has been a story, first and foremost, of tremendous growth. While the field had already changed substantively throughout the second half of the century,

[1] Richard L. Kagan, "Prescott's Paradigm: American Historical Scholarship and the Decline of Spain," *The American Historical Review* 101(2) (April 1996): 423–46.

[2] Kagan, "Prescott's Paradigm," 443.

[3] Richard L. Kagan and Geoffrey Parker, eds., *Spain, Europe and the Atlantic World: Essays in Honour of John H. Elliott* (Cambridge: Cambridge University Press, 1995).

the twenty years since "Prescott's Paradigm" and *Spain, Europe, and the Atlantic World* have seen an ever-increasing rethinking of how to study the early modern Hispanic world produced by scholars throughout the Americas and Europe. One small token of the field's growth might be seen in the list of scholars who self-identify as "historians of early modern Spain" on one frequently consulted blog, now numbering more than 700.[4]

Iberia stands at the center of key trends in current historiography, one of the most important of which is an interest in bodies of water, instead of masses of land. Turning to the seas – Atlantic, Pacific, Mediterranean – has made it possible for scholars to find creative ways of breaking free from national paradigms and rethinking the effects of mobility on pre-modern societies.[5] It has also facilitated the recent explosion of interest in transnational and global histories, which is now overtaking European historiography in all time periods. The position of Portugal and Spain as the first European kingdoms to "go global" – as well as the birthplace of the first global missionaries, the Jesuits – has fueled the exponential expansion of the study of their intertwined histories in recent years.

This volume aims to take stock of key areas of current scholarship that are in the process of again transforming our understanding of the Hispanic world. The Prescottian vision of the Spanish past tended to generate excessively monolithic, static, or moribund understandings of people and institutions of the global Spanish monarchy. More recent revisionist scholarship, on the other hand, has rediscovered the dynamism of the early modern world. In this volume, we argue that one of the most important recent scholarly contributions is awareness that the inhabitants of the far-flung societies of the early modern Spanish monarchy lived in worlds marked by change. What these inhabitants thought they knew about the features and boundaries of their worlds – geographic, literary, social, philosophical, religious, and political – altered repeatedly across the early modern era, as did the ways in which they

[4] Counted on January 25, 2015. The blog "Early Modern Spanish History Notes," is the work of Scott K. Taylor. "Historians of Early Modern Spain," https://emspanishhistorynotes.wordpress.com/historians-of-em-spain/.

[5] Nicolás Canny and Philip D. Morgan, eds., *Oxford Handbook of the Atlantic World* (New York: Oxford University Press, 2011); David Abulafia, *The Great Sea: A Human History of the Mediterranean* (Oxford: Oxford University Press, 2011); Matt K. Matsuda, *Pacific Worlds: A History of Seas, Peoples, and Cultures* (Cambridge: Cambridge University Press, 2012). The most influential text on thinking with the sea is Fernand Braudel, *The Mediterranean and the Mediterranean World in the Age of Philip II*, 2 vols. (New York: Harper Collins, 1972; original published in French in 1949).

represented and transmitted that knowledge. As fluidity and dynamism replace uniformity and fixedness, this volume takes inspiration from the current transnational historiography, defined by Isabel Hofmeyer in 2006 as a methodological approach that claims "not simply that historical processes are made in different places but that they are constructed in the movement between places, sites, and regions."[6] We offer multiple accounts of the ways in which the early modern Hispanic world was constructed through movement – of people, ideas, and texts. While doing so, we continue to remain sensitive to the historical processes that developed out of deeply local contexts and retain our awareness that movement did not mark every historical process or every person's life. Hence, we have taken transnational history as an important concept to think with, rather than the defining paradigm of the early modern Hispanic world.

Thinking with transnational history led us to consider the often-neglected relationship between movement and fixedness, which in turn evolved into a deeper engagement with the concept of circulation and circuitry – what we have termed here "mapping." The book argues that the mapping of the Hispanic world was a fluid and dynamic process, in which circulation of many kinds played a critical role in shaping communities both at the center and the peripheries. Yet the metaphor of mapping permits us to integrate fixed points alongside movement. It further allows a wide geographical and chronological swath, precisely because such disparities open up some of the possibilities found in comparative history – that is, awareness of what are key differences between specific towns, regions, or time periods, and how such awareness challenges or confirms our understanding of historical processes. Our understanding of "mapping" encourages the reader to draw connections between disparate phenomena. Even when, as in the examples in this volume, the individual studies do not center their analysis on transnational circulation or global history, reading these articles in concert with one another allows us to see how disparate times and places related to one another, permitting a more complete understanding of the whole. The essays in this volume consider the broad geographical dimensions of the Catholic Monarchy, examining in various ways what tied the Iberian Peninsula to other regions of Europe, the Mediterranean, and Atlantic worlds. They offer perspectives on distinct aspects of the process of circulation as authors grapple with

[6] C. A. Bayly, Sven Beckert, Matthew Connelly, Isabel Hofmeyer, Wendy Kozol, and Patricia Seed, "*AHR* Conversation: On Transnational History," *The American Historical Review* 111(5) (December 2006): 1444. Hofmeyer's definition of transnational should not be considered definitive; see the essay for alternate viewpoints.

the micro and the macro, spiritual movements and cultural controversies, economic problems and encounters with terrains both familiar and unfamiliar. Together, the essays chart how signal institutions of the Spanish monarchy were made and remade across the early modern era. And they explore how individuals and communities navigated within the frameworks of monarchy and Church, adapting their identities, initiatives, ideologies, and modes of communication to changing circumstances and varying norms of thought and expression. Taken together, the essays demonstrate adaptation and negotiation (themselves types of movement) were central processes through which the early modern Hispanic world evolved.

The essays in this volume make significant contributions to the field of early modern Spanish history as it has developed over the course of the past two decades. The Spanish monarchy, as a complex, global, political entity has become a testing ground for emerging historiographical trends. It figures prominently in attempts to write new histories of the problem of empire.[7] Scholars of the Iberian world have been especially engaged in rethinking its place in debates both about how to interpret the early modern world and the origins of crucial modern phenomena.[8] New approaches to questions of space and place as historical frameworks have propelled not just attention to oceans, but also to thinking about shifting terrestrial boundaries in new ways as well. In this arena, borderlands studies have been driven by experts and archives of Iberia and have in turn pushed investigations of the Hispanic world in fresh directions. While the borderlands of the American southwest might

[7] See the reflections in Barbara Fuchs, "Imperium Studies: Theorizing Early Modern Expansion," in *Postcolonial Moves: Medieval through Modern*, eds. Patricia Clare Ingham and Michelle R. Warren (New York: Palgrave Macmillan, 2003), 71–90; John H. Elliott, *Empires of the Atlantic World: Britain and Spain in America, 1492–1830* (New Haven, CT: Yale University Press, 2006); Jane Burbank and Frederick Cooper, *Empires in World History: Power and the Politics of Difference* (Princeton, NJ: Princeton University Press, 2010); Thomas James Dandelet, *The Renaissance of Empire in Early Modern Europe* (New York: Cambridge University Press, 2014). On the political structure of the monarchy, see the signal essay John H. Elliott, "A Europe of Composite Monarchies," *Past & Present* 132(1) (1992): 48–71.

[8] For example, Francisco Bethencourt, *Racisms: From the Crusades to the Twentieth Century* (Princeton, NJ: Princeton University Press, 2013); David Nirenberg, *Anti-Judaism: The Western Tradition* (New York: W. W. Norton & Co., 2013); Lauren A. Benton, *Law and Colonial Cultures: Legal Regimes in World History, 1400–1900* (Cambridge and New York: Cambridge University Press, 2002); Stuart Schwartz, *All Can Be Saved: Religious Tolerance and Salvation in the Iberian Atlantic World* (New Haven, CT: Yale University Press, 2008); Irene Silverblatt, *Modern Inquisitions: Peru and the Colonial Origins of the Civilized World* (Durham, NC: Duke University Press, 2004).

spring first to mind, this set of approaches has combined with other historiographical shifts to create a new sensitivity to the flexibility and particular dynamics of frontier spaces and societies – from the conquest and Christianization of the kingdom of Granada to the long-term warfare along the Araucanian frontier of the viceroyalty of Peru to the ways in which the frontier zones of colonial Mexico were made and remade across the sixteenth century.[9]

The expanse and complexity of the early modern Spanish monarchy also requires continuing reflection on how far-flung places, tied together by the claims and aspirations of the crown, related to one another. It requires simultaneous engagement with multiple geographic scales.[10] One fruitful approach to this quandary has been to consider the dynamics that operated between centers and peripheries. Scholars have increasingly examined how one-time centers were decentered and, sometimes, reinvigorated. They are exposing how seeming peripheries might be re-read as centers of other networks, and how new circuits were created between peripheries of the Hispanic colonial world.[11] Studying the Spanish monarchy also requires continual reflection on the dynamic interaction of the local and the universal.[12]

[9] One pathbreaking approach to thinking about frontiers and national histories was Peter Sahlins, *Boundaries: The Making of France and Spain in the Pyrenees* (Berkeley, CA: University of California Press, 1989). For rethinking frontier spaces, Ida Altman, *The War for Mexico's West: Indians and Spaniards in New Galicia, 1524–1550* (Albuquerque: University of New Mexico Press, 2010); David Coleman, *Creating Christian Granada: Society and Religious Culture in an Old-World Frontier City, 1492–1600* (Ithaca, NY: Cornell University Press, 2003); Laura Matthews, *Memories of Conquest: Becoming Mexicano in Colonial Guatemala* (Chapel Hill, NC: University of North Carolina Press, 2012). These lines of scholarship are also being used to interrogate the concept of empire and its tendency to Eurocentrism, Pekka Hämäläinen, *The Comanche Empire* (New Haven, CT: Yale University Press, 2008); Paul A. Kramer, "Power and Connection: Imperial Histories of the United States in the World," *The American Historical Review* 116(5) (December 2011): 1348–92.

[10] Francesca Trivellato, "Is There a Future for Italian Microhistory in the Age of Global History?," *California Italian Studies* 2(1) (2011), online at: http://escholarship.org/uc/item/0z94n9hq (accessed February 22, 2015); John-Paul Ghobrial, "The Secret Life of Elias of Babylon and the Uses of Global Microhistory," *Past & Present* 222(1) (February 2014): 51–93; Tonio Andrade, "A Chinese Farmer, Two African Boys, and Warlord: Toward a Global Microhistory," *Journal of World History* 21(4) (December 2010): 573–91.

[11] Kagan and Parker organized their Festschrift in honor of John Elliott along this axis: *Spain, Europe, and the Atlantic World.*

[12] The dynamic between the local and the universal has been studied with particular sensitivity in the context of religious practice and institutions, often overturning earlier assumptions about a a top-down model. One of the most influential works on this theme

As new frameworks of analysis have caused a widespread rethinking of histories centered on nation-states, two other approaches have served to further deconstruct the paradigm of Spanish exceptionalism, another key feature of the Prescottian view of Spanish history. The first technique is that of comparative history. Scholars are increasingly recovering what the early modern Spanish monarchy shared with other contemporary European states, while not losing sight of the ways in which institutions, practices, and ideologies were based in varied origins, responded to distinctive local conditions, and diverged across the early modern era. This approach has been especially meaningful in disrupting the view of Spain and England as mirror images of one another, itself rooted in the polemics and propaganda of the sixteenth century.[13] Intertwined with the effort to look across early modern European states, societies, literary or scientific cultures has been the impetus to bridge the divide between the historiographies of the Age of Reformations. There is increasing effort to compare the Catholicisms and Protestantisms of the early modern world, and the religious underpinnings of various political, social, and imperial enterprises; likewise, the diversity within early modern global Catholicism is being rethought.[14] These historiographical moves are also

has been William A. Christian, *Local Religion in Sixteenth-Century Spain* (Princeton, NJ: Princeton University Press, 1981). See also important work by Sara T. Nalle, *God in La Mancha: Religious Reform and the People of Cuenca, 1500–1650* (Baltimore, MD: Johns Hopkins University Press, 1992); Simon Ditchfield, *Liturgy, Sanctity, and History in Tridentine Italy: Pietro Maria Campi and the Perservation of the Particular* (Cambridge: Cambridge University Press, 1995); and recently, Martin Nesvig, ed., *Local Religion in Colonial Mexico* (Albuquerque: University of New Mexico Press, 2006); Erin Kathleen Rowe, *Saint and Nation: Santiago, Teresa of Avila, and Plural Identities in Early Modern Spain* (University Park: Penn State University Press, 2011).

[13] J. N. Hillgarth, *The Mirror of Spain, 1500–1700: The Formation of a Myth* (Ann Arbor, MI: University of Michigan Press, 2000); Elliott, *Empires of the Atlantic World*; and Glyn Redworth, *The Prince and the Infanta: The Cultural Politics of the Spanish Match* (New Haven, CT: Yale University Press, 2003). See also other frameworks in which Spain is brought into comparison: Thomas Dandelet and John Marino, eds., *Spain in Italy: Politics, Society, and Religion, 1500–1700* (Leiden: Brill, 2007); and Daniel Aznar, Guillaume Hanotin, and Niels F. May, eds., *À la place du roi: Vice-rois, gouverneurs et ambassadeurs dans les monarchies française et espagnole (XVIe–XVIIIe siècles)* (Madrid: Casa de Velázquez, 2014).

[14] In ways also that complicate the older theses of confessionalization and the rise of the modern state. Charles H. Parker and Gretchen Starr-LeBeau, eds., *Judging Faith, Punishing Sin: Inquisitions and Consistories in the Early Modern World* (New York: Cambridge University Press, in press); Carla Gardina Pestana, *Protestant Empire: Religion and the Making of the British Atlantic World* (Philadelphia, PA: University of Pennsylvania Press, 2009);Linda Gregerson and Susan Juster, eds., *Empires of God: Religious Encounters in the Early Modern Atlantic* (Phildadelphia, PA: University of Pennsylvania Press, 2010);

intimately related to various efforts to rethink the nexus between religion and politics in the early modern world – from reopening consideration of religious dissent to the varied status and identities of religious minorities, looking with fresh eyes at religious syncretism and at the remarkable complexities of early modern global Catholicism as it developed in myriad local environments. And breaking the binaries of the mid-twentieth century is having even further interpretive effects, for example in various attempts to disrupt the continuing influence of the Weberian thesis of modernity's alignment with Protestantism.[15]

A second technique to combat Spanish exceptionalism is more concerned with integration. There is new interest in the translocal or transregional within the territories of the Spanish monarchy. Thus the political relations and sociocultural connections between Spain and Spanish Italy are increasingly the focus of scholarly attention.[16] And there are renewed calls to write across longstanding divides in the field – most strikingly, to bring the historiographies of Latin America and Spain into closer conversation with one another.[17] Another variation of these approaches has been to look especially at the circulation of non-Spaniards within the early modern Spanish monarchy.[18] Circulation and mobility – of goods

Alexandra Bamji, Geert H. Janssen, and Mary Laven, eds., *The Ashgate Research Companion to the Counter-Reformation* (Farnham and Burlington, VT: Ashgate Press, 2013); Francisco Bethencourt, *The Inquisition: A Global History, 1478–1834* (Cambridge: Cambridge University Press, 2009). See also the stimulating comparative effort of Julie Hardwick, Sarah M. S. Pearsall, and Karin Wulf, eds., *Centering Families in Atlantic Histories*, special issue of *The William and Mary Quarterly* 70(2) (April 2013).

15 For two recent examples, see Larry Siedentop, *Inventing the Individual: The Origins of Western Liberalism* (Cambridge, MA: Belknap Press, 2014); Brad S. Gregory, *The Unintended Reformation: How a Religious Revolution Secularized Society* (Cambridge, MA: The Belknap Press, 2012).

16 Dandelet and Marino, eds., *Spain in Italy*; John Marino, *Becoming Neapolitan: Citizen Culture in Baroque Naples* (Baltimore, MD: Johns Hopkins University Press, 2010); and Gabriel Guarino, *Representing the King's Splendour: Communication and Reception of Symbolic Forms of Power in Viceregal Naples* (Manchester: Manchester University Press, 2010).

17 Evonne Levy and Kenneth Mills, eds., *Lexikon of the Hispanic Baroque* (Austin, TX: University of Texas Press, 2014); Allan Greer and Jodi Bilinkoff, eds., *Colonial Saints: Discovering the Holy in the Americas* (New York: Routledge, 2002); Tamar Herzog, *Frontiers of Possession: Spain and Portugal in Europe and the Americas* (Cambridge, MA: Harvard University Press, 2015).

18 Céline Dauverd, *Imperial Ambition in the Early Modern Mediterranean: Genoese Merchants and the Spanish Crown* (Cambridge: Cambridge University Press, 2014); Robert Descimon and José Javier Ruiz Ibáñez, *Les ligueurs de l'exil: Le refuge catholique français après 1589* (Seyssel: Champ Vallon, 2005); Richard L. Kagan and Philip D. Morgan, eds., *Atlantic Diasporas: Jews, Conversos, and Crypto-Jews in the Age of Mercantilism, 1500–1800* (Baltimore, MD: Johns Hopkins University Press).

and ideas, of devotional, artistic, or political cultures – characterized the early modern Hispanic world. And recapturing that motion also sheds light on meaningful early modern dynamics, such as the effects that news had on early modern political culture.[19] Thus, in the early twenty-first century, there is a wealth of fresh approaches to how to locate the early modern Hispanic world. Yet these bring new challenges, which mirror the difficulties of the surge in oceanic frames of reference, and of global histories.[20] The challenge in overturning Spanish exceptionalism is to preserve the distinctive features of the Hispanic world, to keep a fine-grained attention to change over time, and to maintain a grasp of the power relations and hierarchies that structured the early modern Spanish monarchy.

There is a confluence between transnational approaches – and the related new imperial histories, oceanic frameworks, concerns with place and space – and interdisciplinarity. In line with the "cultural turn" in historical study, scholars of the Hispanic world have also pioneered approaches to interdisciplinary collaboration.[21] There has been an ongoing cross-pollination of literary and historical scholarship, and collaboration among historians and specialists in art and architectural history and the history of science is bringing an interest in materiality and visual culture ever more into the field, as, for example, in the multiple ways in which urban spaces are being re-read.[22] The newly appeared *Lexikon*

[19] Heinz Schilling, Istvan Toth, Donatella Calabi, et al., eds., *Cultural Exchange in Early Modern Europe*, 4 vols. (Cambridge: Cambridge University Press, 2007); M. Olivari, *Fra trono e opinione. La vita politica castigliana nel Cinque e Seicento* (Venice, 2002); Filippo de Vivo, *Information and Communication in Early Modern Venice: Rethinking Early Modern Politics* (Oxford: Oxford University Press, 2007).

[20] David A. Bell, "This is What Happens When Historians Overuse the Idea of Networks," *The New Republic* (2013), online at: www.newrepublic.com/article/114709/world-connecting-reviewed-historians-overuse-network-metaphor (accessed February 18, 2015).

[21] For influential examples of collaborations between historians and art historians in the field, see John H. Elliott and Jonathan Brown, *A Palace for a King: The Buen Retiro and the Court of Philip IV*, 2nd ed. (New Haven, CT: Yale University Press, 2003); Richard L. Kagan with Fernando Marías, *Urban Images of the Hispanic World, 1493–1793* (New Haven, CT: Yale University Press, 2000); and María Cruz de Carlos Varona, Pierre Civil, Felipe Pereda, and Cécile Vincent-Cassy, eds., *La imagen religiosa en la Monarquía hispánica: Usos y espacios* (Madrid: Casa de Velázquez, 2008).

[22] In addition to the works cited above, the field abounds with interdisciplinary efforts in the past ten years, including works by: Barbara Fuchs, Ricardo Padrón, Laura R. Bass, Felipe Pereda, Amanda J. Wunder, Guy Lazure, Charlene Villaseñor-Black, Daniela Bleichmar, Benjamin Schmidt, Juan Pimentel, Tanya J. Tiffany, and Jesús Escobar. See also the interdisciplinary thematic volume of the *Hispanic Review*, edited by Barbara Fuchs, *Re-Envisioning Early Modern Iberia: Visuality, Materiality, History*, issue of *Hispanic Review* 77(1) (Winter 2009).

of the Hispanic Baroque – a collection including scholars from multiple disciplines and which pairs thematic essays on Spain with corollaries on Spanish America – suggests the field's potential to reinvent itself in ways that transcend old divides and invest categories that once tended to further create Spain as an outlier in early modernity, like the Baroque, with new possibilities of meaning.[23]

Such work suggests another set of challenges on the horizon – that of re-interrogating the chronological boundaries that have customarily circumscribed the field. To borrow a recent acute observation by Robert Vinson: "Moving away from nation-state periodization is the next challenge in transnational history, a still developing field that features audacious geographical expansion but often remains saddled by conventional nation-bound time frames."[24] Scholars of the Hispanic world are increasingly exploring how the use of 1492 as a breaking point – a fundamental year in empire-building and in ending earlier patterns of religious toleration – has been overdrawn. A prime driver of this critique of periodization has resulted from studying the Mediterranean and Atlantic worlds, and, with them, topics such as the fifteenth-century Iberian struggles for control of the Canaries and the Iberian engagement with North Africa, places in which the events of 1492 can be read as part of much longer-range processes.[25] At the other end of early modern periodization, there is increasingly scholarly interest in the Hispanic Enlightenment and its place in larger eighteenth-century global trends.[26] There has long been a

[23] Levy and Mills, *Lexikon.*

[24] Robert Trent Vinson, "Review of Brenda Gayle Plummer, *In Search of Power: African Americans in the Era of Decolonization, 1956–1974*," *American Historical Review* 119(3) (June 2014): 828–30 (p. 828).

[25] Barbara Fuchs, "1492 and the Cleaving of Hispanism," *Medieval/Renaissance: After Periodization, Journal of Medieval and Early Modern Studies* 37(3) (Fall 2007): 493–510. The work of Teofilo F. Ruiz has often pushed against the medieval–early modern divide, as in *A King Travels: Festive Traditions in Late Medieval and Early Modern Spain* (Princeton, NJ: Princeton University Press, 2012). For the engagement between Spain and North Africa, see especially the many important works by Mercedes García-Arenal, including her most recent collaboration with Fernando Rodríguez Mediano, *The Orient in Spain: Converted Muslims, the Forged Lead Books of Granada, and the Rise of Orientalism*, trans. Consuelo López-Morillas (Leiden: Brill, 2013). Also the current research directions represented in "The Spain-North Africa Project" (http://snap.ucsc.edu/Home.html) and Yuen-Gen Liang, Abigail Krasner Balbale, Andrew Devereux, and Camilo Gómez-Rivas, eds., *Unity and Disunity across the Strait of Gibraltar*, special issue of *Medieval Encounters* 19(1–2) (2013).

[26] For a few works on the Spanish Enlightenment and the eighteenth century more broadly, see Jorge Cañizares-Esguerra, *How to Write the History of the New World: Histories, Epistemologies, and Identities in the Eighteenth-Century Atlantic World* (Stanford,

tendency to break studies of the early modern Hispanic world at 1700. While the year marked important divisions – the end of the Habsburg Spanish monarchy, the war that followed, and the advent of the Bourbon dynasty with its far-reaching administrative and political change – it can also mask processes that extended across a moment of only partial rupture. More studies are needed which increase interest in the still understudied later seventeenth century, and which look across the seventeenth and eighteenth centuries. Not so long ago, the first two decades of the seventeenth century – the era of Philip III – were dwarfed by scholarly consideration of the presumably more meaningful reigns of Philip II and Philip IV. Sophisticated studies of that era suggest what might be learned through further attention to other periods in the Hispanic past that have long been perceived as less meaningful.[27]

***

The fourteen essays which follow map some of the major ways that current scholarship is rethinking the boundaries of the early modern Hispanic past and how to analyze the extent of continuity and discontinuity among the places and populations which made up that world. We have organized the volume along four interconnected but distinct themes, highlighting how boundaries were drawn and redrawn and the fortunes of localities, individuals, and communities rose and fell through legal,

CA: Stanford University Press, 2001); Daniela Bleichmar, *Visible Empire: Botanical Expeditions and Visual Culture in the Hispanic Enlightenment* (Chicago, IL: University of Chicago Press, 2012); Catherine M. Jaffe and Elizabeth Franklin Lewis, eds., *Eve's Enlightenment: Women's Experience in Spain and Spanish America, 1726–1839* (Baton Rouge: Louisiana State University Press, 2009); Rebecca Haidt, *Embodying Enlightenment: Knowing the Body in Eighteenth-Century Spanish Literature and Culture* (New York: St. Martin's Press, 1998); and Regina Grafe, *Distant Tyranny: Markets, Power, and Backwardness in Spain, 1650–1800* (Princeton, NJ: Princeton University Press, 2012).

27 See for example Antonio Feros, *Kingship and Favoritism in the Spain of Philip III, 1598–1621* (Cambridge and New York: Cambridge University Press, 2000); Magdalena Sánchez, *The Empress, the Queen, and the Nun: Women and Power at the Court of Philip III of Spain* (Baltimore, MD: Johns Hopkins University Press, 1998); Francesco Benigno, *L'ombra del re: ministry e lotta politica nella Spagna del Seicento* (Venice: Marsilio, 1992); Laura Bass, *The Drama of the Portrait: Theater and Visual Culture in Early Modern Spain* (University Park, PA: Pennsylvania State University Press, 2008); Elizabeth Wright, *Pilgrimage to Patronage: Lope de Vega and the Court of Philip III, 1598–1621* (Lewisburg, PA: Bucknell University Press, 2001). One example of new work on the later seventeenth century is the work of Silvia Z. Mitchell, such as her "Habsburg Motherhood: The Power of Mariana of Austria, Mother and Regent for Carlos II of Spain," in *Early Modern Habsburg Women: Transnational Contexts, Cultural Conflicts, Dynastic Continuities*, eds. Anne J. Cruz and Maria Galli Stampino (Farnham and Burlington, VT: Ashgate, 2013), 175–94.

political, economic, and intellectual contests across the Hispanic world. The essays demonstrate the central role played by urbanism in Spanish imperial projects, while political authority – both in the peninsula and throughout the empire – was established and reinforced through religious festivals and the creation and recreation of fictive ideals of religious and ethnic purity. They further explore how ideas and modes of arguing, representation, and performance circulated throughout the empire and were remade to suit ever-evolving places and circumstances. The book is predominantly interested in questions of how people in the early modern Hispanic world understood the rapidly changing societies in which they lived – how they defined, visualized, and constructed communities from family and city to kingdom and empire.

In order to engage the current historiographical challenge to grapple with multiple scales of analysis, the essays in this volume deliberately take a variety of distinct approaches. Some are historiographical surveys proposing key changes in the field, while others are focused case studies that aim to illuminate wider possibilities. The chronological span of the volume is broad, moving between the late fifteenth and the late eighteenth centuries. Taken together, the essays both demonstrate the insights that continue to be yielded by specific disciplinary perspectives, by close engagement with particular times and places, and point to how reading such studies together illuminates new elements of each and points the way towards new visions of the field. We incorporate voices from across the Hispanic world in order to think more critically about the interdependence between center and periphery, Castile and the Americas. Thus the essays range from Madrid, seat of the monarchy, to some of its less studied arenas – Sardinia, the early Caribbean, the Río de la Plata region. The way early modern Hispanic people "located" themselves in their immediate environment and the larger empire reflects tensions at the heart of early modern life.

CITY AND SOCIETY

The first section addresses urban spaces and the manifold ways in which cities were a key force in shaping societies and political culture across the early modern Hispanic world. Within what has been characterized as an "empire of towns," malleable urban landscapes were a focal point of political, economic, social, and religious contests.[28] As such, the study of

[28] Kagan, *Urban Images*, 28.

the Spanish monarchy's cities continues to offer new ways to explore the inseparable knot of religion and politics in the early modern world, from the monastic institutions which were crucial engines of urban development to the multiple ways through which cities constructed and fortified their "spiritual walls," to use Richard Kagan's phrase.[29] The increasing body of urban histories likewise soon promises to make new kinds of comparative work possible.[30] Responding to these trends, the three essays in this section bring social history together with newer methods of understanding the ways in which the urban environment shaped, undermined, or reinvented early modern societies. Space and place take center stage as the essays in this section examine social networks, mobility, gender, family, and religious identity. Ida Altman (Chapter 1) examines towns in the Spanish Caribbean and their crucial role in the evolution of the early transatlantic empire. Her analysis calls attention to some of the first town foundations in the newly Spanish Caribbean – urban centers that were integral to early efforts at colonization – and with it to the significance of smaller urban units and to the varied fortunes and frequent impermanence of newly founded places.

The next two chapters further address the role that the urban environment had in constructing identity and community. James Amelang (Chapter 2) explores a literary genre that mirrored early modern artistic interest in capturing the "view" of European cities on the ground – that is, early modern urban descriptions that took the reader on a written walking tour through the urban space. His analysis calls attention to ways of narrativizing the experience of Iberian cities and demonstrates that forms of urban writing utilized in other parts of the early modern world – and that have been most studied in the case of early modern England – had corollaries across the Iberian empires, from the sixteenth century down through the eighteenth. It also serves as a vivid reminder of the variety of early modern people who wrote about their environments and of the texts of varying forms still to be uncovered and analyzed. Erin

[29] Karen Melvin, *Building Colonial Cities of God: Mendicant Orders and Urban Culture in New Spain* (Palo Alto, CA: Stanford University Press, 2012); Richard L. Kagan, "A World Without Walls: City and Town in Colonial Spanish America," in *City Walls: The Urban Enceinte in Global Perspective*, ed. James Tracy (Cambridge and New York: Cambridge University Press, 2000), 117–52.

[30] The classical example is Bartolomé Bennassar, *Valladolid au siècle d'or, une ville de Castille et sa campagne au XVIe siècle* (Paris: La Haye, Mouton et Cie, 1967). Recent additions include Stefano D'Amico, *Spanish Milan: A City Within the Empire, 1535–1706* (New York: Palgrave Macmillan, 2012); Alejandra Osorio, *Inventing Lima: Baroque Modernity in Peru's South Sea Metropolis* (New York: Palgrave Macmillan, 2008).

Rowe (Chapter 3), in turn, examines the movement of religious processions through Madrid in the 1620s, paying close attention to the way space was mobilized to sacralize the monarchy's capital. She argues for the inseparableness of the religious and political elements of royal ritual in the city by following a religious procession along the streets of Madrid. Despite their diverse points of focus and chronological range, each of these three essays demonstrates how analyzing little studied source materials – records of the early Caribbean, a literary genre of urban walks, printed accounts of processions – can illuminate new aspects to the complex and various processes that made up the always-shifting social fabric of the city and the different ways that members of urban communities understood and redefined themselves – socially, spatially, and religiously.

RELIGION, RACE, AND COMMUNITY

As a result of the particular religious diversity of medieval Iberia and the polyglot and multiethnic nature of the early modern Iberian empires, the study of the Hispanic past has become a focal point for efforts to reassess the emergence of modern notions of, and attitudes towards, ethnicity, race, and religious identity; it has provided a set of materials through which to rethink citizenship, belonging, and the ongoing process of narrating collective experience.[31] The second section continues the volume's attention to circulation, while turning the focus to questions of identity: how people defined themselves and others as individuals and in terms of larger nested identities.[32] Examining the two together helps throw light on how people clustered into communities and how they attempted to fashion or

[31] See note 8 above for some examples. Also, Mercedes García-Arenal, "Religious Dissent and Minorities: The Morisco Age," *Journal of Modern History* 81(4) (2009): 888–920; María Elena Martínez, *Genealogical Fictions: Limpieza de Sangre, Religion, and Gender in Colonial Mexico* (Palo Alto, CA: Stanford University Press, 2008). A host of recent symposia and grant projects are also spurring large-scale rethinking of the religious history of the early modern Spanish world. Mercedes García-Arenal is, for example, overseeing an EU-funded grant that aims to remake our understanding of religious dissidence, minorities, and mass conversion in early modern Spain. See the project description for "Conversion, Overlapping Religiosities, Polemics, Interaction: Early Modern Iberia and Beyond," http://proyectos.cchs.csic.es/corpi/. See also the Canadian-funded initiative, "Early Modern Cities as Theatres of Conversion," http://earlymodernconversions.com/research-groups/early-modern-cities-as-theaters-of-conversion/.

[32] We use the term identity as a useful way to think about a variety of dynamic historical processes, well advised by the seminal critique in Frederick Cooper, "Identity," with Rogers Brubaker, in *Colonialism in Question: Theory, Knowledge, History* (Berkeley, CA and Los Angeles: University of California Press, 2005), 59–90.

re-fashion themselves to fit into dominant paradigms. Thus, the essays in this section focus on the multifaceted ways social and religious categories were made and remade in the early modern Hispanic world. They do so through consideration of a wide variety of places across the Spanish colonial world and the ways in which boundaries were drawn and redrawn over time in those far-flung localities. Sara Nalle's essay (Chapter 4) sheds new light on a key problem in the historiography of early modern Spain. Offering an innovative example of how to use Spanish Inquisition records, she demonstrates how *judeoconverso* communities evolved in the several decades after the expulsion of 1492, and successfully rebuts approaches that have treated that group as monolithic. She reflects on the emergence of what she refers to as a "minority within a minority" – new groups of new converts from Judaism in early modern Castile and Aragon following the 1492 expulsion. Her close study of one region's families and their preservation of group boundaries across several decades suggests the need for greater sensitivity to the variegation among *converso* populations as well as to the subtleties of how meaningful terms of self-description could change over time.

Next, Mercedes García-Arenal and Felipe Pereda (Chapter 5) provide a transformative look at one of the most vexing intellectual-spiritual movements of sixteenth-century Spain, *alumbradismo* (Illuminism). They reassess a key issue in early modern historiography: how to categorize and understand the religious dissidents known as *alumbrados*. In so doing, they explore that ongoing tension within the study of the Hispanic world of exceptionalism, that tension between the distinctiveness of Spain's multireligious medieval history and its consonance with the religious history of other parts of Europe in the Age of Reformations. The following chapter turns to look at the frontier – ideological, political, and military – with Islam and how a variety of individuals navigated that frontier in the eastern Iberian region of Valencia. Benjamin Ehlers (Chapter 6) highlights the disruption and anxiety created by the ongoing confrontation with Islam on the Mediterranean. Yet his study also reads beyond that rhetoric to capture a range of forms of engagement with Muslim and Morisco populations across the second half of the sixteenth century; it reinforces the importance of locating the Spanish monarchy – the concerns of both the crown and regional nobility – within a dynamic Mediterranean framework. Finally, this section pushes the chronological and geographical boundaries of the volume by considering the ways in which late eighteenth-century colonists around Buenos Aires adapted to new social roles as owners of slaves. Allyson Poska (Chapter 7) focuses on the lower orders of society and explores the dramatic shifts that occurred in status and identity when Spanish peasants crossed the Atlantic, and made efforts to reinvent themselves as landowners and slave owners. She signals the importance of complex engagement with the study of slavery and its intersection with other ways of thinking about the history of labor and transatlantic mobility. Together, the essays in this section draw particular attention to religious minorities, race, and strategic claims for inclusion and exclusion. They continue the emphases of the first section on examining place and mobility, the politics of religion, and drawing upon less studied historical sources. And they delve more deeply into how Spanish identities evolved during a period when rapid territorial expansion and nation-building began to slip into financial crises and violent conflicts. The volatility of the early modern period gave rise to tremendous creativity as various communities attempted to reinvent themselves and the Hispanic world itself.

LAW AND LETTERS

The third section shifts the primary focus to early modern Spanish intellectual culture and how it can be written into broader considerations of European intellectual history.[33] In doing so, it is tied to the previous section in contributing to the deep rethinking of early modern Catholicism that is underway in current scholarship.[34] The essays indicate an effort

[33] Early modern intellectual history is another sub-field which is becoming increasingly dynamic, yet one into which the Spanish world could still be more strongly integrated, as for example into such magisterial arguments as John Marshall, *John Locke, Toleration, and Early Enlightenment Culture: Religious Intolerance and Arguments for Religious Toleration in Early Modern and 'Early Enlightenment' Europe* (Cambridge: Cambridge University Press, 2006). There are indications of how to better integrate Spanish intellectuals into this scholarship in such important recent work as Stefania Tutino, *Shadows of Doubt: Language and Truth in Post-Reformation Catholic Culture* (Oxford: Oxford University Press, 2014); Susan Schreiner, *Are You Alone Wise? The Search for Certainty in the Early Modern Era* (Oxford: Oxford University Press, 2011), in which the sixteenth-century *alumbrados* appear; or Katherine Elliott Van Liere, Simon Ditchfield, and Howard Louthan, eds., *Sacred History: Uses of the Christian Past in the Renaissance World* (Oxford: Oxford University Press, 2012).

[34] Circulation and persistent attention to space and place as frameworks for analysis are central to the reevaluation of early modern Catholicisms in the essays in Bamji, Janssen, and Laven, eds., *The Ashgate Research Companion to the Counter-Reformation*. See also the insightful survey in Lu Ann Homza, "The Merits of Disruption and Tumult: New Scholarship on Religion and Spirituality in Spain during the Sixteenth Century," *Archiv für Reformationsgeschichte/Archive for Reformation History* 100 (2009): 212–28. Scholars are also fruitfully re-reading theological texts alongside varied other kinds

to more thoroughly investigate the connections between the varied kinds of writing and reasoning that circulated in the early modern Hispanic world and how those changed over time. To that end, this section turns towards the intellectual world of the *letrado*, university-educated and legally trained men who played instrumental roles in the circulation of ideas and texts that shaped Hispanic cultural life.

Complementing the historiographic reflection of the second section's chapter on the *alumbrados* and religious dissidence, this section begins with another chapter that rethinks a key area of the historiography of the early modern Spanish world. María Portuondo (Chapter 8) contests the model of the decline of Hispanic science in this period, and reassesses how we study the history of science in early modern Spain and what that has to say to the prevailing master narratives in that field, and to any understanding of the crucial intellectual movements of the era. She proposes a new paradigm that replaces a focus on decline with one on variegated responses to a "crisis of knowledge." The next two essays look at applied case studies of intellectual culture in the later sixteenth and earlier seventeenth centuries. Kimberly Lynn (Chapter 9) returns to another persisting theme of the scholarly field: the Spanish Inquisition. She looks at the problems and opportunities posed by print to one of the Spanish monarchy's most influential bodies – the Holy Office of the Inquisition – and reconsiders the attitudes of Spanish inquisitors towards the written word. The essay calls attention to how early modern Spaniards weighed and deployed the varied forms of communication at their disposal as well as to the climate of contestation that surrounded inquisitorial action.

Katie Harris (Chapter 10) turns to an especially fruitful area of inquiry in recent years: the study of early modern historiography. Her chapter rethinks some of the core epistemological issues of sacred history, one of the most important and ideologically charged genres of writing in sixteenth- and seventeenth-century Europe. Adding to the insights of Portuondo's chapter, Harris explores the application of empirical observation and scientific experimentation to the determination of the authenticity of relics. Her study, along with Lynn's, points to the important roles played by Rome in the Spanish world, while the conflict it charts also relates to the questions of civic and regional authority raised in the book's first section. Finally, Xavier Gil (Chapter 11) offers a way to rethink the political culture of the

of writings: Jessica Boon, *The Mystical Science of the Soul: Medieval Cognition in Bernardino de Laredo's Recollection Method* (Toronto: University of Toronto Press, 2012); Hilaire Kallendorf, *Sins of the Fathers: Moral Economies in Early Modern Spain* (Toronto: University of Toronto Press, 2013).

Hispanic world at the turn of the seventeenth century in a reflection on the widespread influence of the key Italian political theorist Giovanni Botero, and in particular Botero's stress on demographic expansion and human capital. The essay forms a fitting end to this section as it chronicles vibrant critiques of legal thinking and its utility in governance that circulated among early modern intellectuals, and as it considers the weight accorded to experiential knowledge, the empirical, and the utilitarian.

Together, this section's essays call attention to the importance of law and legal reasoning to early modern Spanish intellectual culture and their relationship to other modes of thought; they propose fresh ways to think about networks of intellectuals and the circulation of writings. They consider the mobility of ideas and the limits on that mobility. And they highlight the politics of communication and to the myriad challenges of managing information in the early modern era. Taken with the essays in the other sections, they point to innovative ways to break down traditional barriers between areas of study by encouraging scholars to look at the breathtaking range of lettered culture in the Hispanic world. In various ways, Gil's and Portuondo's essays, like Amelang's in the first section and García-Arenal's and Pereda's in the second, draw comparisons to other parts of Europe, suggesting how the place of early modern Spain in a multiplicity of prevailing historical narratives is ripe for revision. And with Sardinia taking center stage in Chapter 10, the section continues the book's emphasis on drawing case studies from across the vast geography of the Spanish monarchy, encouraging scholars to engage more deeply with Spain's Mediterranean holdings.

PERFORMANCE AND PLACE

The fourth and final section of the book amplifies the interdisciplinary approach of the volume. It draws attention to the importance of material and visual culture as among the most dynamic areas of current research.[35] Here we see three experts address visual, literary, and

[35] For example, Fuchs, ed., *Re-Envisioning Early Modern Iberia*. A suggestion of how early modern Hispanic material and visual culture of varying kinds is being reconsidered could also be gleaned from such varied works as Felipe Pereda, *Las imágenes de la discordia. Política y poética de la imagen sagrada en la España del 400* (Madrid: Marcial Pons, 2007); Xavier Bray, ed., *The Sacred Made Real: Spanish Painting and Sculpture, 1600–1700* (London: National Gallery, 2009); Byron Ellsworth Hamann, "Interventions: The Mirrors of *Las Meninas*: Cochineal, Silver, and Clay" (and the five Responses), *The Art Bulletin* 92(1–2) (2010): 6–60; Noelia Cirnigliaro and John Beusterien, eds., *Touching the Ground: Female Footwear in the Early Modern Hispanic World*, special issue of *Journal of Spanish Cultural Studies* 14(2) (June 2013).

performative aspects of the early modern Hispanic world in a variety of innovative ways, challenging us to push forward in new methodologies. They offer another vantage point from which to view important trends in the field and suggest the possibilities to be found in revisiting figures and institutions of towering importance with new questions and turning to images and written texts that have often been ignored.

Eminent art historian Fernando Marías (Chapter 12) reconsiders the role of images in the mid-sixteenth-century Inquisition trial of an artisan, providing a means to broaden our understanding of religious dissidence and calling attention to visual materials that were once at the margins of art historical study. He examines the role of the Inquisition in censoring printed engravings they deemed "Lutheran." His essay calls for further conversation between art history and other kinds of social and religious historical inquiry; it adds to other chapters in the volume in showing how inquisitorial archives can still be used to produce new insights. The section then moves to another cultural product that had previously received little sustained study – a drama that belies easy generic classification by the most prolific playwright of the Spanish Golden Age. Elizabeth Wright (Chapter 13) explores the relationship between poetry, performance, and politics surrounding the memory of a key naval victory against the Muslim Ottoman Turks, and so makes a work long treated dismissively a site of productive cultural historical analysis. She explores literary performance of place and event through an analysis of Lope de Vega's turn of the seventeenth-century depiction of the 1571 Battle of Lepanto, fought in the eastern Mediterranean. Thus it complements the concerns of various other chapters of the volume – whether about the complexities of the frontier with the Islamic world or about how to productively read genres of texts (and images) once largely ignored or unknown.

The section then concludes with another piece that delves into the understudied eighteenth century and brings an important component of gender studies to the volume. Marta Vicente (Chapter 14) explores how femininity was expressed on stage, how the theater was regulated, the social position of actresses, and how controversies around these issues can shed light on broader Enlightenment-era debates about morality and the family. She examines women's appearance and the audience gaze by investigating eighteenth-century actresses known as *tonadilleras*, whose enormous popularity stemmed from their elaborate and sexual dress. With Poska's essay earlier in the volume – likewise focused on eighteenth-century material – it highlights women's work and its possible social meanings.

Together, the section re-thinks the ways that sensitive examination of the visual and performative – image, gaze, and representation of place – can provide previously unmined approaches to gaining a deeper understanding of early modern conceptions of community and self, city and empire. In many ways, the fourth section brings the volume full circle, returning to examine the uses of urban space, questions of performance (in trials, on the theater stage, or on the field of epic poetry), and representation. It again draws attention to relatively little studied genres and offers other suggestive avenues by which to produce new insights about the most prominent institutions of the early modern Spanish world (the Inquisition, the monarchy, Madrid as capital).

* * *

The distance traveled in Spanish historiography over the past two decades has been the result of a complex mixture of forces – larger historiographical trends, a new, more urgent interest in understanding the history of Islam, and a push to explore interdisciplinary connections. Yet it also owes a significant debt to the scholars who have been deeply engaged with Spanish historiography throughout this period. The contributors of this volume are a mixture of up-and-coming scholars just beginning to play their role in the continuing evolution of their field, and established scholars who have made and continue to make substantial contributions to it. We continue to be indebted to and inspired by each other, and above all, by the dynamic and indefatigable career of the polymath Richard L. Kagan, to whom this volume is dedicated.

PART I

CITY AND SOCIETY

I

Towns and the Forging of the Spanish Caribbean

Ida Altman

The establishment of towns was fundamental to the Spanish imperial enterprise in the Caribbean as it would be throughout early Spanish America.[1] As was true elsewhere, some early town sites in the large islands of Española, Puerto Rico, and Cuba proved unworkable or even disastrous and were vacated. Yet creating municipalities remained crucial, as Spaniards equated effective control over territories and peoples with the founding of towns and establishment of civilized life. Towns were vital to the extension of Spanish settlement, control over indigenous labor, the development of mining, ranching, and sugar production, and expansion into adjacent territories. They became principal centers for the formation of Spanish American society in the islands, sites for contact among Europeans, Africans, and indigenous inhabitants, and keys to maintaining contact with the larger Iberian world.

The departure from the islands of many Spanish residents in the 1520s and 1530s in pursuit of opportunities elsewhere and the drastic reduction in indigenous population meant a sharp contraction in the size, or even the disappearance, of some early towns. The main port cities – Santo Domingo and San Juan – along with some secondary ports retained their importance, while other places intended primarily

[1] For a comparative overview, see Richard L. Kagan, "People and Places in the Americas," in *The Oxford Handbook of the Atlantic World, 1450–1850*, ed. Nicholas Canny and Philip Morgan (New York: Oxford University Press, 2011), 341–61. For a discussion of urban networks in Spain and Spanish America, see Ida Altman, "Reconsidering the Center. Puebla and Mexico City, 1550–1650," in *Negotiated Empires. Centers and Peripheries in the Americas, 1500–1820*, eds. Christine Daniels and Michael V. Kennedy (New York: Routledge, 2002), especially 43–6.

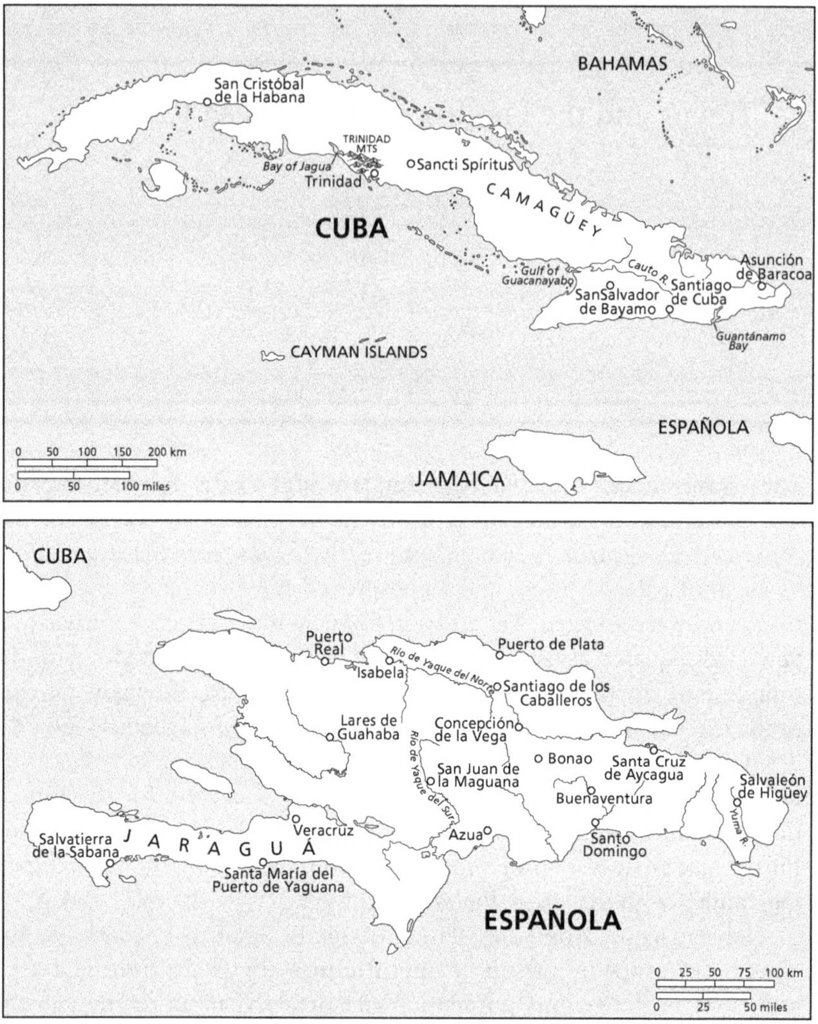

Map 1.1. Early Spanish towns in the Caribbean

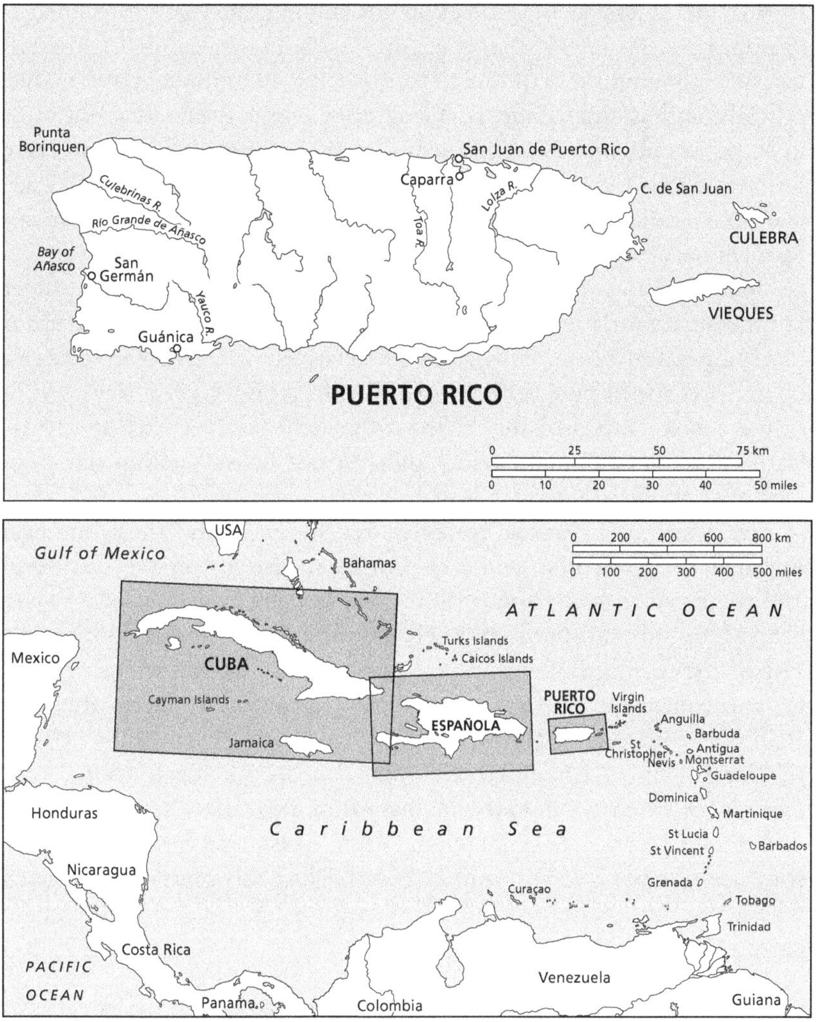

Map 1.1 (continued)

to provide access to or protection for mining enterprises declined or disappeared.[2] As late as the 1520s and 1530s, however, plans to establish new communities continued to draw the attention of both Spanish officials and island residents. Geography – relatively long distances in large islands like Española and Cuba, the difficulties of traversing heavily wooded or mountainous terrain, and the advantages of water transport – and the need for settlements of even modest size to anchor Spanish occupation, together with the early emergence of relatively stable local societies that were increasingly mixed in composition, underlay the survival of many smaller towns. The early towns and cities of the Spanish Caribbean have attracted relatively little scholarly interest, owing perhaps in part to their volatility. In a larger sense their neglect in the scholarship probably reflects a general lack of attention to the islands and the assumption that they do not fit well within the mainstream of Spanish American history.[3]

The island of Española, today shared by Haiti and the Dominican Republic, was the first center of European activity in the Caribbean and remained the only site of European settlement for close to twenty years. Not only did the island become the base for expansion into the rest of the region, but specific towns served as the launching grounds for subsequent moves, such as the 1509 expedition to Puerto Rico that Juan Ponce de León led from Salvaleón de Higüey, founded in 1503 on the eastern tip of Española, where he had been living for some time. During the extensive military campaigns of 1503–04 after frey Nicolás de Ovando became royal governor, Diego Velásquez's forces went to the southwestern part of the island and established Salvatierra de la Sabana, where Velásquez recruited some 300 men for his expedition to Cuba in early 1511.

[2] In the first half of the sixteenth century Havana was not a significant port. The capital of early Cuba was Santiago de Cuba in the southeastern part of the island.

[3] For discussion of this point, see Ida Altman, "The Revolt of Enriquillo and the Historiography of Spanish America," *The Americas* 63(4) (2007): 587–614. Work that has addressed aspects of urban life in the early islands includes Genaro Rodríguez Morel, *Cartas del Cabildo de la ciudad de Santo Domingo en el Siglo XVI* (Santo Domingo: Centro de Altos Estudios Humanísticos y del Idioma Español, 1999); Enrique Otte, *Las perlas del Caribe. Nueva Cádiz de Cubagua* (Caracas: Fundación John Boulton, 1977); Esteban Mira Caballos, "Los cabildos antillanos: aspectos institucionales (1492–1542)" and "Urbanismo y arquitectura en los primeros asentamientos antillanas (1492–1550)," in *La Española, epicentro del Caribe en el siglo XVI* (Santo Domingo: Academia Dominicana de la Historia, 2010); and Alejandro de la Fuente, *Havana and the Atlantic in the Sixteenth Century* (Chapel Hill, NC: University of North Carolina Press, 2008), although the focus is on the later sixteenth century.

FOUNDING TOWNS

The first European effort to establish a base on the northern coast of Española, La Navidad, ended in disaster. During the course of Columbus' first visit to the island, his ship the *Santa María* went aground on Christmas eve 1492 near the village headed by a friendly indigenous chief or *cacique*, Guacanagarí. With his help, Europeans salvaged what they could from the wreck and built a small blockhouse, using planks from the deck and hull of the ship. Thirty-eight men remained there when the rest of the newcomers departed. When Columbus returned to the island in 1493 he found La Navidad destroyed and all the men dead.[4]

The decision to create an outpost at La Navidad was dictated by the happenstance of the shipwreck, although the benefits of locating the colony close to the village of the friendly cacique Guacanagarí very likely played a role in his decision as well.[5] While subsequently throughout the Americas Spaniards demonstrated a strong preference for establishing their towns and cities close to or even at the very sites of existing indigenous communities (including very large urban places, such as Tenochtitlan in central Mexico or Cuzco in highland Peru), Columbus' second attempt at town-founding took place on Española's northern coast at the mouth of the Río Bahabonico in a fairly sparsely populated area. Although many modern scholars have emphasized the disadvantages of La Isabela's site, contemporary observers praised the fertility of the land, proximity to rivers, availability of fish, and access to good building materials.[6]

In January 1494 Columbus initiated construction of a walled town following the natural contours of the terrain. It included a fortified tower, storehouse, church, and house for himself, all built of stone with

[4] See Samuel M. Wilson, *Hispaniola. Caribbean Chiefdoms in the Age of Columbus* (Tuscaloosa, AL: University of Alabama Press, 1990), 68–71, 75. For information on what many consider to be the probable site of a large indigenous community located close to where La Navidad was built at En Bas Saline in Haiti, see www.flmnh.ufl.edu/histarch/ebs_archaeology.htm.

[5] Carl O. Sauer, *The Early Spanish Main* (Los Angeles and Berkeley, CA: University of California Press, 1966), 33.

[6] See Kathleen Deagan and José María Cruxent, *Columbus's Outpost among the Taínos* (New Haven, CT: Yale University Press, 2002), 50–2. They argue that the discovery of a "second, satellite community in the river drainage area" and existence of a "stone quarry located between the two settlements [that] provided building materials and a source of lime ... considerably alters the traditional view that Columbus made a misguided and poorly informed choice of site that led ultimately to the colony's demise" (98–9). See Sauer, *Early Spanish Main*, 75, for a negative view.

the exception of the earthen walls, and featured an open plaza. Housing for most residents was modest, utilizing materials favored by the local inhabitants, wood and thatch. The problems that led to La Isabela's abandonment were as much human as natural. The colony suffered from internal social and political conflict as well as illness and food shortages. Columbus' harsh rule and imposition of a strict work regime to which he subjected virtually all the colonists in hopes of completing construction as quickly as possible fostered deep resentment, especially in face of illness and privation. Many gave up and returned to Spain. Supplies arrived sporadically, hurricanes struck the bay and destroyed ships, efforts to cultivate crops met with little success in the first couple of years, and most of the local indigenous inhabitants fled or died. Most important, perhaps, La Isabela was too distant from sources of gold and major centers of indigenous population to be ideal for the purposes of overseeing and protecting gold mines and organizing native labor. By 1496, under the leadership of Columbus' brother Bartolomé, the settlement was in the process of being relocated to Santo Domingo on the southern coast of the island. By 1498 La Isabela lay empty, never to be reoccupied.[7]

More typically early urban sites shifted rather than being abandoned altogether. Following the fiery destruction in 1513 of the first Spanish settlement in Puerto Rico, Caparra, the new capital (then called Puerto Rico and later San Juan) took shape about ten miles away on the coast and subsequently moved again to its present site.[8] After a major hurricane and a plague of ants, in 1502 Santo Domingo moved from the left to the right side of the Ozama River under the orders of governor Ovando. Over the next six or seven years an intense building campaign resulted in the construction in stone of the city's fortress, houses for Ovando and several other leading citizens, and the hospital of San Nicolás. Work on a House of Trade (*Casa de Contratación*) also commenced.

By the end of Ovando's term as governor and the arrival in the summer of 1509 of Columbus' son, Don Diego Colón, as second governor with his high-born wife, Doña María de Toledo, a number of other towns had taken shape in Española. Originating as a fort in the Vega valley, by 1510 Concepción de la Vega (today La Vega) became a leading city and seat of one of three bishoprics in the islands (Santo Domingo and San Juan in Puerto Rico were the other two). Other early settlements had mixed experiences, but many survived, as did a number of Cuba's early

7 Deagan and Cruxent, *Columbus's Outpost*, 54–70.

8 In the early years the island was known as San Juan and its capital as Puerto Rico.

towns.[9] Probably because of its smaller size and conflict in the southern part of the island, there were only two towns in early Puerto Rico, the capital city of San Juan and San Germán, founded in 1512 on the island's western end.

BUILDING TOWNS

Shifting the site of Santo Domingo from one side of the river to the other gave rise to a bitter legal battle between governor Ovando and Cristóbal de Tapia, who for a time served as the warden (*alcaide*) of the fortress.[10] The testimony of witnesses in the lawsuit offers insights into early urban life and construction in the Spanish Caribbean. At the heart of the dispute were two properties that Tapia had been granted, one of agricultural land producing yucca and *ajes* (sweet potatoes) that the governor appropriated when the city was moved, the second a *solar* (urban lot) close to where Ovando was having the *Casa de Contratación* built. At the latter site, located near a fountain, Tapia owned a building or buildings made of wood and thatch that witnesses referred to as a *bohío* (or bohíos) housing several shops that yielded a monthly rental income. When depositions for the case were taken in 1509, according to a couple of witnesses, these structures had collapsed during a hurricane about a year before.[11] The men in charge of building the Casa de Contratación claimed that even though Tapia's lot was on the other side of a narrow street from the construction site, they needed it as a staging ground and storage area for materials and equipment. Perhaps most telling was a comment from the city's *alcalde ordinario* (magistrate), Gerónimo de Mendoza, who testified that because Tapia had rented his shops to tavernkeepers and they were located close to the construction site, there were concerns that "the blacks and other people" who were supposed to be working there instead would while away their time in one of the taverns.[12]

Twenty-six-year-old Francisco de la Barrera stated that he had been Tapia's *criado* or *mayordomo* (retainer, steward) for more than two

[9] Emilio Rodríguez Demorizi, *El pleito Ovando-Tapia. Comienzos de la vida urbana en América* (Santo Domingo: Editora del Caribe, 1978), 108–09, lists Española's towns with the dates of their establishment.

[10] Rodríguez Demorizi published the 1509 depositions, with a lengthy introduction, in *El pleito*. On the ants, see p. 51.

[11] The term *bohío* was adopted from the Taino language. Spaniards mainly used it for modest indigenous-style structures.

[12] For the quote from Mendoza, see Rodríguez Demorizi, *El pleito*, 256–7; see also 266.

years. He was in charge of renting out the *tiendas* (shops) for four pesos per month each. When the buildings collapsed in the hurricane he had cleaned up the site in order to rebuild. Another witness stated that some people actually lived in the buildings. Juan Martínez de la Palma said that he had been renting one of the shops at the time of the hurricane, while another was rented to a carpenter, a third to tavernkeeper, and the fourth to a woman known as La Cordobesa. Martínez had rebuilt his hut with Tapia's approval, but a man named Ramos who was in charge of royal construction projects demanded that all work should cease. Several days later, having left his *bohío* in the hands of a friend, Martínez returned to the site to find that it had been knocked down and the *solar* surrounded by an enclosure to prevent unauthorized entry.[13]

Given Santo Domingo's importance and its concentration of royal officials and wealthy *vecinos*, it is not surprising that a campaign promoting construction in stone, brick, and tile over other materials yielded some success. Structures of wood and thatch were vulnerable to fire as well as tropical storms, although packed earth (*tapia*) survived those hazards far more successfully. Stone houses also were markers of social distinction, wealth, and permanence, and desirable as such. By Ovando's time, carpenters, masons, and other specialists in construction probably were fairly numerous in the city, but some years passed before locally produced bricks, roofing tile, or lime were available in any quantity. In the early years the islands imported bricks – not an ideal transatlantic cargo – as well as specialists in construction, who might return to Seville when their contracts were up.[14]

A 1528 report to the crown described Concepción as having "a fort and stone church and a Franciscan monastery of stone and a city hall and another 25 or 30 very good stone buildings."[15] Elsewhere progress towards converting to stone construction was slower. The 1530 report prepared by Francisco Manuel de Lando, lieutenant governor of Puerto

[13] Rodríguez Demorizi, *El pleito*, 203–05, 210.

[14] On importing bricks from Spain, see *ibid.*, 100. See testimony that Juan Ponce de León's house and the fort in Caparra were built of imported bricks in Aurelio Tió, *Nuevas Fuentes para la historia de Puerto Rico* (San Germán: Ediciones de la Universidad Interamericana de Puerto Rico, 1961), 282, 287, 288 (petition of Juan Troche Ponce de León, 1564). On the need for workers in early Puerto Rico, see Ricardo E. Alegría, ed., *Documentos Históricos de Puerto Rico. Vol. 1: 1493–1516* (San Juan: Centro de Estudios Avanzados de Puerto Rico y el Caribe, 2009), 248, 266, 402.

[15] Report from *oidores* Espinosa and Zuazo, in Roberto Marte, comp., *Santo Domingo en los manuscritos de Juan Bautista Muñoz* (Santo Domingo: Ediciones Fundación García Arévalo, 1981), 277–8.

Rico, noted that San Juan had a only few stone buildings, including those for the royal treasurer and accountant, the smelting house, and the house of the bishop. Only two other vecinos had stone houses, while another twenty were constructed of wood and tile and twenty of thatch. There was a church on which construction in stone had begun and a hospital "half fallen down because of a storm and in bad condition." In addition there were thirty-five shops of wood and thatch or tile. San Germán entirely lacked stone buildings. It had only two wooden houses and a church built of packed earth with a tile roof.[16] In Santiago de Cuba, the island's first capital, fire destroyed many houses. Despite royal orders to build in stone, however, not only did the city lack master masons and carpenters to undertake such work, some *vecinos* were convinced that stone houses were unhealthily humid, preventing the entrance of sun and breezes. Some argued that these houses were responsible for illnesses and even deaths, as in the case of governor Gonzalo de Guzmán and his wife.[17]

TOWN LIFE

Early urban centers were diverse milieus where indigenous workers and African and Indian slaves from all over the region not only worked on construction projects and performed virtually all other menial labor but formed an essential part of Spanish-headed households as well. Racial or ethnic diversity was not the only source of socioeconomic variability, however. Some witnesses for the Tapia–Ovando lawsuit made explicit distinctions between the powerful and well-to-do who held grants of indigenous labor (*repartimientos*), servants, and slaves, and whose profits from gold mining, commercial agriculture, and trade allowed them to live in some luxury, and poorer members of Spanish society who had little access to the benefits to be derived from access to indigenous labor and political connections.[18] Ironically, Cristóbal de Tapia, who eventually became embittered towards Ovando, at one time enjoyed close ties with

[16] Archivo General de Indias (hereafter AGI) Justicia 106, no. 3.

[17] Leví Marrero, *Cuba: Economía y sociedad, vol. I: Antecedentes. Siglo XVI: la presencia europea* (Río Piedras: Editorial San Juan, 1972), 372–3.

[18] Pero Gallego testified that "it's well known that all the magistrates and officials and [their?] criados are rich, and for the most part of those who are not, married men as well as other vecinos, are lost"; in Rodríguez Demorizi, *El pleito*, 195. Rodríguez Demorizi, 30–2, discusses the implications of the split between wealthy and poor for Cristóbal de Tapia's case.

the governor, which no doubt contributed to his accumulation of wealth. Ovando dined with Tapia and his wife in their home, apparently a rare sign of favor, and was godfather to Tapia's children.[19]

With the departure of many Spaniards for destinations on the mainland following the success of Hernando Cortés' campaign in central Mexico (completed 1521) and Francisco Pizarro's in Peru (1532), many of the early towns became far more modest places than originally projected. The 1528 report from the judges of the *audiencia* included an account of Española's struggling towns. Concepción "used to have ... 200 vecinos of whom 100 were horsemen, hidalgos and principal people, [but] now there are no more than 20 vecinos ... old and unmarried and without children." Of Santiago's former 100 *vecinos*, also "principal people," there remained only eight who were living mainly on their estates "and not in the manner of a town." Puerto Real, which had been joined to Lares de Guahava and also at one time had 100 *vecinos*, had no more than fifteen *vecinos*; and Salvaleón de Higüey, many of whose 100 or more *vecinos* were married to women from Castile, also was down to fifteen *vecinos*. The litany of decline continued, Azua, San Juan de la Maguana, Santa María del Puerto, Salvatierra de la Sabana all having lost significant numbers of *vecinos*.[20] The observation that many of Santiago's *vecinos* actually were living outside of the town proper probably applied to many other towns, especially given that the principal economic activities of the period – gold mining and agriculture or cattle raising – were based in the countryside. Francisco Manuel de Lando's 1530 description and census of Puerto Rico's two towns reported that there were fifty or sixty married men living in houses in San Juan, while another twenty or thirty either lived in other people's households or in the countryside "because of their poverty."[21]

Of particular interest in terms of what happened to these towns as Spaniards departed in search of other opportunities, indigenous populations contracted, and the gold economy faltered is the testimony of *vecinos* of the closely connected towns of Trinidad and Sancti Spíritus in Cuba in 1533–4.[22] The testimony was elicited during an inquiry by lieutenant governor Manuel de Rojas into the viability of the two towns,

[19] Rodríguez Demorizi, *El pleito*, 252, 270.

[20] Marte, *Santo Domingo*, 277–8.

[21] AGI Justicia 106, no. 3.

[22] AGI Patronato 177, no. 1, ramo 18. On the founding of the towns, see Oswaldo Morales Patiño, "Los indígenas en los primeros municipios cubanos," *Revista de arqueología y etnología* (Havana) VII(13–14) (1951): 379–80.

which officials proposed merging into one. Most of the *vecinos* who testified had been living there for years, some since the time of towns' founding in 1514, and they all reported their steadily worsening economic condition. Many had overseen substantial mining and agricultural operations that at one time employed many Spaniards as well as Indian and African workers and slaves. By the 1530s most of these enterprises barely functioned and the former *encomiendas* had few if any remaining Indians.[23]

A plan to merge the two towns had been proposed five years before, as a result of which some *vecinos* of Trinidad had gone to live in Sancti Spíritus, taking with them the Indians they held in *encomienda*, an act resented by those who stayed behind since they saw it as further diminishing the already dwindling labor resources of their town.[24] Some of those who relocated had returned to Trinidad, however, and at the time of Rojas' visit the issue of merging the towns at one locale remained unresolved. *Vecinos* of Trinidad argued its superiority over its sister and rival town, pointing out that it possessed a port where ships in trouble often stopped for repairs. Juan de Espinosa acknowledged that at one time Trinidad had forty-two "vecinos ricos y prósperos," while two other *vecinos* claimed that formerly there had been as many as seventy; but all agreed that their number had shrunk to somewhere between ten and twelve. Notwithstanding this contraction, Espinosa and others contended that because of its port, fertile land, nearby mines, and generally healthier situation the site of Trinidad was superior. The wealthy and powerful Vasco Porcallo de Figueroa, unquestionably the most prominent man in the region and a *vecino* of both towns, argued in favor of Sancti Spíritus as the preferable site because of its mines, agricultural production, and location on a river. He also pointed out that it had a larger population and its *vecinos* were "mas arraigados" (more rooted, although the term also could imply that more of them held land in the district around the town). Manuel de Rojas concluded his report with a statement that while he felt it would be advantageous to join the two towns into one, it seemed unlikely to happen because of opposition on both sides.[25]

[23] Originally a grant of indigenous labor was *repartimiento*. By the 1530s the term *encomienda* was being used as well although the term *encomendero* for the holder of such a grant was in common use much earlier.

[24] The report lists five men from Trinidad who held *encomiendas* but actually were living in Sancti Spíritus and two women, the wives of Alonso Mendes and Antonio de Paredes.

[25] Rojas said he would refer the issue to the Crown to be resolved. Both towns still exist.

Altogether seventeen men testified regarding their personal circumstances: seven from Trinidad, eight from Sancti Spíritus, and two who called themselves *vecinos* of both. In Trinidad two of the men were married to native women of Cuba and had children, one was married to a Castilian woman with whom he had two unmarried daughters, and the others were unmarried, although one unmarried *vecino* said he had a daughter (most likely mestiza).[26] In Sancti Spíritus four of the men were married to Castilian women, all with at least one child, and five were or had been married to women they referred to as "de esta tierra"; most of the latter had children. The wives of two of the latter group actually were mestizas, both described as "hija de español." Vasco Porcallo de Figueroa, *vecino* of both towns and a dominant figure in early Cuba, testified that he was not married but had a son and two daughters. By the time of his death in 1550 he probably had more children.

If we include the two unmarried men in the group that had children and the seven men who were married to either indigenous or mestiza women, a slight majority of the seventeen men who testified were part of couples or families that were ethnically mixed. Possibly these were the families that were most rooted in the towns, even though nearly all the witnesses stated that they would like to go elsewhere, Peru not surprisingly being a favored destination at this date. Alonso de Cepeda, for example, had been a *vecino* of Sancti Spíritus for nineteen years and had a son and a married daughter with his indigenous wife. Juan Rodríguez de Córdoba, who had been a *vecino* of both towns for twenty years, was a widower who had been married to an indigenous woman and had a daughter who had married in Sancti Spíritus as well as two illegitimate children. Sebastián de la Fuente, also a *vecino* for twenty years, was married to a native woman and also had a daughter who had married in the town. Alonso de Oviedo, married to a mestiza woman, had one legitimate daughter and another who was illegitimate. He said that he wanted to go to live somewhere else and that if he did not do so it was because he did not want to relocate his wife and children.

Vasco Porcallo de Figueroa, who participated in the founding of both towns, was unmarried. Notwithstanding his importance in early Cuba and a wide-ranging career that took him to Mexico and briefly to Florida as part of Hernando de Soto's expedition, he was exceptional among prominent Spaniards in that he apparently never married or established

[26] The term *mestiza* was not used by any of the *vecinos* who testified.

a legitimate household. He made efforts to ensure that his mestizo children were well placed in Cuban society, however. Porcallo's mestizo son, Gómez Suárez de Figueroa, accompanied him on Hernando de Soto's expedition to Florida; another mestizo son, Vasco Porcallo, became *regidor* (town councilman) and *alcalde* (magistrate) in Puerto Príncipe.[27] In the 1530s Porcallo still commanded a substantial labor force that included some 150 Indians in *encomienda*, fifteen or sixteen male and female African slaves, and as many as thirty indigenous slaves from the island and elsewhere. He also had half a dozen Spaniards attached to his household.

By contrast, among the other *vecinos* of the two towns who testified, the largest *repartimientos* were held by a man in Trinidad (with thirty Indians) and an *alcalde* of Sancti Spíritus who held twenty-two. Nearly all the other *vecinos* of the towns who testified had no more than a handful of Indians or none at all, although many still owned African or Indian slaves, many of the latter from New Spain (probably Pánuco).[28] Although in the 1533 deposition Vasco Porcallo, like the other Spaniards who testified, suggested that he might want to leave Cuba, when the bishop conducted his *visita* of the island ten years later, in 1544, he reported that in La Sabana, which he called "the pueblo of Vasco Porcallo, where he has his estates and settlement ... there are twenty houses and huts, lodging for Indians and Spaniards. There is a church and an educated man [*letrado*] who for twenty years has been his chaplain and teaches the Indians and slaves with zeal and diligence ... There are 80 and 120 [*sic*] black slaves. He has ten Spaniards apart from another ten who personally serve Porcallo."[29]

The marriages, families, and fortunes of the long-time *vecinos* of these two small Cuban towns in the 1530s suggest some features of urban development in the large islands of the Caribbean. Española, Puerto Rico, and Cuba moved through roughly comparable phases of expansion and economic boom that attracted large numbers of immigrants, followed by years of retrenchment and the departure for other destinations of many people who had been attracted by the early gold rushes. Because of the relative lateness of the Spanish occupation of Puerto Rico and Cuba (around 1509 and 1511, respectively) compared to Española, in the 1530s the situation of those islands in some ways resembled that of

27 Marrero, *Cuba*, vol. I, 203–10.

28 AGI Patronato 177, no. 1, ramo 18.

29 Quoted in Marrero, *Cuba*, vol. I, 210.

Española at the time of the *repartimiento* of 1514.[30] The *repartimiento* was an attempt to redistribute the already diminished indigenous labor force to the island's Spanish *vecinos*. Information included in the report on the marital status of *vecinos* of different towns suggests a good deal of variation in patterns of marriage partners for Spanish men. Not surprisingly, Santo Domingo had the largest concentration of Castilian women and wives, with larger percentages of Spanish-indigenous couples in the smaller towns. It is reasonable to conclude that by the 1520s and 1530s in all three islands Spaniards who were living in towns other than the port cities and had settled down rather than moving on to other destinations were as likely to be married to or living with indigenous, mestiza, or possibly African women as Castilians, fostering extensive interethnic mixing in Spanish households and towns.

The presence of women in some ways defined urban life.[31] Households managed and overseen by women were crucial to the existence of towns; together they formed the basis for civilized society as Spaniards understood it. There is little doubt that Spanish women were living in the islands from an early time and were playing essential and visible roles in cities and towns. As seen, "La Cordobesa" rented one of Cristóbal de Tapia's stores or taverns in Santo Domingo in the early years of the sixteenth century. While the witnesses who testified in Tapia's suit with Ovando were uniformly male, they frequently mentioned women as their sources of information on local life, suggesting that they played a more public role in urban society than often is assumed.

Information about other aspects of town life for the early years is fairly limited. Santo Domingo led the way as a center for institutions and in acquiring the trappings of urban life. During Ovando's term as governor there were at least two *cofradías* (confraternities), one connected to the hospital of San Nicolás. Even in Santo Domingo, however, progress towards a full set of the institutions associated with a capital city was fairly slow. A convent for women was founded in 1535 and a

[30] For a transcription and discussion of the repartimiento of 1514, see Emilio Rodríguez Demorizi, *Los Dominicos y las Encomiendas de Indios de la Isla Española* (Santo Domingo: Editora del Caribe, 1971). On the presence of Spanish women Española's towns as reflected in the *repartimiento*, see Ida Altman, "Spanish Women in the Caribbean, 1493–1540," in *Women of the Iberian Atlantic*, eds. Sarah E. Owens and Jane E. Mangan (Baton Rouge, LA: Louisiana State University Press, 2012), 62–3.

[31] On Spanish women and marriage, family, their implications for interethnic relations, and royal policies regarding these issues, see Altman, "Spanish Women," 57–81 and Ida Altman, "Marriage, Family and Ethnicity in the Early Spanish Caribbean," *William and Mary Quarterly*, 3d ser., 70(2) (April 2013): 225–50.

university chartered in 1538. The port cities had cathedral chapters and monasteries, and most of the smaller towns had churches and at least some clerical presence, although often only sporadically.[32] Residents frequently mentioned the poor state of churches. When Manuel de Rojas visited Trinidad in 1534 he found that a small hut that had belonged to a "poor vecino" was being used for a church, although there was a hospital. Even in larger places like San Juan funds to build or maintain churches and public works were scarce. Most towns had municipal councils,[33] although in Trinidad Rojas had to arrange for the election of two *regidores* and an *alcalde*.[34]

Many smaller places lacked basic amenities. *Vecinos* of San Germán in Puerto Rico complained that there was no doctor or pharmacist and that they had to go to San Juan for medical treatment.[35] They also probably offered little in the way of social life. Rojas' tour of Cuba's towns provided a relatively rare opportunity for residents to participate in the ceremonial and social aspects of such a visit and to associate themselves with a representative of royal authority. The notary of Santa María del Puerto de Príncipe wrote on May 27, 1534 that Rojas arrived, bringing with him:

> the royal notary Cristóbal de Torres ... and other persons, Spaniards, slaves and Indians in his service, all of whom stayed in the house of Antón de Alaminos, vecino and regidor of this town who went to stay in another lodging and left his house ... for the said señor Manuel de Rojas and his entourage, and I saw that on the first holiday that occurred after they came to the town the alcaldes, regidores and other honorable persons of the town left mass with señor Manuel de Rojas and accompanied him to his lodging.

32 Much research on the presence and activities of the regular and secular clergy in the islands in these years remains to be done. See Genaro Rodríguez Morel, *Cartas de los cabildos eclesiásticos de Santo Domingo y Concepción de la Vega en el siglo XVI* (Santo Domingo: Patronato de la Ciudad Colonial de Santo Domingo; Centro de Altos Estudios Humanísticos y del Idioma Español, 2000).

33 See Genaro Rodríguez Morel's introductory essay to his compilation *Cartas del Cabildo de Santo Domingo* and Esteban Mira Caballos, "Los cabildos antillanos" in *La Española*, 469–92.

34 AGI Patronato 177, no. 1, ramo 17.

35 In 1537 Doña Juana de Villasante, heir of Blas de Villasante, declared that she wanted to assume management of her father's estate, especially his *encomienda* and sugar mill. She mentioned that there was no pharmacy in San Germán and that the *ingenio* provided sugar for medicines and other things as well as carpenters and blacksmiths and other tradesmen that San Germán lacked. Gaspar de Bruselas, a *vecino* of San Germán, testified in December 1536 that he had become seriously ill and had to go to San Juan because of the lack of a physician, pharmacist, or medications in San Germán. AGI Santo Domingo 2280, legajo 2.

On another Sunday:

> All the married and honorable women of the town got together in the house of Francisco Vásquez to enjoy themselves and eat as the ladies of that town sometimes do with one another, and señor Manuel de Rojas was present there, which they [the women] did to have a celebration as for someone recently arrived to that town; and they asked him to eat there with them, which he accepted to please them. After that I saw how many honorable and principal people of the pueblo ate ... many times with señor Manuel de Rojas, sometimes some and sometimes others.[36]

The celebratory reception of Rojas and his entourage suggests the isolation (and, probably, boredom) that both men and women experienced living in these small towns.

PLANS AND POLICIES

The crown and its officials promoted colonization efforts that emphasized the establishment of new communities. Luis de Arriaga, a *vecino* of Seville, undertook an ambitious charge to recruit fifty settlers for each of four towns in Española as early as 1501. He was able to send only seventy-three married men, but with their wives and families they accounted for a still fairly substantial group of some 200 settlers.[37] In 1520 a group of thirty-seven families from the town of Antequera (near Malaga) arrived in Española after stopping in Puerto Rico, where officials and some of the emigrants themselves hoped they would remain. The newcomers were promised their own settlement in Española with the standard elements of a municipality.[38]

These efforts continued for some time, notwithstanding – or perhaps because of – the continued departure of Spaniards from the islands for other destinations. In 1528 *oidores* (judges) of the *audiencia* Licenciados Zuazo and Espinosa solicited royal approval for a plan to sponsor new settlements in Española. These new communities would consist of "at least fifty married *vecinos* with their wives, provided that half be Spaniards and the other half blacks and that all should be brought from Spain" (or Portugal). These new *vecinos* would receive land and livestock and other forms of assistance, and each new community would have its own church and priest. In addition, each settler could bring

36 Both quotations from AGI Patronato 177, no. 1, ramo 17.

37 Rodríguez Demorizi, *El pleito*, 73 and Marte, *Santo Domingo*, 150.

38 See Marte, *Santo Domingo*, 324 and AGI Patronato 176, ramo 5 for the reaction of officials in Puerto Rico.

up to 100 African slaves to the island free of any duty, suggesting that the officials might have envisioned the proposed settlements as a means to expand sugar production. The prominent men who sponsored the new pueblos would receive a portion of the royal revenues generated in perpetuity.[39]

Although it is not clear if anyone in Española responded directly to this proposal, which received the king's approval, in Puerto Rico in 1532 Asencio de Villanueva proposed to implement this scheme and to establish a new settlement – to be named Villanueva – on lands that he held about halfway between San Juan and San Germán which he claimed promised high gold yields. Villanueva was a long-time resident of the island and well-known figure in local society. He stated that he had arrived there in 1510, lived in both San Juan and San Germán, and served first as *alguacil* (constable) and then *receptor* for the Inquisition.[40] On his lands he grazed cattle and had planted wheat, barley, vegetables, grapevines, and fig and other trees. One of the witnesses in his *probanza* testified that he had received seeds from Villanueva as had others. He had hired a priest to instruct his workers in Christianity.[41]

In early 1534 Villanueva was at court in Spain where he requested permission to return to Puerto Rico and "to take with him fifty vecinos, twenty-five of them married Spaniards and twenty-five slaves," also married. The inclusion of African slaves in the category of *vecinos* is ambiguous, as the royal license conceded to Villanueva stipulated that each of the twenty-five Spaniards could take with him two slaves, a man and a woman.[42] If the colonists remained on the island for at least five years they would not have to pay duty on the slaves or other necessities. Villanueva asked the king to pay the passage for two Franciscans who "will assist the priest who will be in that town" and for financial assistance in building a church. Villanueva himself asked for permission to return to Puerto Rico and to take seeds and cattle free of any duty.[43] He apparently tarried for some time in Spain, perhaps encountering difficulties in recruiting the new colonists, and his plans became increasingly ambitious. In May 1535 he also asked for a license to take a horse that would be the sire for a herd

39 Marte, *Santo Domingo*, 284–6, 290.

40 Villanueva was in Spain in 1527–8 acting as representative for Puerto Rico's bishop Alonso Manso and planning to return to the island with materials for the construction of a church as well as arms; see AGI Indiferente General 421, legajo 13.

41 The *probanza* is in AGI Santo Domingo 9, no. 35.

42 On African slaves as potential *vecinos*, see Altman, "Marriage, Family," 237–8.

43 AGI Santo Domingo 2280, legajo 2. See also AGI Indiferente General 737, no. 33 (1533).

of mares. The next month he received permission to take with him "some miners and people who know how to discover mines of silver and copper and other metals."

Villanueva probably did not succeed in establishing the new town, but in those years other prominent men also proposed recruiting colonists, if not actually creating new towns. Around the same time that Villanueva was pursuing his colonization scheme Puerto Rico's treasurer Juan de Castellanos in May 1535 obtained a license to take "his wife and children and all his household to that island with the desire to live and remain in it" and also offered to take with him "thirty people with their wives and households." The king agreed to pay their passage if they would stay on the island for at least five years. Castellanos also proposed taking an additional fifty single men, all of them to be allowed to take slaves with them for their personal service. He apparently was a successful recruiter, as a year later he took fifty men to Puerto Rico, and another twenty-five had contracted to join them.[44]

Whether or not these recruitment efforts resulted in the establishment of new towns, the estates of some of the largest landowners, such as that of Vasco Porcallo de Figueroa at La Sabana described by the bishop, could become significant nuclei of population. Diego Caballero, the longtime and influential secretary of the *audiencia*, owned an estate and sugar mill that in a deposition of 1538 he described as including a settlement with some sixty houses. In 1545 there were more than 300 African slaves and fifty Indians working there.[45]

After the arrival of Europeans, towns in the islands were Spanish almost by definition. The island peoples did not have the opportunity to reorganize their polities as quasi-autonomous units under Spanish rule as occurred elsewhere. The dismal story of the rapid reduction in indigenous island populations due to disease, famine, military conflict, and flight is well known, while perhaps the massive relocation of captives from other parts of the Caribbean and the surrounding mainland is less familiar. These phenomena, together with the arrival of substantial numbers of Europeans and Africans, contributed to the extreme demographic transformation of the region.

[44] AGI Santo Domingo 2280, legajo 2. Like Villanueva Castellanos was returning to the island. In 1558 he testified that he was sixty-five years old, had known Juan Ponce since 1505 in Española, and had participated in the conquest of Puerto Rico. See Tió, *Nuevas fuentes*, 293.

[45] See Altman, "Marriage, Family and Ethnicity," 236, 248.

Very little is known about indigenous communities after contact. They certainly did not disappear all at once,[46] but populations diminished and the chiefly group suffered significant losses, especially as a result of the military campaigns that took place during Ovando's term as governor in the first decade of the sixteenth century. The *encomienda* system also undermined native villages. In theory, Indians assigned to *encomiendas* remained in their home communities except during the periods they were working, but in reality they were relocated both temporarily and permanently. Only belatedly did the deterioration of the indigenous communities become a concern to the crown and its officials, not because of the extremely negative impact of Spanish labor demands and relocations on the welfare of the indigenous population but rather because the Spaniards were losing their labor supply. The ostensible reform efforts of the Hieronymites (1517–19) accomplished little beyond occasioning further dislocations and relocations.[47] As *encomiendas* were vacated and the increasingly tenuous situation of the surviving indigenous inhabitants became ever clearer, the Spanish crown and its officials promoted the establishment of communities for "free" Indians. They referred to these as *experiencias*, that is, experiments in creating quasi-independent communities for once-sovereign people who now had to prove themselves capable and worthy of the opportunity to live on their own.

Cuban officials attempted to implement the *experiencia*, in at least one instance soliciting and recording indigenous testimony.[48] In late April 1531 the lieutenant governor of Cuba, Gonzalo de Guzmán, announced a plan to create an experimental community with the Indians of an *encomienda* vacated by Pedro de Morón, who had been a *vecino* and *regidor* of San Salvador (Bayamo). Guzmán said that he had arranged for the priest Francisco Guerrero to live with and instruct them in the Catholic faith.

[46] See Kathleen Deagan, "Reconsidering Taíno Social Dynamics after Spanish Conquest: Gender and Class in Culture Contact Studies," *American Antiquity* 69(4) (2004): 597–626 on the archaeological study of pre- and post-contact En Bas Saline, located near Puerto Real and possibly the site of Guacanagarí's town.

[47] For the 1517 inquiry made by the Hieronymites after they arrived in Española regarding whether the island's native residents would be better off if they were released from the *encomiendas*, see the transcription of the *Interrogatorio Jeronimiano* in Rodríguez Demorizi, *Los domenicos*, 273–354. For a discussion of the reform efforts of the Hieronymites and arguments made by Dominicans and Franciscans, see Floyd, *Columbus Dynasty*, 173–4 and 176–7.

[48] The documentation on the experiencias in Cuba is fairly extensive. See Esteban Mira Caballos, *El Indio Antillano: repartimiento, encomienda y esclavitud (1492–1542)* (Seville: Múñoz Moya Editor, 1997), 180–4.

Guerrero wanted to ascertain if the Indians in question understood the terms of their "freedom," so Guzmán summoned some of the leading people from the pueblos that had been part of the *encomienda* and arranged for a Spaniard named Pedro de Ribadeneyra to act as interpreter. Ribadeneyra agreed to explain to the Indians that they should live like "labradores de Castilla" following Christian ways and "come to live near the town of San Salvador and there cultivate their fields for themselves and have their pueblo for themselves." The additional instructions cast doubt on the likelihood of their achieving meaningful independence, since they also were expected "to raise cattle and extract gold and give to His Majesty what belongs to them as his vassals." Once Ribadeneyra had communicated these expectations, the Indians would confer with the people of their pueblos and come back with an answer. Diego Ramirez returned to say that he and the rest of his pueblo wanted "to come [to live] close to the town of Bayamo where the Spaniards are and to establish their pueblo and fields and extract gold for the king and to pray to God and Saint Mary and to pay their tithes." Testimony of other indigenous men summoned for the deposition echoed these affirmations, although one man expressed his willingness to provide lodging to travelers, which the Spanish officials quickly corrected, saying that was not part of their charge.[49]

The choice to opt for "freedom" was far from free. Indians could be reassigned to *encomiendas* if they refused to participate in the experiment, as part of which they would cultivate yucca, corn, and cotton and raise poultry and pigs and also agree to give up their dances, fiestas, "idolatries," and custom of painting themselves and donning masks – suggesting that in twenty years Spaniards had made little progress in Christianizing the Indians. The priest would reside with them every other month, which would allow him to judge if they were capable of upholding their end of the agreement while living on their own.

In July 1534 Manuel de Rojas reported that the Indians who had belonged to Pedro de Morón's *encomienda* had "scattered and strayed" ("*han andado derramados y descarriados*") and were not cultivating their lands. Apparently the man designated to oversee them during the previous year fell ill and went to Santiago to be treated where he remained for five months.[50] Once again a number of "principal" Indians were summoned

49 AGI Patronato 177, no. 1, ramo 12.

50 According to Mira Caballos, *El Indio Antillano*, 182–3, the priest assigned to the community proved highly venal and exploitative and was accused of having taken an Indian woman from her husband to live with him instead.

and the interpreter Pedro de Ribadeneyra deputized to explain to them their failure to comply with the conditions of their "liberty." Only one, a man named Gaspar, responded, explaining that if they lacked money and cultivated fields it was because they had been "summoned to work on their properties by certain Christians who gave them nothing more than a shirt." Although Rojas considered making a second attempt at the experiment, by this time it seems clear that Spaniards were ready to abandon it in favor of settlements that essentially would be attached to Spanish towns.[51]

The results of this effort to resettle Cuba's dwindling indigenous population near Spanish towns perhaps accounted in part for the Indians reported by the bishop Juan del Castillo in 1570 to be living in or near a number of Cuban towns, although there might not have been any direct relationship to the "experiment." Puerto Príncipe (today Camagüey) had twenty-five "poor Spaniards" and forty married Indians, Trinidad fifty married Indians and no Spaniards, while Sancti Spíritus's forty residents were evenly split between Spaniards and Indians. Guanabacoa was a pueblo of sixty married Indians about one league from Havana.[52] This report suggests that many towns that originated as centers for Spanish economic and political control over resources and the native population had become home to an ethnically mixed group of survivors of economic decline and demographic disaster.

* * *

This essay focuses on the establishment, decline, or survival of towns and the lives and experiences of those who inhabited them, suggesting the vital role urban centers played in the early development of the islands. Towns were diverse, both internally and in comparison with one another, but they also they shared characteristics that would typify urban centers throughout Spanish America: ethnically and socioeconomically diverse populations, a combination of economic and administrative functions, and tensions between geographic realities on the one hand and strategic and symbolic considerations on the other. For Spaniards, towns represented permanence, civilization, and control over the surrounding countryside and its people. Since many were ports, they also were vehicles for contact and communication with the outside world.

As in Spain, towns provided a focus and matrix for individual and collective identity and loyalty. Despite the turbulence and rapid change

[51] AGI Patronato 177, no. 1, ramo 17.
[52] Marrero, *Cuba*, vol. I, p. 329.

that characterized the society and economy of the Spanish Caribbean in its first fifty or so years, some Spaniards remained in the places that they adopted as their new homes after they crossed the Atlantic. Witness after witness in depositions, such as that of Asencio de Villanueva, who hoped to found a new town in Puerto Rico, or in the inquiry into the potential merger of Trinidad and Sancti Spíritus in Cuba, testified to having lived in the same towns for many years, often since their founding, despite economic downturn and population loss. Why did they stay? Family considerations, time and labor already invested, and the lack of viable alternatives probably were paramount. As seen among the long-time *vecinos* of Trinidad and Sancti Spíritus, and surely was true for others who remained in smaller towns especially, many men who settled in these places not only had married in the islands but frequently into the fast-fading remnants of island communities. Their wives and mistresses often were native to these places, as would be their children. Even as indigenous communities deteriorated and disappeared, very likely a sense of connection to and rootedness in place fostered by indigenous wives and mothers persisted, contributing to the stability and continuity, and also to the transformation, of the towns.

2

The Walk of the Town: Modeling the Early Modern City[1]

James S. Amelang

In 1716, the young English poet and playwright John Gay – now famous for his authorship of the 1728 satirical "ballad opera" *The Beggar's Opera* – published a poem with the curious name "Trivia." As its subtitle makes clear, it deals with "the art of walking the streets of London." The narrator pulls the reader along with him as they wind their way through the labyrinth of the city by day and night. As they dodge dangers and eye opportunities amid the sights, smells, and sensations of urban life, the poet treats his companion to a lively and topographically explicit evocation of a crowded, confusing, but inviting metropolis.

"Trivia" was widely read in its day, and scholars have pored over it ever since. Literary studies of this work, while recognizing its novelty, have until recently focused on a fairly narrow range of topics. These include a search for antecedents in classical literature, particularly poetic recreations of Rome such as Horace's stroll down the Via Sacra in *Satires* I.9.[2] The Latin corpus is a natural place to look for sources; after all,

[1] I have previously discussed this material twice. The first time was at a roundtable held at the August 2012 Prague meeting of the European Association of Urban History; I am indebted to Roey Sweet and John Gallagher for their comments and suggestions. The second occasion was a paper given at a conference organized by Peter Arnade on "Cities and Empire in the Early Modern Spanish Habsburg World," held in March 2013 at the Clark Library, UCLA. Funding for this project has been provided by HAR2011-27898-C02-02 and HAR2011-27898-C02-00 of Spain's Plan Nacional I+D+i (MINECON), 2011–14 (PI: Prof. Fernando Andrés).

[2] As in Max Byrd, *London Transformed: Images of the City in the Eighteenth Century* (New Haven, CT: Yale University Press, 1978), 62–3. For more recent approaches, along with a fully annotated edition of Gay's poem, see Clare Brant and Susan E. Whyman, eds., *Walking the Streets of London: John Gay's Trivia (1716)* (Oxford: Oxford University Press, 2007); Susanna Morton Braund's "Gay's *Trivia*: Walking the Streets of Rome,"

Gay wrote at the height of England's Augustan age, during which men of letters in a growing world empire worked obsessively to import and adapt classical models into their own maturing national literature. Still, this essay takes a different tack. In it I will suggest that Gay and others like him had a considerably broader array of models to draw on, and that these included relatively recent works written in prose as opposed to verse, and in vernacular instead of the classical languages. I will also go on to argue that walks constituted one of the more compelling literary forms early modern citizens drew upon in order to describe their cities. Writers throughout Europe introduced readers to individual towns by taking them on textual tours of their streets, main buildings, and distinctive topographical features. Hoping to draw attention to this important subgenre, I will briefly discuss the varieties, degree of innovativeness, stylistic characteristics, sources, and intentions of this intriguing if hitherto little explored means of representing early modern urban space, and much else along with it.

One might start by remarking the overall trajectory of these texts (see the very tentative Checklist below). While the literary tradition of presenting cities in the form of walks through them stretched back to Antiquity, it did not reach fruition until the early modern era. These depictions moreover appeared in various forms. By far the best known of these pioneering works was an early exercise in urban history: John Stow's "perambulation" of London's twenty-six wards (neighborhood districts) in his *Survey of London*, first published in 1598 and periodically updated and reprinted thereafter. Beginning with Portsoken Ward in the eastern end of the city, and working his way westward and southward until he finally crossed the Bridge and reached Southwark, Stow patiently listed every lane he tramped and building he entered. The result is an extraordinary still-life rendition of a city undergoing numerous changes, many of which – especially those which originated in the dissolution of monasteries under Henry VIII – Stow did not approve of. The sheer accumulation of fact overwhelms the occasional touch of grace, as Stow was determined to list all the benefactors, patrons, and protagonists of the city's past, as reflected in its monuments, documents, and streets. His tome was the closest the sixteenth century got to Google Street View, and

pp. 149–66, provides a thorough assessment of the deployment of classical motifs in this poem. For these literary models in general, see also her "City and Country in Roman Satire," in *Satire and Society in Ancient Rome*, ed. Susanna Morton Braund (Exeter: University of Exeter Press, 1989), 23–47, and Diane Favro, *The Urban Image of Augustan Rome* (Cambridge: Cambridge University Press, 1996).

that it is still available in paperback today bespeaks its popularity and authority as a guide to Central London's multiple topographies.[3]

Still, Stow's "ambulatory writing" was far from unique. Nor was he the first to portray London from the vantage-point of a stroller. Very little is known about Isabella Whitney, apart from the fact that she was apparently the first Englishwoman to publish a collection of her own poetry, in 1567. Her second book, printed in 1573, contains a literally unprecedented work, a Nashe-like last "Will and Testament" in which the narrator takes her leave of London, street by street, place by place. The effect is somewhat chaotic, a jumble mixed with ramble, and its topographic progression is far from continuous. But it is highly effective as a lively invocation of a city presented above all as an ensemble of streets designated individually and occasionally marked by the author with sentimental significance.[4] Moreover, it was merely the first in what would become a rich tradition of written urban walks. One unusually well-known instance is the presentation of London in Letter Five of Daniel Defoe's *A Tour through the Whole Island of Great Britain*, originally published in the mid-1720s. The eighteenth century in particular saw a proliferation of such texts both in the capital and in the rest of Britain itself – and we are acquainted with these thanks to a characteristically lucid and learned article by Penelope Corfield, who was to my knowledge the first scholar to draw systematic attention to this source for urban history.[5]

[3] Admittedly, I read Stow differently from Cynthia Wall, who in her stimulating *The Literary and Cultural Spaces of Restoration London* (Cambridge: Cambridge University Press, 1999) stresses the relatively sedentary character of his writing in comparison with later London topographies. I see him from the other end, as being the author of the first extensive description of London to adopt a dynamic, ambling approach.

[4] The text is reproduced with helpful identifications in Betty S. Travitsky, "'The Wyll and Testament' of Isabella Whitney," *English Language Review* 10(1) (1980): 76–94, and has since appeared in several anthologies of early modern women's writing. For more on this work, see Ann Rosalind Jones, "Apostrophes to Cities: Urban Rhetorics in Isabella Whitney and Moderata Fonte," in *Attending to Early Modern Women*, eds. Susan D. Amussen and Adele Seeff (Newark: University of Delaware Press, 1998), 155–75; Jean A. Howard, "Textualizing an Urban Life: The Case of Isabella Whitney," in *Early Modern Autobiography: Theories, Genres, Practices*, eds. Ronald Bedford, Lloyd Davis, and Philippa Kelly (Ann Arbor, MI: University of Michigan Press, 2006), 217–33; and Karen Newman, *Cultural Capitals: Early Modern London and Paris* (Princeton, NJ: Princeton University Press, 2007), 70–2.

[5] Penelope Corfield, "Walking the City Streets: The Urban Odyssey in Eighteenth-Century England," *Journal of Urban History* 16(2) (1990), 132–74. For the separate but related phenomenon of the formal promenade – that is, a walk (or carriage ride) within a segregated space in or on the outskirts of a city, which started as a form of public display by royalty and aristocrats and broadened its social base beginning in the seventeenth century – see Marcel Poëte, *La Promenade à Paris au XVIIe siècle. L'art de se promener.*

Yet England was not the only country to produce such texts. Depicting cities through walks also found favor in the early modern Iberian world, and on both sides of the Atlantic. Feet-first itineraries through Hispanic towns constituted a sort of paper parallel to the visual riches that Richard Kagan unfolded in a splendidly illustrated overview in 2000.[6] He did not limit his attention to the wide range of artistic representations featured in this globe-trotting work. Instead, he interpreted urban imagery alongside urban discourse. Bringing text and image together, Kagan intimated that how artists and mapmakers painted or drew the civic form or landscape was closely interwoven with how an even wider range of bureaucrats, travelers, poets, social critics, and a host of others wrote about cities. The following brief essay seeks to develop his insight into the significance of urban discourse. Reporting on a book in progress which bears the tentative title of *Writing Cities*, it probes how people from different social and cultural backgrounds wrote about towns during the early modern era. More specifically, it examines textual wandering through the city as a literary form endowed with several specific resonances. At the same time it suggests that "Trivia" and its counterparts in England met their match on the other side of the Channel, as Spaniards devised a rich ambulatory discourse of their own.

Our point of departure is a highly provisional checklist of nine Hispanic city walks:

Granada:

- In 1550 one Sebastián Martínez printed a poem centering around a walk through the city, whose overriding theme was to trumpet how the former Islamic capital had now been triumphantly transformed into a visibly Christian one.
- In a dialogue written by the scholarly priest Luis de la Cueva in 1603, Cecilio – a native of the city named for its patron saint – guides the outsider César around it, singling out and explaining its more imposing buildings and sites.

Les lieux de promenade dans Paris et aux environs (Paris: Armand Colin, 1913); Peter N. Borsay, "The Rise of the Promenade: The Social and Cultural Use of Space in the English Provincial Town, c. 1660–1800," *British Journal for Eighteenth Century Studies* 9 (1986): 125–40; Olivier Dautresme, "La promenade, un loisir urbain universel? L'exemple du Palais-Royal à Paris à la fin du XVIIIe siècle," *Histoire Urbaine* 1(3) (2001): 83–102; and Laurent Turcot, *Le promeneur à Paris au XVIIIe siècle* (Paris: Gallimard, 2007).

[6] Richard L. Kagan, with Fernando Marías, *Urban Images of the Hispanic World, 1493–1780* (New Haven, CT: Yale University Press, 2000).

- In a 1764 dialogue by Juan Velázquez de Echevarría, a local historian, theologian, and notorious scholarly forger, once again "Granada" instructs a "Foreigner" on its past and present by taking the latter on a lengthy series of walks in and around the town.

Lisbon:

- In 1554 the eminent humanist Damião de Góis published a formal eulogy in Latin of Lisbon. He devotes part II to walking around the city walls, and lists each of the main gates and different quarters of the city.

Mexico City:

- The social-climbing professor-priest Francisco Cervantes de Salazar published in 1554 a series of dialogues in Latin which included a street-by-street tour of the capital of Mexico on horseback ("that we may talk as we ride without becoming tired").
- In 1777 a priest named Juan de Viera wrote a walking tour of Mexico City, as part of a cluster of texts which document the capital in a moment of considerable expansion.[7]

Toledo:

- One of the most ambitious descriptions of an early modern Spanish city, Luis Hurtado de Toledo's 1576 leisurely survey on foot of each of Toledo's parishes originated as an singularly lengthy reply to a bureaucratic questionnaire from the king known as the *Relaciones topográficas*.

Valladolid:

- Capital of the Spanish monarchy from 1601 to 1606, Valladolid attracted numerous visitors, including the Portuguese noble Tomé Pinheiro da Veiga. The second part of his *Fastiginia* reproduces conversation and jokes among courtiers during their *paseo*, or afternoon walk. Its name – *Pratología* – derives from the Prado, the large square on the city's outskirts where these gossip-laden promenades took place.

[7] For a list of these texts, see Rogelio Jiménez Marce, "Entre Roma y Jerusalén. Conciencia criolla y elogio a la ciudad de México en la obra de Juan de Viera," *Fronteras de la Historia* 13(1) (2008): 37–65 (p. 40).

Barcelona:

- A 1777 manuscript (now lost) by the Catalan noble Don Rafel d'Amat i de Cortada, better known as the Baron of Maldà, presents Barcelona in the form of a walk that starts and ends in the cathedral located in the city's center.[8]

In short: if one moves beyond the traditional literary focus on classical (and later) poetry and the relatively well-studied case of England, it is not difficult to find a substantially broader range of texts. Even more noteworthy is the degree of innovation many of these texts exhibit in terms of the urban features they cover and the techniques they deploy to describe them. In regard to the former, it is true that the Roman prototypes similarly combined (brief) allusion to monumental buildings and sites with evocation of the hustle and bustle of crowded streets. Yet a fair number of the early modern texts linger far more on the latter, portraying the city as a site of daily life literally pushed and pulled by different sorts of commerce, bestowing on the urban sphere an air of activity and abundance. There are few if any precedents for the Cervantes de Salazar's showing the Spanish overlords picking their way through all the plants and foodstuffs sold by the natives in Mexico City's marketplaces – and fewer still for the care with which the author lists the Indian names of these exotic products. In a similar vein, Hurtado de Toledo's detailing the socio-economic composition of each of the parishes he visits in Toledo also marks a new departure in comparison with earlier literary walks. (For example, he notes that San Nicolás in the city center boasts "very noble people," and has the "best houses" and "largest shops," whereas the much less distinguished San Isidro on the outskirts is inhabited by bakers, potters, and above all Moriscos, or converts from Islam.) Clearly, the angle of vision of these texts broadens as the corpus expands. All sorts of new subject matter rub up against the static, even antiquarian drift that the traditional fixation on monumental architecture and the like wound up encouraging.

This expanded corpus encouraged stylistic innovation as well. By far the most significant change in this regard is the fact that the vast majority of these texts are written in prose instead of verse (more on this below). And it is far from accidental that many of these walks showcase

[8] References in his *Viles i ciutats de Catalunya*, ed. Margarida Aritzeta (Barcelona: Ed. Barcino, 1994), 72 and 82n. The lengthy description of Barcelona published on pp. 81–140 derives from a later version which, while very attentive to topographical detail, is not structured around an itinerary.

spoken dialogues. This gambit was not unknown to the ancients; after all, Horace's satire famously featured the author making his way through the city while desperately trying to shake a particularly annoying companion. What was new in the early modern texts, however, is how they heed the urban surroundings. Horace's poem contains only one topographical reference, to the Temple of Vesta at the entry to the Forum. Many if not most of the later walks, on the other hand, resort to dialogue in order to enhance and extend their micro-description of the city. Typically, a native introduces an outsider to the city by taking him on a walk and answering the questions the naive newcomer poses. These and other features confer an informal, even conversational quality on the texts. At the same time, they make them more accessible to more diverse publics, although in truth we know all too little about the readers and consumers of this specific discursive tradition.[9]

Another relevant concern is that of influence, and more specifically, the extent to which one can conceive of this form of writing as a connected whole, with authors finding inspiration in earlier textual models. It is true that some walks mention previous ones. Such is the case of, for example, Viera's 1777 paean to Mexico City, which cites the earlier effort by Cervantes de Salazar. Note also that towards the beginning of his letter on London, Defoe demurs that "to enter here ... into a particular description of the city of London, its antiquities, monuments, etc. would be only to make an abridgment of Stow and his continuators ..."[10] But on the whole, one senses that direct reference of this sort is rare. While influence undoubtedly exists, it is usually indirect. Such was indeed the case of the impact of Gay's "Trivia" on later poetic renderings of walking in London, such as Samuel Johnson's "London" (1738), which suggests that instead of searching for explicit borrowings from one text to another, it may be more rewarding to try to reconstruct a fuller array of available models, along with the uses to which they were put.

Here variety wins the day. Early modern authors could draw on a broad swath of sources for their efforts at ambulatory writing – far broader, in fact, than those traditional literary scholarship has identified. Three points are especially germane here. First, most of the available

[9] For more on the strong appeal of the dialogue as a form of urban discourse, see my "Artisans Discuss the City: Urban Dialogues in Early Modern Europe," in *Storia sociale e politica: Omaggio a Rosario Villari*, eds. Alberto Merola, Giovanni Muto, Elena Valeri, and Maria Antonietta Visceglia (Milan: Franco Angeli, 2007), 408–24.

[10] Daniel Defoe, *A Tour through the Whole Island of Great Britain*, abridged and ed. Pat Rogers (London: Penguin, 1986), 295.

models were written nearer in time than the classical precedents alluded to above. Moreover, most of this enlarged corpus takes the form of prose instead of poetry. Finally, many of these models have recognizably popular as well as elite origins and tones.

The point of departure here is fairly predictable. Literary scholars – thanks above all to their ability to detect traces of classical culture in the works they study – have understandably focused on these references largely to the exclusion of others. That students of poems such as "Trivia" would look backward (and upward?) towards classical verse is equally comprehensible. In fact, it clearly was what the author hoped readers would do, and helps explain why he gave the poem a Latin title and opened with an epigraph and motto from Virgil. Still, a more inclusive approach to the question of antecedents would wind up expanding the field of possible predecessors beyond this fairly narrow catchment. It would turn up, first, the fact that Antiquity produced walking portraits of cities in prose as well as poetry. One obvious candidate is Pausanias' *Description of Greece*, written in the second century AD. Its most revelant passage in this regard is the section on Athens with which the book opens. This takes the form of a circular walk within the confines of the city, beginning with entry through the Diplyon Gate where the road from Piraeus reaches the walls, and ending at the same spot, which allows the author to finish his tour with a description of the Academy and other outlying buildings.[11] One can moreover locate at least one early modern instance in which this walk served as a model for later urban discourse. The Italian traveler and diplomat Cornelio Magni directly patterned the depiction of Athens he published in 1688 on Pausanias' visit – just as Jane Ellen Harrison and countless other classicists have tried to do since then.[12]

I bring up this widely read text in order to make clear that I am *not* trying to suggest that the classics were unimportant for the sort of early

[11] For more on Pausanias' tour around the Athenian *agora*, see for example Christian Habicht, *Pausanias' Guide to Ancient Greece* (Berkeley, CA: University of California Press, 1998), 77–83.

[12] Cornelio Magni, *Relazione della città di Atene, colle provincie dell'Attica, Focia, Beozia e Negroponte* (Parma, 1688). I have consulted the version in Marziano Guglielminetti, ed., *Viaggiatori del Seicento* (Turin: UTET, 2007), 649–57 (his walks in Athens in 1674); for the citation of Pausanias, see p. 656. For a wry account of Harrison's use of the Greek text in her book on the monuments of Athens (co-authored with Margaret Verrall), see Mary Beard, "'Pausanias in Petticoats', or *The Blue Jane*," in *Pausanias: Travel and Memory in Ancient Greece*, eds. Susan E. Alcock and Arthur F. Thurnau (Oxford: Oxford University Press, 2001), 224–39.

modern urban description being studied here.[13] If anything, the urban walks themselves amply demonstrate the ability of ancient models to stimulate literary invention and critical reflection. Perhaps the most influential exemplar in this regard was Petrarch. When recreating his walks in Rome in his *Familiar Letters*, he remarks the dissonance between what he actually sees and the images based on his earlier reading about the classical city. It has been suggested that this experience triggered a crucial change in his thinking, one that led him towards the discovery of history itself, that is, the recognition that the past differed from the present, and that it was indeed more admirable.[14] His text in turn made a deep impression on later walkers in Rome, especially humanists hailing from throughout Europe, from Poggio to Montaigne and Peiresc, and reached even further, to the mass (but elite) tourism of the Grand Tour.

Yet even if the influence of the ancients remained intact throughout the early modern era, something did change in the meantime. Paperbound city walks proved to be a fertile site for various sorts of invention. Especially visible was the widening of the base of models authors drew upon. Here other sources now joined the classics, and began to challenge their predominance. Some of these moreover deal explicitly with popular as opposed to privileged experience in the urban environment. This is certainly the case of the "Cries" of London, Paris, Bologna, and other cities, which present salespersons and hucksters of all sorts making the streets their own as they advertise their goods and services. (Interestingly, this subgenre also took on aural and visual expression, as distinguished composers including Clément Janequin and Henry Purcell set these cries to music, and artists such as Annibale Carracci depicted their performers in series of engravings.)[15] Pamphlets featuring walks also appeared during the early modern era. One frequently reprinted example is the 1728

[13] For reliance on classical (Roman) models for the description of the urban landscape of early modern London, see Lawrence Manley, *Literature and Culture in Early Modern London* (Cambridge: Cambridge University Press, 1995), esp. pts. II–IV.

[14] Kenneth Gouwens, "Perceiving the Past: Renaissance Humanism after the 'Cognitive Turn'," *American Historical Review* 103(1) (1998): 55–82 (p. 69). See also David Galbraith, "Petrarch and the Broken City," in *Antiquity and its Interpreters*, eds. Alina Payne, Ann Kuttner, and Rebekah Smick (Cambridge: Cambridge University Press, 2000), 17–26.

[15] The most detailed study of this genre I am familiar with is Vincent Milliot, *Les Cris de Paris ou le peuple travesti: Les représentations des petits métiers parisiens, XVIe–XVIIIe siècles*, preface by Daniel Roche (Paris: Sorbonne, 1995), although see also Charles Hindley, *A History of the Cries of London, Ancient and Modern* (London: Reeves and Turner, 1881), and [Robert] Massin, *Les cris de la ville: commerces ambulants et petits métiers de la rue* (Paris: Albin Michel, 1985).

miscellany *A Trip through London*, attributed to one Erasmus Jones, and which complains of (among many other things) the state of repair of the capital's streets.[16] Naturally, the guidebooks to cities which began to proliferate in this era similarly use walks as templates for orienting visitors (as well as natives) through civic space.[17] Finally, urban diaries often offer a bottom-up view of the city thanks to their habit of registering not only walking in general but also specific itineraries both inside and outside the city. Perhaps the best-known case here is that of Samuel Pepys, a university-educated and socially mobile bureaucrat, to be sure, but endowed with enough popular roots to feel embarrassed about them. And the close attention to his own movement through urban space he shows in his diaries provides the modern-day reader with a remarkably vivid sense of a street-level topography that managed to survive (as did the author) the Great Fire of 1666, if not later mishandling by the Luftwaffe and more recent urban planners.[18]

When discussing the question of sources of this new and yet not so new mode of writing, one must also keep in sight a number of variants that, while not full-scale urban walks themselves, nevertheless hover around closely enough to serve as a possible stimulus to this type of writing. I am thinking for example of walks through single civic buildings and monuments, such as the foot tour through the cathedral of Toledo registered in an anonymous early seventeenth-century manuscript account of a journey through the Iberian Peninsula known as the "Floresta española."[19]

[16] [Anon.], *A Trip Through London: Containing Observations on Men and Things ...* (London: J. Roberts, etc., 1728 [2nd ed.]). I first saw this referred to in Robert B. Shoemaker, "Gendered Spaces: Patterns of Mobility and Perceptions of London's Geography, 1660–1750," in *Imagining Early Modern London: Perceptions and Portrayals of the City from Stow to Strype, 1598–1720*, ed. J. F. Merritt (Cambridge: Cambridge University Press, 2001), 144–65 (p. 151); see p. 164 for a list of other such pamphlets.

[17] One example would be the architect Andrea Palladio's guidebook to Rome, published in numerous editions from 1554 on. See in general Gilles Chabaud, Évelyne Cohen, Natacha Coquery, and Jérôme Penez, eds., *Les guides imprimés du XVIe au XXe siècle. Villes, paysages, voyages* (Paris: Belin, 2000).

[18] For the record, Pepys' father was a Merchant Taylor and his mother an illiterate washerwomen. Their son was characteristically frank about his discomfort when dealing with his poorer relations outside the city; distance was probably not the only reason why he visited them in a coach. For reconstructions of his walks, see John S. Pipkin, "Space and the Social Order in Pepys' Diary," *Urban Geography* 11(2) (1990): 153–75, and Ian Archer, "Social Networks in Restoration London: The Evidence from Samuel Pepys' Diary," in *Communities in Early Modern England: Networks, Place, Rhetoric*, eds. Alexandra Shepard and Phil Withington (Manchester: University of Manchester Press, 2000), 76–94. Note that Pepys recorded not only his own walks, but also those of others, above all King Charles II.

[19] "La península a principios del siglo XVII," ed. Luis Sánchez Costa, *Revue Hispanique* 34 (1915): 300–542 (pp. 308–54).

One can also point to in this regard the specific motif of viewing a city while walking along its walls. The merchant Jean Pagès included a lengthy account of such a bird's-eye promenade in his late seventeenth-century chronicle of Amiens, and other instances could be cited. A third example is the textual trudge down a single street, as when John Hall inserted into his 1565 English translation of Lanfranco of Milan's *Chirurgia parva* a walk along Bucklersbury Street in London, a thoroughfare famous for the sale of spices, drugs, and medicinal herbs.[20] Finally, one might also mention written walks whose aim was not so much urban description as something else. One example would be the antiquarian who is doing his best *not* to notice the city which houses his itinerary. Thus the French humanist Pierre Gilles' *De topographia Constantinopoleos*, published posthumously in Lyon in 1561, takes his reader walking through Istanbul, but refuses to let the modern city distract him from his single-minded pursuit of classical remnants and ruins.[21] The same indifference to the cityscape marks the *Discursos medicinales* the Spanish physician Juan Méndez Nieto published in 1607. This book includes a first-person account of his walking through the Castilian city of Arévalo during an epidemic of plague in 1558. In the company of an alderman Méndez traverses the entire town while attending to his professional duties. He then lists one by one the different measures he ordered to be taken to try to stop the contagion, which include filling in ditches cluttered with dead animals, and having the streets cleaned, garbage removed, fires lit, and the sale of foodstuffs closely overseen.[22]

One final ensemble of urban textual walks that could be seen as possible models are those which took place only in the author's imagination. I refer here above all to city walks that appear within literary or philosophical texts. Many of these are set in real cities, and their authors take the same pride in displaying minute topographical knowledge that we find in the rest of the genre. Examples include the pedagogical dialogue

[20] Cited in Deborah E. Harkness, "'Strange' Ideas and 'English Knowledge': Natural Science Exchange in Elizabethan London," in *Merchants and Marvels: Commerce, Science, and Art in Early Modern Europe*, eds. Pamela Smith and Paula Findlen (New York: Routledge, 2002), 137–60 (p. 144).

[21] *Pierre Gilles' Constantinople: A Modern English Translation*, ed. and trans. Kimberly Byrd (New York: Italica Press, 2008–09), 3 vols. Details in Amanda J. Wunder, "Western Travelers, Eastern Antiquities, and the Image of the Turk in Early Modern Europe," *Journal of Early Modern History* 7(1–2) (2003), esp. pp. 99–103.

[22] "Sobre cierta pestilencia del año 1558 y sus remedios médicos," in *Los siglos XVI–XVII: cultura y vida cotidiana*, eds. Luis Enrique Rodríguez-San Pedro Bezares and José Luis Sánchez Lora (Madrid: Síntesis, 2000), 265–8.

the humanist Juan Luis Vives structured around a street-by-street stroll through his native city of Valencia (1538); the walk through Rome undertaken by the recently arrived prostitute Aldonza in Francisco Delicado's risqué colloquy *The Luscious Andalusian* of 1528; or the phantasmagoric dialogue (once again!) in which the Italian philosopher Giordano Bruno recounts his stumbling through the streets of London at night-time known as *The Ash Wednesday Supper* (1584).[23] Finally, one even comes across walks taken in wholly imaginary cities. Admittedly, Christian in *The Pilgrim's Progress* constantly keeps before him two cities ("of Destruction" and "Celestial") during his foot travel, but in the end says precious little about them as places. Slightly more detailed is the Spanish satirist Francisco de Quevedo's reverie in which an old man accompanies the author down the high road of Hypocrisy – "the main street of the world"– during which (rather like his English counterpart) they run into a parade of avatars of immorality, disbelief, and the like.[24]

Perhaps the easiest question to pose – and the hardest to answer – is the final one in my list: why? Why did early modern writers on cities resort to the motif of walking? Once again, diversity must be given its due. The texts studied here were characteristically generated with multiple intentions in mind, and no single purpose dominates the field. That said, one can certainly think of several obvious attractions of this form of writing. First, there is the lure of the local, the deep pleasure of recognition afforded to those privileged with enough proximity to navigate successfully through this particular type of spatial discourse. Then there is the special appeal of the concreteness that derives from attention to topography and other forms of familiarity that are built on close knowledge. Localist lore has always benefited from its ability to generate positive formulas by combining insiderdom and a sense of place. And it surely is no accident that so many of the walking dialogues involve instruction of an ignorant but interested outsider by a generous indigene whose authority is rooted in his experiential command of the smallest details. This tilt towards an insider perspective may also explain why relatively

[23] Juan Luis Vives, *Diálogos. Fábula del hombre. Templo de las leyes*, ed. and trans. Juan Francisco Alcina (Barcelona: Planeta, 1988), esp. pp. 9–13; Francisco Delicado, *La lozana andaluza*, ed. Bruno M. Damiani (Madrid: Castalia, 1969), esp. pp. 61–91; Giordano Bruno, *The Ash Wednesday Supper*, ed. and trans. Edward A. Gosselin and Lawrence S. Lerner (Hamden CT: Archon, 1977), 110–27.

[24] Francisco de Quevedo, "El mundo por de dentro," in his *Los sueños*, ed. Ignacio Arellano (Madrid: Cátedra, 2003 [orig. ed. 1627]), 276–306. Recall that Hypocrisy also makes a brief appearance in Bunyan's book.

few among the abundant travel accounts of the early modern era adopted the walk as a literary model. Presumably it required too much local knowledge – and loyalty? – for a short-term visitor to acquire on the run.

But detail can exist for its own sake only up to a certain point. The texts evoked here are in the end *not* primarily about walking. Instead, these are texts in which walking serves a larger purpose, that of exploring and revealing a city.[25] Seen in this light, "Trivia" and its companions may be approached as pieces within a series of larger puzzles of urban knowledge whose key is topography. And when in the early modern era the mode in which these puzzles were written shifted from poetry to prose, they took on a demotic quality that allowed new sorts of authors to reach new sorts of readers – a possibility which has often been overlooked in standard surveys of belle-lettristic writing about early modern cities.

In conclusion, this very tentative overview has focused on one specific form within the large and heterogenous sphere of early modern urban discourse: the walk as a means of presenting and explicating a city. It is a literary form with a long history, dating back to the ancient world and reaching the present. To a certain degree what I have wound up doing is to locate further back in time, in the early modern period, a source which historians of nineteenth- and twentieth-century cities have found especially valuable. With this in mind I would suggest that a closer look at this perambulatory prose may serve to help measure continuities and changes in civic prose (and poetry) over the long haul. Looking back over the more recent stages in this trajectory, one can detect several turning points and shifts in emphasis. Yet on the whole, the walk has retained intact its importance as a discrete form of knowledge, as well as a distinctive literary exercise. If anything, its significance and visibility has increased over time.

The nineteenth century witnessed the consolidation of the walker not only as a crucial figure within the urban landscape, but also as the author of some of the most stimulating reflections about the city in terms of urban life and culture, its ever more rapid rhythms of change, and

[25] In fact, in a few of these texts the actual reference to strolling is muffled altogether. This is the case of Whitney's "Wyll and Testament," which mentions place-names without explicitly connecting them by walking. The author also reverses the flow of classical influence found in Gay and others. Thus, while her previous works made sustained reference to ancient Roman literature – indeed, her earliest poems were modeled directly on Ovid's *Heroides* – her London itinerary contains no visible classical allusions, although one could perhaps argue that its general thrust has echoes of Juvenal's lingering farewell to Rome in his third *Satire*.

the like. It is well known how the re-creation in Baudelaire's poetry of modern Paris from the perspective of the *flâneur* was later taken up in what was perhaps the most ambitious and complex project of urban cultural history essayed in the twentieth century, the *Passagenwerk* book that the German critic Walter Benjamin left unfinished at his death in 1940.[26] Slightly less well known is the fact that the poetic tradition of city-walking also reached the twentieth century alive and well; think, for example, of the reflections of a foreigner ambling through Paris in Rilke's *The Notebooks of Malte Laurids Brigge* (1910). Finally, one should also keep in mind the prominence walks achieved in the representation of the modern city in prose literature, as in the London of the Sherlock Holmes stories, or the way the reader comes to know Dublin by following the footsteps of Stephen Dedalus in *Ulysses*.

My aim has not been to surprise the *flâneur* in the early modern city – other studies do a good job of that.[27] Instead, it has been to get a sense of how certain individuals wrote about early modern cities, as a means of responding to a broader query: what people from all walks of life (so to speak) thought about the cities in which they lived, and how they reflected on their urban experience. This essay has placed emphasis on the impressive diversity of the texts they wrote. It has paid particular attention to their sources, in order to offer some glimpse of the interpenetration between high and low culture that marked a complex and intriguing form of writing. Finally, it offers the hope that further research and analysis will bring to light other aspects of a world of urban discourse about which we know all too little. And there is no better place in which to venture these reflections than in a volume which offers homage to a historian who has taught us so much about Hispanic cities, and about the texts and images to which they gave rise.

[26] Available in English as Walter Benjamin, *The Arcades Project*, ed. Roy Tiedemann, trans. Howard Eiland and Kevin McLaughlin, prepared on the basis of the German volume ed. by Rolf Tiedemann (Cambridge MA: Harvard University Press, 1999). For studies of walks as a means of portraying modern cities, see Judith R. Walkowitz, *City of Dreadful Delight: Narratives of Sexual Danger in Late-Victorian London* (Chicago, IL: University of Chicago Press, 1992); Deborah Epstein Nord, *Walking the Victorian Streets: Women, Representation, and the City* (Ithaca, NY: Cornell University Press, 1995); and Dana Arnold, *Re-presenting the Metropolis: Architecture, Urban Experience and Social Life in London, 1800–1840* (Aldershot: Ashgate, 2000), esp. ch. II, "The Art of Walking the Streets."

[27] See for example Miles Ogborn, *Spaces of Modernity: London Geographies, 1680–1780* (London: Guilford Press, 1998), 75–115, and Newman, *Cultural Capitals*, 60–75.

URBAN TEXTUAL WALKS: A PARTIAL CHECKLIST

Granada, 1550, Sebastián Martínez

"Las partidas de la gran Ciudad de Granada en metro, o en manera de Perqué, hecho por Sebastián Martínez, vezino de las Mesas, que es en el marquesado de Villena. Con un villancico. Año de M.D.L.," in Jesús Luque Moreno, *Granada en el siglo XVI: Testimonios de la época*, Granada: Universidad de Granada, 2013, pp. 401–412, along with pp. 39–46

Lisbon, 1554, Damião de Góis, humanist/courtier

Lisbon in the Renaissance: A New Translation of the Urbis Olisiponis Descriptio, trans. and ed. Jeffrey S. Ruth, New York: Italica Press, 1996

Mexico City, 1564, Francisco Cervantes de Salazar, educator/priest

Life in the Imperial and Loyal City of Mexico in New Spain and the Royal and Pontifical University of Mexico as Described in the Dialogues for the Study of the Latin Language Prepared by Francisco Cervantes de Salazar for Use in His Classes and Printed in 1554 by Juan Pablos, trans. Minnie Lee Barrett Shepard and ed. Carlos Eduardo Castañeda, Austin, TX: University of Texas Press, 1953, pp. 37–67

London, 1573, Isabella Whitney, poet

Betty S. Travitsky, "'The Wyll and Testament' of Isabella Whitney," *English Language Review* 10(1) (1980): 76–94

Toledo, 1576, Luis Hurtado de Toledo, parish priest

"Toledo: Memorial de algunas cosas notables que tiene la Imperial Ciudad de Toledo (1576)," in Carmelo Viñas and Ramón Paz (eds.), *Relaciones histórico-geográfico-estadísticas de los pueblos de España hechas por iniciativa de Felipe II. Reino de Toledo*, Madrid: Instituto Balmes de Sociología, 1951, vol. III, pp. 481–576

London, 1598, John Stow, chandler/chronicler

The Survey of London, reprinted from the text of 1603, ed. Charles Lethbridge Kingsford, 2 vols. Oxford: Clarendon Press, 1971

Granada, 1603, Luis de la Cueva, priest

Diálogos de las cosas notables de Granada y lengua española y algunas cosas curiosas, ed. José Mondéjar, Granada: Colección Archivum, Universidad de Granada, 1993 [facs. ed.]

Valladolid (Spain), *c.* 1605, Tomé Pinheiro da Veiga, Portuguese courtier

Fastiginia: vida cotidiana en la corte de Valladolid, ed. Narciso Alonso Cortés, Valladolid: Ambito-Ayuntamiento de Valladolid, 1989, pp. 173–285

Livorno, undated, early 17th century, "Don Antonio de Contreras," Spanish nobleman (the likely real author was "Fernando Texeda," aka Tomás Carrascón, a Spanish monk who converted to Protestantism and lived in England and then Amsterdam in the 1620s–1630s)

Biblioteca Nacional, Madrid/Ms. 18247; also M. Caplan, *Danielillo ó Respuestas á los Cristianos. Escrito en Amsterdam por Isaac Mendes en el año de 1738*, Brussels: B. G. van Gelder, 1868

Athens, 1674, Cornelio Magni, traveler

Marziano Guglielminetti (ed.), *Viaggiatori del Seicento*, Turin: UTET, 2007, pp. 649–57

Amiens, *c.* 1680, Jean Pagès, merchant

Manuscrits de Pagès, marchand d'Amiens, écrits à la fin du 17e et au commencement du 18e siècle, sur Amiens et la Picardie (continués par Achille Machart), ed. Louis Douchet, 6 vols. Amiens: Alfred Caron, 1856–64

London, 1685, Guy Miège, Huguenot language teacher

Nouvelle Methode Pour Apprendre l'Anglois. Avec une Nomenclature Françoise & Angloise, London, 1685[28]

London, 1690, Thomas D'Urfey, poet/playwright

Collin's Walk through London and Westminster: a Poem in Burlesque, London: John Bullord, 1690

London, 1698–1700, Ned Ward, satirist/tavernkeeper

The London Spy: The Vanities and Vices of the Town Exposed to View, ed. Arthur L. Hayward, London: Cassell, 1927

London, 1716, John Gay, poet/playwright

Clare Brant and Susan E. Whyman (eds.), *Walking the Streets of London: John Gay's Trivia (1716)*, Oxford: Oxford University Press, 2007, pp. 169–245

London, 1724–6, Daniel Defoe, novelist/journalist

A Tour through the Whole Island of Great Britain, abridged and ed. Pat Rogers, London: Penguin, 1986, pp. 286–336

Granada, 1764, Juan Velázquez de Echevarría, priest/antiquarian

[28] My thanks to John Gallagher of Cambridge University for this reference.

Paseos por Granada y sus contornos, ed. Cristina Viñes Millet, Granada: Universidad de Granada, 1993 [facs. ed.]

Canterbury, 1774, William Gostling, cathedral canon

A Walk in and about the City of Canterbury, With many Observations not to be found in any Description hitherto published, Canterbury: Printed and sold by Simmons and Kirkby, and W. Flackton, 1774

Mexico City, 1777, Juan de Viera, priest

Breve compendiosa narración de la Ciudad de México, corte y cabeza de toda la América Septentrional, Mexico City: Guarania, 1952

Barcelona, 1777, Don Rafel d'Amat i de Cortada, baró de Maldà, noble

3

The King, the City, and the Saints: Performing Sacred Kingship in the Royal Capital

Erin Kathleen Rowe

It was a balmy June day in Madrid. The sun beat down on the newly renovated Plaza Mayor, an architectural marvel at the center of the growing city. Crowds poured into the plaza, standing around the entrances and inside the residences surrounding the square, leaning down from balconies to witness the scene below. In the plaza stood enormous structures built for the celebration in progress, decorated with allegorical figures representing Madrid and Spain. Also depicted were the city's civic patron saint, Isidore, and his wife María de la Cabeza, along with another notable son, the fourth-century Pope Saint Damasus. Among these saints stood the figure of the king himself, Philip IV. Philip, Isidore, Damasus – king, patron saint, pope – clustered into the plaza that stood at the most important space of the capital city located in the center of the Spanish kingdoms, the metaphorical heart of the vast Spanish empire. The Plaza Mayor had been converted into a stage to view these images in close relationship to each other. Suddenly, the living king appeared in the middle of this space from the royal apartments near the Panadería, along with members meof the most powerful councils of his government. They joined the crowd and moved along the procession route out of the Plaza Mayor and towards the royal palace on the western edge of the city.

Those who planned Baroque entertainments did so with great sophistication and a canny eye for the layers of meaning they invested into space, place, and participants; we must therefore read this moment as the result of deliberate staging. The scene described above occurred during the festivals honoring the canonization of five individuals in 1622, four of whom were Spaniards. One of those honored was the civic patron of Madrid, Isidore; hence, the celebrations took on a heightened intensity in Madrid, though the saints were honored with festivals throughout the Spanish kingdoms. Javier Portús has remarked that the "purpose of these festivals was very clear: they set out to exalt not only the four new Spanish saints, but also the monarchy itself and the social and religious order that it defended."[1] What Portús observed, and what I argue here, is that close study of religious festivals in the seventeenth century reveals that they provided an ideal medium for communicating a sacred vision of king and kingdom. Such a vision played a crucial role in the Habsburg monarchy, as kings vanished behind palace doors, abandoning for the most part the long-held tradition of royal travel, and requiring innovative new ways of performing royal power and augmenting political authority.[2]

It is a truism to suggest that the Habsburg monarchs were deeply pious and they assiduously promoted Catholicism both at home and abroad. This truism often elides the differences between the Habsburg monarchs, who did not maintain uniform models of representing kingship and monarchy. Such differences are most easily observed in royal portraiture. While much emphasis has been placed on similarities – and particularly the Habsburg penchant for posing in stark black with no royal adornments – this style of portraiture evolved over time. Charles V's (1516–56) portraits (most famously, the many done by Titian) projected a quite different style of kingship, one that emphasized strength and military prowess. A distinct shift took place at the end of Charles V's reign and continued into Philip II's (1556–98), signaled by the increasing number of allegorical paintings representing the Habsburg monarchs and monarchy in ways that centered Catholicism as key to their legitimation. Two

[1] (Translation mine.) Javier Portús Pérez, "Público, públicos e imágenes en el Madrid de 1622," *Cuadernos de Arte e Iconografía* IV-8 (1991): 8: "El propósito de estas fiestas era muy claro: se trataba de exaltar no sólo a los cuatro nuevos santos españoles, sino también a la propia Monarquía y al orden social y religioso que defendía ésta."

[2] Civic festivals such as royal entrances have recently received a wealth of scholarly attention. Examining religious festivals (and specifically, their political and civic dimensions), however, has lagged behind. One of the most important works on festive culture in Spain is María José del Río Barredo, *Madrid, Urbs Regia: La capital ceremonial de la Monarquía Católica* (Madrid: Marcial Pons, 2000). For a few examples in addition to Del Río of the large body of work on festivals in Spain, see José Jaime García Bernal, *Fasto Público en la España de los Austrias* (Seville: Universidad de Sevilla, 2006); León Carlos Albarez Santaló, "El espectáculo religioso barroco," *Manuscrits* 13 (1995): 157–83; Teofilo Ruiz, *A King Travels: Festive Traditions in Late Medieval and Early Modern Spain* (Princeton, NJ: Princeton University Press, 2012); and José Antonio Mateos Royo, "All the Town is a Stage: Civic Ceremonies and Religious Festivities in Spain during the Golden Age," *Urban History* 26(2) (1999): 165–89. Mateos argues: "Finally, the religious festivities and civic ceremonies in Spain became elevated ... by their splendour and magnificence to the category of art" (189).

early examples include Titian's *The Trinity in Glory* (1552–4), featuring Charles V and Philip II kneeling at the feet of the holy Trinity, and Titian's *Religion Succored by Spain* (1572–5), in which the allegorical figure of Spain is shown coming to the rescue of nearly defeated Religion.[3] Such allegorical works were an important feature of Philip II's representation of the monarchy; they reflect the necessity of continuing older models of Spanish monarchy as the defender of Christianity, in an era where the king was no longer a military hero.[4]

While Charles V brought Burgundian court etiquette to Castile in the early sixteenth century, it was not until the reign of Philip II that we can discern the patterns of court life we associate most closely with the Habsburg monarchs – that is to say, the tightly restricted access to the physical person of the king, which reduced opportunities for public demonstrations of kingship. The extension and maintenance of authority greatly preoccupied Habsburg Spanish rulers, particularly because the early modern period marked a time of transition to the impersonal, or bureaucratic, state. While medieval kings shored up their authority through the charisma of their physical presence and traveling their kingdoms extensively, monarchs became increasingly stationary over the course of the sixteenth century, eschewing the peripatetic model for fixed capitals. They continued to travel, but the decreasing frequency of such travel eroded its central role in the pageantry of kingship.[5] The difference between Charles V, who spent little time in his Spanish kingdoms, and Philip IV, who never left Iberia, could not be starker.[6] Even within Iberia, late sixteenth- and seventeenth-century kings

3 Jason McCloskey argues: "The iconographic religiosity of Hispania suggests a fusion of the identities of Religion and Spain from Titian's painting into a single personification": "Spain Succored by Religion: Titian and Lope de Vega's *La Dragontea*," in *Signs of Power in Habsburg Spain and the New World*, eds. Jason McCloskey and Ignacio López Alemany (Lewisburg, PA: Bucknell University Press, 2013), 205.

4 Fernando Checa Cremades discusses the significant changes Philip II made to court life and royal representation: "Monarchic Liturgies and the 'Hidden King': The Function and Meaning of Spanish Royal Portraiture in the Sixteenth and Seventeenth Centuries," in *Iconography, Propaganda and Legitimation*, ed. Allan Ellenius (Oxford: Clarendon Press, 1999), 89–104. See also Fernando Bouza, "Ardides del arte: Cultura de corte, acción política y artes visuales en tiempos de Felipe II," in *Felipe II: Un monarca y su época* (Madrid: Museo del Prado, 1998), 57–81.

5 On the travels of medieval Spanish kings through Philip II, see Ruiz, *A King Travels*, which not surprisingly ends in the latter sixteenth century as Habsburg monarchs became increasingly stationary.

6 Aurelio Espinosa provides a useful table detailing Charles' itinerary throughout his life; one can clearly see how often he was on the move, and how little time he spent in his Spanish kingdoms. The longest block of time that Charles spent in Iberia was seven years (1522–9) in the immediate aftermath of the *comunero* revolt: *The Empire of the*

covered less territory. As their charismatic presence became less necessary for the maintenance of power, the Spanish Habsburgs substituted the power of absence, deliberately taking on the mantle of invisibility behind a nearly impenetrable court etiquette.[7] The retreat of the king into the palace was balanced in part through the continuance of festivals, where the king could be present (in Madrid, for the most part) or represented figuratively (elsewhere in the Spanish monarchy).

The main way that later Habsburg monarchs, particularly King Philip IV (1621–64), appeared in public was at religious events – walking to mass, accompanying the Eucharist, and attending religious festivals.[8] These moments highlighted both the deep personal piety of the king and a specific vision of the Spanish monarchy as chosen by God. Examining two high-profile religious festivals of the first decade of Philip's reign – the 1620s – reveals the ways in which the new king actively sought to promote sacred kingship as an integral part of shoring up royal authority in Castile and beyond. The 1620s marked a critical moment in Philip's reign; the decade was his first as king, heir to a vast empire that had fallen prey to grave economic and political crises. The monarchy was not only vast, but politically disparate – that is, nothing unified the various parts of the kingdom (no laws, language, currencies) except loyalty to the king himself. In their composite monarchy, Habsburgs employed religion as crucial means to create (or to attempt to create) cohesive community in the absence of a politically unified state.[9] Philip IV, like most seventeenth-century monarchs, had a keen eye for propaganda and the opportunities it provided for augmenting authority.[10] By taking a careful

Cities: Emperor Charles V, the Comunero Revolt, and the Transformation of the Spanish System (Leiden: Brill, 2008), 289.

[7] John H. Elliott, "The Habsburg Court: A Peculiar Institution," in *Spain and its World* (New Haven, CT: Yale University Press, 1989), 142–61. On the dangers of absent kings, see María de los Ángeles Pérez Samper, "El rey ausente," in *Monarquía, imperio y pueblos en la España moderna*, ed. Pablo Fernández Albaladejo (Alicante: Universidad de Alicante, 1997), 379–93.

[8] On the devotion of Philip IV to one Marian shrine and how it changed the devotional and civic landscape of Madrid, see Jeffrey Schrader, *La Virgen de Atocha: Los Austrias y las imágenes milagrosas*, trans. Teresa Sans and Fabián Chuenca (Madrid: Ayuntamiento de Madrid, 2006).

[9] John Elliott argues: "Church and king were the two common elements in the disparate and fragmented Spanish monarchy; and religious uniformity, enforced by king, church, and inquisition, was the guarantee of continuing political order and stability": "Power and Propaganda in the Spain of Philip IV," in *Spain and its World*, 169.

[10] We can see this in Philip IV's portraiture, architectural projects, and perhaps most importantly, sponsorship of history-writing. For an important discussion of Philip and the count-duke of Olivares' patronage of histories, see Richard L. Kagan, *Clio and*

look at religious festivals – routes, decorations, and festival books – we can see how they acted as key sites for the production and dissemination of a specific vision of a holy Catholic monarchy, in which disparate people were brought together through the witnessing of and participation in religious events. As Philip IV's reign continued, Spanish saints played increasingly significant roles as representatives of royal authority and national community. Rather than mediating between heaven and earth, the saints mediated between the king and his people, a transposition that both secularized the sacred and sacralized the civic.

***

Royal entrances had been an important part of Habsburg court culture in the sixteenth century, as in most other European kingdoms.[11] Such events were not only local celebrations, but culminated in the publication of a festival book – an account of the festivities often written in lavish detail.[12] While on the decline in seventeenth-century peninsular Spain, such festivals continued throughout the seventeenth century in the Americas in the entrances of viceroys, who acted as the representative of the king. Scholars of colonial Latin America have turned a keen eye to the representation of the viceroy and royal authority in imperial capitals as part of the process of colonization. The entrances of the viceroys into colonial cities, in particular, have been shown to be a key moment in the solidification of viceregal and royal authority in cities thousands of miles from the monarchy's capital.[13] As Alejandra Osorio remarks in her work

the Crown: The Politics of History in Medieval and Early Modern Spain (Baltimore, MD: Johns Hopkins University Press, 2012); see especially ch. 6, "'To Mortify Our Enemies': History and Propaganda at the Court of Philip IV."

11 Most of the recent historiography on festival focuses on civic festival, and sixteenth-century royal entrances in particular: J. R. Mulryne and Elizabeth Goldring, eds., *Court Festivals of the European Renaissance: Art, Politics, and Performance* (Aldershot: Ashgate, 2002); Helen Watanabe-O'Kelly and Pierre Béhar, eds., *Spectaculum Europaeum: Theatre and Spectacle in Europe (1580–1750)* (Wiesbaden: Harrassowitz, 1999); and John A. Marino, *Becoming Neapolitan: Citizen Culture in Baroque Naples* (Baltimore, MD: Johns Hopkins University Press, 2011).

12 It is worth noting that while we refer to such events as "entrances," the Spanish festival books almost always called them "recibimientos" (receptions), which placed the emphasis on the city receiving the king, rather than on the king's arrival.

13 Mulryne, Watanabe-O'Kelly, Showing, eds., *Europa Triumphans*, includes a section entitled "The New World: Seventeenth-Century Festivals in the Viceroyalties of Mexico and Peru," which contains work by D. A. Brading, Linda A. Curcio-Nagy, and Peter Davidson (ii, 345–434). See also Linda A. Curcio-Nagy, *The Great Festivals of Colonial Mexico City: Performing Power and Identity* (Albuquerque: University of New Mexico Press, 2004); and Alejandro Cañeque, *The King's Living Image: The Culture and Politics of Viceregal Power in Colonial Mexico* (New York: Routledge, 2004).

on Lima, "The majesty and power of the king was, therefore, established every time his alter ego appeared in public surrounded by magnificence and splendor."[14] Similarly, Linda Curcio-Nagy has argued that festivals in early modern Mexico City served a colonizing function.[15] Such authors locate the entrance processions and public presence of the viceroys as a pivotal moment for displaying and augmenting colonial authority. Their studies provide a template for understanding the ways in which royal power could be represented and enforced in festivals even when the king was absent.[16] Since the king was absent from Iberian towns and villages almost as completely as he was from Lima and Mexico City, studies of such festivals offer a parallel example to the meaning and uses of religious festival discussed here, where the saints acted as viceroys.

Alejandro Cañeque asserts that absence, like presence, carried with it its own form of power. The power of absence derived largely from the sacred. Attendants at Mass, for example, were not permitted to view the act of transubstantiation, considered too sacred for lay eyes.[17] Thus, monarchs took lessons from the Church on how to cultivate awe and power, which had the additional advantage of associating the king with the sacred. The saints themselves could act as representatives, as viceroys for the kings, representing his authority and working to foster royal authority and national identity. Recent work by Cornelius Conover on the cult of the saints in colonial Mexico City has called attention to what Conover dubs "imperial religion." He points out that "Spanish imperialism guided the long-established process of devotional change in Catholicism."[18] Spanish saints were uniquely situated to serve imperial needs – at home with God in Heaven, they maintained ties to the earthly

[14] Alejandra B. Osorio, *Inventing Lima: Baroque Modernity in Peru's South Sea Metropolis* (New York: Palgrave Macmillan, 2008), 62. Alejandro Cañeque also asserts, "On his [the viceory's] displayed body, exhibited in processions ... royal authority was legible for all": *The King's Living Image*, 121.

[15] Curcio-Nagy, *The Great Festivals*, 3–6.

[16] I am inspired here by the work of Barbara Fuchs: "Imperium Studies: Theorizing Early Modern Expansion," in *Postcolonial Moves: Medieval through Modern*, eds. Patricia Clare Ingham and Michelle R. Warren (New York: Palgrave Macmillan, 2003), 71–92. Fuchs argues: "[B]y reconceptualizing its object of study to focus on both colonies and metropole, this approach alerts the critic to the continuities and interdependence between the formation of early modern nations and their imperial aspirations" (72).

[17] Cañeque, *The King's Living Image*, 37. He connects the rhetoric of royal invisibility with religious etiquette surrounding the Eucharist (pp. 36–8).

[18] Cornelius Conover, "Catholic Saints in Spain's Atlantic Empire," in *Empires of God: Religious Encounters in the Early Modern Atlantic* (Philadelphia, PA: University of Pennsylvania Press, 2011), 87.

community and to the monarchy, to which they were bound by loyalty and birth. These ties of allegiance assured they would continue to work on behalf of its well-being as advocates at the heavenly court. The ways in which the sacred was intertwined with the projects of colonialism, nation-building, and community-identity formation in the Americas clearly speaks to how the sacred functioned back in peninsular Spain as well.

The Habsburgs put much focus on the cult of the saints – not merely veneration of saints generally, but in the promotion and spread of Spanish saints in particular. The seventeenth century marked an apogee for Spanish sanctity. A sixty-year break in canonizations (1523–88) brought about by the institutional crisis of the sixteenth-century Church ended with the 1588 canonization of Spaniard Diego de Alcalá (d. 1463). Forty years later, in 1622, four Spanish saints were canonized. Throughout the seventeenth century, roughly half the holy people beatified or canonized were born in the Spanish monarchy, a feat unparalleled in the history of canonizations.[19] The numbers reflect both the transformative role that Spain played in early modern Catholicism and the lobbying power of the Habsburgs monarchs at a critical juncture for the papacy.[20]

Monarchs had multi-faceted reasons for promoting the cults of specific holy people. One, of course, was personal piety. Villalon points out that Philip II remained devoted to the cult of San Diego after the holy man's relics miraculously cured his son, Carlos.[21] Aggressive promotion of Diego's status as a saint would serve to reinforce the validity of the royal miracle, and also act as a way for Philip to express gratitude for saving his heir's life. In addition to private devotion and prestige, cultic devotions played powerful roles in the extension, development, and

[19] The majority of saints in every century except the twentieth were from the Italian Peninsula. For the period of 1588–1767, Peter Burke has determined that seventeen out of fifty-five canonizations were of Spaniards, which is about a third, coming in second to the Italians, although the French were close behind with fourteen. Peter Burke, "How to Become a Counter Reformation Saint (1984)," in *The Counter-Reformation: Essential Readings*, ed. David Luebke (Oxford: Blackwell, 1999), 134–7. Twenty Spanish saints were beatified and fifteen canonized in the seventeenth century alone: Philippe Jansen and Dominique Le Tourneau, "Causes of Canonization," in *The Papacy: An Encyclopedia*, ed. Philippe Levillain, vol. I (New York: Routledge, 2002), 271.

[20] L. J. Andrew Villalon, "San Diego de Alcalá and the Politics of Saint-Making in Counter-Reformation Europe," *Catholic Historical Review* 83(4) (1997): 713–14. Villalon argues that Diego de Alcalá's canonization resulted from papal gratitude to Philip II for his role in taking on the English with the attempted Armada invasion that same year.

[21] Villalon, "San Diego de Alcalá," 695–702.

maintenance of political authority.[22] My purpose here is not to present an instrumentalist view of religious belief or practice – for example, that the Spanish Habsburgs used or manipulated the sacred for cynical political ends. The argument that saints could be deployed as agents for political authority does not negate the sincerity of the message, especially for Philip IV, who was not only deeply pious but felt himself personally responsible for the moral and spiritual well-being of his subjects. Such a sense of responsibility cannot and should not be passed over lightly, since it had the potential to shape royal policies. At the same time, however, we must not ignore the aggressiveness with which the Habsburg monarchs pursued the consolidation and expansion of royal authority throughout the monarchy, including the Spanish Church, often at the expense of the papacy when the papal will conflicted with theirs.

Political messages found a broad audience in the performance of public festivals.[23] Unlike commissioning a work of art, or the development of elaborate court rituals, staging festivities in the royal capital could mobilize thousands of people as participants and spectators, including members of religious orders, the royal family, courtiers, nobility, foreign visitors, and people from all walks of life. Festivals were more than liturgical solemnities; they were parties. Since a feast day was by definition a holiday, it marked a time when laborers took rest, alms and other prizes of money were distributed, and public entertainment was performed, such as bull fights, literary contests, and firework displays. The wide array of entertainments and a general atmosphere of celebration ensured that such events would draw a large audience. Audience members were more than passive observers – they were participants who could walk in

[22] Augustine Thompson discusses this process in the context of late medieval Italian cities in *Cities of God: The Religion of the Italian Communes, 1125–1325* (University Park, PA: Pennsylvania State University Press, 2006). For examples of patron saints and the state, see Erin Kathleen Rowe, *Saint and Nation: Santiago, Teresa of Avila, and Plural Identities in Early Modern Spain* (University Park, PA: Pennsylvania State University Press, 2011); and Florence Buttay-Jutier and Axelle Guillausseau, eds., *Des saints d'État? Politique et sainteté au temps du concile de Trente* (Paris: Presses de l'Université Paris-Sorbonne, 2012).

[23] Edward Muir, "The Eye of the Procession: Ritual Ways of Seeing in the Renaissance," in *Ceremonial Culture in Pre-Modern Europe*, ed. Nicholas Howe (South Bend, IN: University of Notre Dame Press, 2007), 129–53. Muir claims: "If this exposition of the radiation principle is correct, then the goal of the civic and church officials ... was to create numerous points for beneficent spiritual emanations that would project piety, passivity, and peacefulness into the souls of the populace ... The procession included besides sacred objects, people, typically ecclesiastics or civic officials whose authority when so displayed also irradiated the city" (140).

the processions holding floats, cram into a church to kiss or view relics, or help build or adorn the decorations that were placed throughout the city.[24] This participation in turn led to a higher degree of emotional connectivity to the events transpiring, and perhaps of the overall efficacy of the propaganda being disseminated.[25] Scholars have demonstrated that a great deal of the persuasive power of rituals stems from their ability to – in the words of Edward Muir – "give access to emotional, psychological, or spiritual states that often resist expression in language."[26]

Rituals used by Habsburg kings underwent a material change as the sixteenth century passed into the seventeenth. As mentioned above, kings began to travel less and to rely less on their physical presence to create authority. Festivals, however, continued and even multiplied throughout the same period, becoming more elaborate as well as more frequent. Festivals that could serve as vehicles for celebrations of the Habsburg dynasty or an aid to solidifying authority took two main forms in Spain: 1) celebrations in honor of events in the royal life cycle (such as baptisms, weddings, and funerals); and 2) the religious festivals of Corpus Christi and Spanish saints. These two categories played complementary roles, though the political significance of religious festivals has been understudied. While the celebration of religious festivals had traditionally been an important element of Catholic piety, the seventeenth century witnessed a dramatic increase in the frequency, intensity, and elaborateness of such events.[27] In this essay, I focus on Spanish saints, since the development of the role of the Spanish saint vis-à-vis the monarchy took a central role in the seventeenth century as the saints became more deeply intertwined with political legitimization and imperial propaganda.

[24] As Nicholas Howe argues, "To observe meant to participate": "Introduction," *Ceremonial Culture*, 4. In this passage, Howe contrasts the kind of participatory observation we see in early modern festivals with the passive observation of twentieth-century television watching.

[25] Political messaging became increasingly important throughout the seventeenth century, as newspapers and pamphlets vied for an ever-expanding public opinion. For recent work on public opinion in early modern Europe, see Judith Pollman and Andrew Spicer, eds., *Public Opinion and Changing Identities in the Early Modern Netherlands* (Leiden: Brill, 2006); and Andrew Pettegree, *Reformation and the Culture of Persuasion* (Cambridge: Cambridge University Press, 2005). José Antonio Maravall viewed the desire to "move" people as a quintessential feature of the Baroque: *Culture of the Baroque: Analysis of a Historical Structure*, trans. Terry Cochran (Minneapolis, MN: University of Minnesota Press, 1986; Spanish ed. *La cultura del Barroco*, 1975), 75–100.

[26] Muir, "The Eye of the Procession," 148. Curcio-Nagy agrees, claiming: "Officials sought to stir up emotions, to affect, attract, persuade the masses to their message": *The Great Festivals*, 6.

[27] This point is made by García Bernal, *El Fasto Público*, 270–4.

Two important festivals celebrated in Madrid during the early years of Philip IV's reign highlight such development: the 1622 events honoring the unprecedented canonization of five saints, four of whom were Spanish; and the feast day of St Teresa of Avila in 1627, following the papal brief proclaiming her co-patron saint of Spain. The festivals celebrated distinct events, yet both reveal the significant ways that religious celebrations could express and communicate a specific vision of community and sacred kingship.[28] They both emphasized the "Spanishness" of their saints in ways that connected the religious celebration to royal authority and power. The public performance of these proceedings in the capital city provided poets, preachers, artists, and those who wrote accounts of the events a unique opportunity to disseminate specific visions of the king and monarchy. By tracing the way festivals depicted the king and the saints, we can understand more clearly how the saints were transformed into conduits of royal authority.

1622

In 1622, Pope Gregory XV (1621–3) canonized five saints, four of whom were Spanish: Teresa of Avila, Ignatius of Loyola, Francis Xavier, and Isidore the Laborer.[29] Celebrations erupted throughout the Catholic world; the Spanish kingdoms celebrated this unprecedented cluster of canonizations with particular energy. But nowhere were the celebrations more elaborate that in the capital city of Madrid. One of the Spanish saints recognized by the pope was the city's beloved patron, Isidore the Laborer, a twelfth-century farmer venerated for his holiness in the area around Madrid since his death. In contrast, the other four were contemporary saints who died at the end of the sixteenth century and whose cults served as the basis for the Tridentine Church's efforts at revitalizing Catholic spirituality. In this illustrious group, Isidore appears as something of an odd man out. While the other saints had the power of weighty religious orders and important men of the Church behind their cults, Isidore was a layman with a small regional cult. The municipal authorities in Madrid had been pushing for their patron's official beatification

[28] García Bernal argues that sacred processions performed the same function as royal entrances: "Sus funciones polisémicas en la fijación de la memoria colectiva, demarcación de identidades y espacios de poder, no han dejado de ser subrayadas por los especialistas": *El Fasto Público*, 265.

[29] The Italian saint Philip Neri (d. 1595), who founded of the Congregation of the Oratory in Rome, was also canonized at this time.

and canonization since the end of the sixteenth century.[30] Their ability to do so successfully is testimony to the power of the Spanish faction at the Holy See. Additionally, Philip III (1598–1621) was purportedly healed by Isidore's relics, a miracle that echoed the one attributed to Diego de Alcalá and raised Isidore to the status of a royal as well as civic patron.

The celebrations that erupted in Madrid in the summer of 1622 were orchestrated (and paid for) by the municipal council, as most festivals were. While the town and court, municipal authorities and the king, were often in conflict, Madrid derived its wealth and status from its position as royal court. Unlike other European capitals that had been thriving cities in their own right before becoming official capitals, Madrid had been a sleepy farming village before Philip II settled the court there in 1561. In spite of Philip's decision, however, Madrid's status as court remained uncertain; after the consecration of El Escorial, Philip II spent much time in his new palace-monastery, while his son, Philip III, under the advice of his chief minister the duke of Lerma, moved the court to Valladolid from 1601 to 1606.[31] The 1622 celebrations provided the city the ideal opportunity to celebrate its solidified prestige, symbolized by the completion of the most important architectural site in the city: the Plaza Mayor, which acted as the central stage for the festivals, as we will see.[32]

As in all feast days celebrated in the seventeenth century, the festivities in honor of the 1622 canonizations included masses, sermons, processions, poetry contests, and theater (including the performance of two plays by Lope de Vega honoring Saint Isidore).[33] The author of one of the *relaciones* (festival books) in honor of the event was the Golden Age poet and playwright Lope de Vega, whose account I follow here, partly because his longstanding quest for a royal office suggests that his account might

[30] On the history of efforts to elevate Isidore, see Elizabeth R. Wright, "Virtuous Labor, Courtly Laborer: Canonization and a Literary Career in Lope de Vega's 'Isidro'," Hispanic Issue of *Modern Language Notes* 114(2) (March 1999): 223–40. See also María José del Río Barredo, "Literatura y ritual en la creación de una identidad urbana: Isidro, patrón de Madrid," *Edad de Oro* 17 (1998): 146–68.

[31] On this decision, see Antonio Feros, *Kingship and Favoritism in the Spain of Philip III* (Cambridge: Cambridge University Press, 2000), 86–7.

[32] The most comprehensive work on the Plaza Mayor is: Jesús Escobar, *The Plaza Mayor and the Shaping of Baroque Madrid* (Cambridge: Cambridge University Press, 2003). On Madrid's architecture, see also *Juan Gómez de Mora (1586–1648): Arquitecto y trazador del rey y maestro mayor de obras de la villa de Madrid* (Madrid: Concejalía de Cultura, 1986).

[33] Reyes Escalera Pérez provides an excellent overview of the components of festivals: *La imagen de la sociedad barroca andaluza* (Málaga: Universidad de Malaga, 1994), 22–33, 271–81.

have been crafted to please the king, hence emphasizing the version of the event that the king wished observers to see.[34] Festival books offered official accounts of the event, including descriptions of processional routes and ephemeral decorations in varying degrees of detail. Because the two accounts I analyze here were officially produced, they reflect idealized, not complete, accounts of what transpired.[35] *Relaciones*, of course, are not only source material for the events, they were part of the events, designed to provide an account that would glorify the city and/or king long after the ephemeral decorations had vanished. The authors of official *relaciones*, therefore, had an eye to the immortalizing aspects of the genre; their job was to transform a moment in time and space into an enduring historical memory.

The bishop of Valladolid, Enrique Pimentel, led the procession in pontifical vestments, followed by the members of the Councils of Castile, the Indies, the Orders, and Finance. While religious festivals in all Spanish towns included members of the municipal government, one unique feature of Madrid's is the participation of members of royal administration. The joint participation of members of the secular government (municipal and royal) along with members of the clergy highlighted the unity of secular and religious authority.[36] The procession passed from the parish church of San Andrés, where Isidore's relics were housed, to the Plaza de la Cebada in the southern part of the city along the Calle de Toledo, one of the city's largest thoroughfares, up to the Plaza Mayor and then to the

[34] I chose Lope de Vega's account because he was deeply involved in many aspects of the festival: writing the commemorative book; overseeing the poetry contest; and writing and performing two plays. See Hugo Albert Rennert, *The Life of Lope de Vega, 1562–1635* (Glasgow: Gowans & Gray, 1904), 277–96; and Elaine M. Canning, *Lope de Vega's Comedias de Tema Religiosos: Re-Creations and Re-Presentations* (Woodbridge: Boydell & Brewer, 2004), 49–84. On Lope's struggles to gain royal patronage, see Elizabeth R. Wright, *Pilgrimage to Patronage: Lope de Vega and the Court of Philip III, 1598–1621* (Lewisburg, PA: Bucknell University Press, 2001) (see especially p. 139 for her discussion of Lope and the 1622 celebrations). Two additional relaciones are: Fernando de Monforte y Herrera, *Relación de las fiestas que ha hecho el Colegio Imperial de la Compañia de Jesus de Madrid en la canonización de San Ignacio de Loyola y S. Francisco Xavier* (Madrid, 1622); and Miguel de Leon, *Fiestas de Madrid, celebrados de XIX de Junio de 1622 años* (Madrid, 1622).

[35] A "complete" account might include alternate perspectives (there were occasionally multiple accounts of *relaciones* produced for one event), informal accounts (from diaries or correspondence), and structural information (more detail on the behind-the-scenes pre-production). Helen Watanabe-O'Kelly cautions against reading festival books as eyewitness accounts or as fully "accurate" in "The Early Modern Festival Book: Function and Form," in *Europa Triumphans*, i, 9.

[36] García Bernal, *El Fasto Público*, 282.

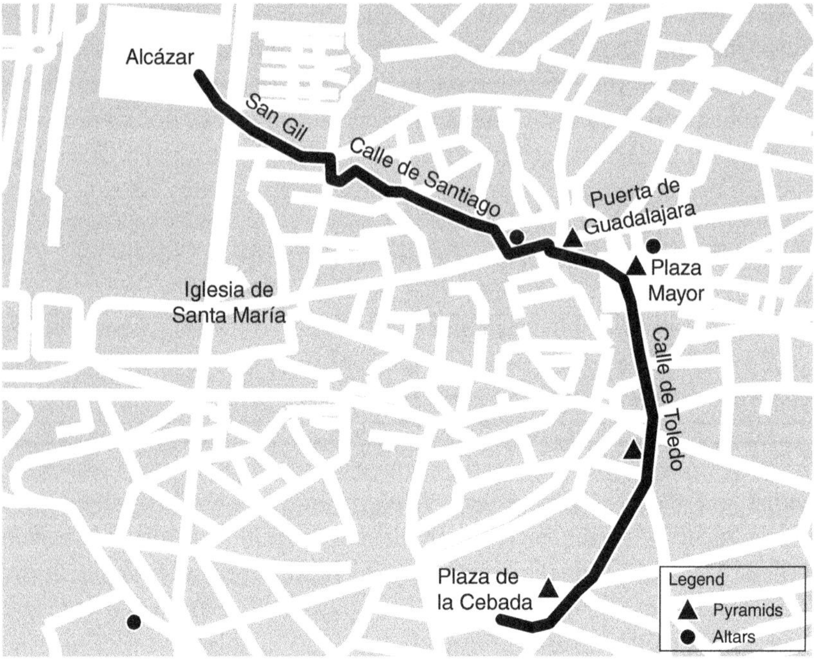

MAP 3.1. Map of procession route, 1622 canonization festivals, with pyramids and altars, Madrid

Puerta de Guadalajara. From there, it most likely moved up the Calle de Santiago to the Alcázar, the royal palace. This last phase of the procession followed the route already established for the major processions associated with the feast of Corpus Christi.[37] The procession thus moved along a route that was both sacred and civic, encompassing the major streets and monuments of Madrid's urban landscape. Like the figures participating in the procession, the space in which they moved echoed the integration of sacred and secular.

The route was marked by ephemeral objects built for the event; these were pyramids and altars that not only guided the walkers but also highlighted the central public spaces of the town.[38] Lope de Vega describes the pyramids as structures seventy-four feet high, sitting on pedestals of

[37] Escobar argues that some of the streets between the Alcázar and the Plaza Mayor had been reformed in the latter decades of the sixteenth century in order to facilitate the Corpus Christi festivals: Escobar, *The Plaza Mayor*, 186.

[38] Our knowledge of ephemeral decorations at public festivals remains incomplete; however, for a preliminary foray, see Fernando Moreno Cuadro, *Artistas y mentores de barroco efímero* (Cordoba: Servicio de Publicaciones de la Universidad de Córdoba, 1985).

twelve and a half feet high and seven feet wide, flanked by smaller pedestals that held statues of eight and half feet decorated in gold. They were adorned with elaborate images and allegorical figures, along with "hieroglyphic" inscriptions in both Latin and Castilian. The Egyptian flare is an odd addition to the Catholic festival, unconnected thematically to the lives of any of the saints being honored. The Egyptian objects could serve multiple functions, including reflecting the spike in popularity of Egyptian styles during the early modern period. They also quietly asserted Spain's Egyptian genealogy (going back to Hercules), and the ancient nobility of the Habsburg dynasty. Hence, in the middle of the celebration of the canonization of four Spanish saints – by Rome – the pyramids claimed Spain's greatness equal to Rome's.[39] The key sites of the city where the pyramids could be found were the Plaza de la Cebada, the Calle de Toledo, the Plaza Mayor, and the Puerta de Guadalajara.

Each pyramid was dedicated to one of the newly canonized saints, clustered together by religious order: the Guadalajara entrance depicted the new Jesuit saints, Ignatius and Francis Xavier; the Plaza Mayor held images of Isidore; the Calle de Toledo represented the Italian saint, Philip Neri; and the Carmelite Teresa of Avila was found in the Plaza de la Cebada. In addition to the saints honored in each location, the sites contained additional figures, generally positioned in holy duos. Ignatius and Francis Xavier appeared alongside Luis de Gonzaga (d. 1591) and a Jesuit martyr (representing all Jesuit martyrs). Gonzaga, beatified in 1605, was a member of the ruling family of the Italian duchy of Mantua, which maintained close political ties with the Spanish monarchy.[40] Philip Neri stood alongside Pedro de Alcántara (d. 1562), a Spanish Franciscan who was beatified in 1622. Aside from biblical figures, Neri and Gonzaga are the only two non-Spaniards depicted during the events, and Gonzaga was connected to one of the Habsburg's key Italian allies. The insertion

[39] The Spanish Habsburgs made the mythical figure of Hercules one of the main symbolic representations of its dynasty; it is less well known that early modern antiquarians often claimed Hercules was the son of the Egyptian god, Osiris. See Brian Curran, *The Egyptian Renaissance: The Afterlife of Ancient Egypt in Early Modern Italy* (Chicago, IL: University of Chicago Press, 2007), 117–24; and R. B. Tate, "Mitología en la historiografía española de la edad media del renacimiento," in *Ensayos sobre la historiografía peninsular del siglo XV*, trans. J. Díaz (Madrid: Gredos, 1970), 13–32. For a discussion of the Habsburgs and their Herculean imagery, see Jonathan Brown and John H. Elliott, *A Palace for a King: The Buen Retiro and the Court of Philip IV* (New Haven, CT: Yale University Press, 2003), 163–70.

[40] The Spanish monarchy became involved in a disastrous and humiliating war in northern Italy over the succession of the duchy of Mantua, 1628–31.

of Alcántara next to Neri can be read as an attempt to neutralize Neri's foreignness by having him appear next to a contemporary Spanish holy person.[41] The emphasis on the Spanish origins of these saints is striking, and suggests that the celebrations projected a specific vision of holy community, anchored by the presence of the king and the civic patron Isidore.

The Plaza Mayor marked the halfway point of the procession, and it is here that the king appeared, as we have seen: "And when the procession arrived at the houses in the Plaza Mayor that are called the Panadería, where the greatest princes and lords live," Lope noted, "the King, our lord, came to accompany it [the procession], along with the Councils of Aragon, the Inquisition, and Italy."[42] The emergence of the king into the Plaza Mayor half-way through the procession underscores the significance of the plaza as the central space for the event. The king and his councilors then joined the procession from the plaza north to its termination at the royal palace. Three nights of fireworks lit the sky in front of the royal palace, and spectators were delighted with a *juego de cañas*, a popular game involving thirty participants on horseback with shields and lances, as well as a hunt of stag and bear.[43] The involvement of both the royal palace as a site for part of the festivities and of the king himself, as well as the participation of the members of the most powerful royal councils, emphasized the civic and royal dimensions to the event.[44]

If the plazas abutting the royal palaces featured the most spectacular events of the festival, the Plaza Mayor was the most symbolically potent locus. The Plaza Mayor hosted the pyramid dedicated to Saint Isidore along with allegorical representations of the king and Madrid. Here too the temporary stage was erected for the performance of plays written in honor of the civic patron's canonization (including two by Lope himself).

[41] Other than being contemporaries, there is no obvious connection between these two figures – Neri was an Italian priest, and Pedro a Spanish Franciscan.

[42] Lope de Vega, "Relación," 454–5.

[43] Lope de Vega, "Relación," 457. For an image of a *juego de cañas* and a description of civic festivals in the Plaza Mayor, see Glyn Redworth, *The Prince and the Infanta: The Cultural Politics of the Spanish Match* (New Haven, CT: Yale University Press, 2003), 113–16.

[44] The king was present for the vespers of the festivals as well: "La vispera deste dia fueron a palacio en alarde con musica de trompetas y chirimias todas las danzas que la Villa tenia prevendias y los carros referidos, como previendo los animos a la esperanza de las fiestas, y alegrando a sus majestades, que favorecieron este dia con su real presencia." Lope de Vega, "Relación," 455. Ronald L. Grimes asserts: "Moving through the civic center of town, the plaza, is a way of declaring symbolically that the event is of citywide, not merely churchwide, significance": *Symbol and Conquest: Public Ritual and Drama in Santa Fe, New Mexico* (Ithaca, NY: Cornell University Press, 1976), 68.

The pedestals in the plaza held statues of Pope Saint Damasus (d. 380), whom the city claimed as a native son, Isidore with his wife, Saint María de la Cabeza, and Philip IV together with Spain and Madrid. The images were accompanied by more hieroglyphics. The allegorical image associated with Pope Damasus was a bear with seven stars, the heraldry of the city, highlighting Damasus' role as an illustrious native of Madrid.[45] Philip was represented as a lion rampant alongside a caduceus and a cornucopia, with a verse proclaiming: "With counsel, royal power augments the good republic and sustains peace and abundance." Spain was depicted as the king's lady, sumptuously dressed and crowned, sitting atop two worlds ("ours and the New World"), representing the far-flung Spanish monarchy and claims to universal sovereignty. Madrid appeared as a royal peacock (*pavón real*) looking at the sun, with the following verse: "The presence of Philip, divine Spanish Caesar, transforms my tail [feathers] in the sunlight."[46] The verse evokes the image of a peacock, tail feathers extended, basking in the sunlight of the king, which in turn reflects onto the feathers, rendering the bird even more beautiful, demonstrating the glory Madrid derived from its proximity to the person of the king.[47]

It is no coincidence that this cluster of images – Damasus, Isidore, Philip, Spain, Madrid – appear together in the heart of the city. The inscriptions and images connected to Philip, Spain, and Madrid proclaim the well-being and majesty of king and monarchy. It is not the saints, but rather the king and town that stood at the center of the festival. Philip IV's presence, in allegory and on the streets of the city, gave shape and meaning to the events; he acted not only as a participant, but also became an object of devotion alongside the saints. Rather than gathering to rejoice in the recognition of new holy people in Heaven, we see

[45] Escobar, *The Plaza Mayor*, 45. Escobar cites a late sixteenth-century author, Juan López de Hoyos, who claimed that the area surrounding Madrid used to be called "the land of the bears."

[46] Lope de Vega, "Relación," 450–1. The verse reads: "La presencia de Felipe,/ César divino español,/ Convierte mi rueda en sol."

[47] Covarrubias repeats the common belief that peacock feathers shine more beautifully in the sun: Sebastián de Covarrubias Orozco, *Tesoro de la Lengua Castellana o Española [1611]*, eds. Felipe C. R. Maldonado and Mauel Camarero (Madrid: Castalia, 1995). Underneath peacock, Covarrubias also provides a verb derived from the root for peacock: *pavonear*, which means "es hacer muestra de su gentileza; pasear por lugares publicos a fin de ser vistos y mirado" (to demonstrate one's gentility; to walk through public places in order to be seen and looked at) (890). This intriguing verb suggests that peacocks were associated with nobility, who, in the allegory of Madrid, cluster around the king as members of his court.

collective acclamation of the power and spiritual wealth of the Spanish monarchy. In this instance, the political did not merely absorb the power of the holy through proximity; rather, it changed the significance of the holy, incorporating it into Spanish and royal greatness.

Lope de Vega's cycle of plays about Saint Isidore commissioned for and performed during the 1622 festival expands on some of the dominant themes of the festival. One play, *La juventud de San Isidro* (*The Youth of Saint Isidore*), begins with a *loa* – a prefatory piece designed to praise the important members of the audience.[48] In this case, the *loa* was spoken to the king, who was present during the performance. Lope declared that the young king, then in the first year of his reign, would bring great things to the monarchy – "the first year begins/[with] honors, crowns, laurels,/ in Heaven and on earth."[49] The honors, crowns, and laurels naturally included the four Spanish saints. Lope addresses the king directly: "And you, said Prince,/ high and divine defender / of the Church, to whom you sent crowns of such precious stones." Each saint was represented by a different jewel: Ignatius a ruby, Francis Xavier an emerald, Teresa a sapphire, and Isidore a diamond. The relationship between sanctity and kingship is made clear by Lope's choice of metaphor – the saints are jewels, both decoration and part of the symbolic accoutrement of kingship. As adornments, they reflect the king's wealth and glory (spiritual and material); furthermore, they are *his*, given as gifts to the Catholic world. Here we see the rhetoric of Spain in service to the Church subverted, as the saints become tools for demonstrating the power and sacredness of the Habsburgs.[50]

After this description of the saints as jewels, Lope plays at some length with the number four in honor of the king. He points out the four Philips and four saints: "You are the fourth of the Philips/the sun in its fourth sphere/shines; four the saints/ that today our Spain celebrates."[51] Here Lope connects the king with the sun, shining on all. Moreover, the four saints share a line with the shining of the sun. The verb Lope used to describe the sun's action, "da su luz," was also an idiom in Spanish for "to

[48] On the tradition of the *loa*, see Jonathan Thacker, *A Companion to Golden Age Theatre* (Woodbridge: Boydell & Brewer, 2007), 153–4.

[49] Lope de Vega, "La juventud de San Isidro," in Marcelino Menédez Pelayo (ed.), *Comedias de vidas de santos* (BAE 178, Madrid: Atlas, 1965), 363. The "piedras" (stones) are described as Isidore the diamond (light and fortitude), Ignatius the ruby (charity), Xavier the emerald (chastity), and Teresa the sapphire (zeal and penitence).

[50] *Ibid.*, 364.

[51] *Ibid.* "Cuarto sois de los Felipes;/el sol en su cuarta esfera/ da su luz; cuatro los santos/ que hoy nuestra España celebra."

proclaim."[52] The structure of the verses leaves the impression that Philip brought forth the saints just as the sun illuminated the world; they were lights that originated in his majesty. The ideal king and his happy subjects here acted as an invincible source of military and spiritual victories, while the enemies who feared the might of the Spanish numbered four as well.[53] Four is proclaimed a whole and even perfect number; Philip embodies the perfection of his status as the fourth. Far from standing on their own as objects of devotion, the saints' significance diminished as accessories to glorify the king.

The presence of the king on the streets and in allegory transformed the celebration of the canonization of new holy people into a triumphant celebration of the might of the Spanish monarchy, its identity deeply rooted in its divine favor and sacred mission. In this way, Madrid acted as a microcosm for the Spanish monarchy. For his part, the king appeared alongside the saints as an equal object of devotion, and in Lope's play, the source of Spanish holiness. The qualities of good kingship exhorted in the Plaza Mayor and praised by Lope's *loa* exhibited similar static qualities that we find in hagiography; thus, enacting an unchanging ideal of good (and sacred) kingship promised to sustain the power and wealth of the monarchy, to overlay prosaic and fickle realities with the immutable and eternal qualities of the Heavenly City. This festival provides a stepping-stone to the ways in which Philip would be incorporated into the cult of the saints as a stand-in for the cult of the monarchy.

CELEBRATING A PATRON SAINT

While the 1622 celebrations in Madrid came together to present an exalted vision of sacred kingship and Spanish greatness, there were limitations to the royal uses to which such a festival could be put. Outside of Madrid, the celebrations took on different shades of meaning, further removed from the king and monarchy.[54] So while festivals in Madrid

[52] Covarrubias, *Tesoro*, 722. (It is found under "luz," not "dar.")

[53] The four enemies listed are metonyms: North Africa (home of pirates lurking in the Mediterranean); England; Ottoman Empire; and most likely, France.

[54] Different cities might emphasize one of the canonized saints over the others, or printed accounts emerged from festivals hosted by a particular religious order. The canonization of its two founders, Ignatius of Loyola and Francis Xavier, meant that these two saints were celebrated with particular glory by the Jesuits in some cities: Francisco Ruiz, *Relación de las fiestas que hizo el Colegio de la Compañia de Jesus de Girona en la Canonizacion de su patriarca San Ignacio y del apostol de la India San Francisco Javier i Beatificacion del angelico Luis Gonzaga. Con el torneo poético* (Barcelona, 1623).

could play a powerful role in celebrating a specific vision of kingship and monarchy within the capital, it was not a message that traveled easily. Yet it provided a fertile start for the possibilities available to the Habsburgs for public devotions to Spanish saints, and the role this could play in helping to create unity and submission to the authority of the king.

A mere five years after the 1622 canonizations, a new opportunity presented itself during the feast of one of Spain's most popular early modern saints: Teresa of Avila (1515–82). Teresa's feast day in 1627 (October 5) was no ordinary festal event – it was a double octave, eight days of celebration at the highest level of liturgical formality, to proclaim Teresa the co-patron saint of Spain. The previous year, Philip and the count-duke of Olivares petitioned the parliament of Castile (the Cortes) to elect Teresa as patron saint of Spain, a title she would hold alongside Spain's traditional patron, St James. After the motion passed, the king asked the pope to issue a brief ratifying their choice of patron, which arrived in Madrid sometime in early September 1627. As soon as the brief arrived, Philip sent out an official request that all major Castilian cities and towns celebrate Teresa's next feast day as an octave in honor of her new patron sainthood.[55] The octave marked an innovative and daring moment of political propaganda for the monarchy. More than a reflection of royal power, Teresa became the representative for Spain and a symbol meant to rally disparate parts of the kingdom into a national community under centralized royal authority. The request that all towns celebrate her feast day provided Philip and Olivares with dozens of stages throughout the kingdom of Castile for the performance of Habsburg propaganda and authority, as the celebration of the new national patron saint required a triangulation of saint, king, and Castile, inextricably locked together. Teresa would act as the king's representative, his alter ego and viceroy; she was paraded around Castilian towns and cities promoting the king's authority and a unified vision of the sacred monarchy.

The anonymous author of the official *relación* of Teresa's feast day in Madrid described the king as the master of events; Philip provided the Discalced Carmelite church with musicians from the royal chapel, silver, and decorations, which included Tunisian linens and a canopy for the main altar. Underneath the king's canopy sat a life-sized image of Teresa

Notice in the title of this piece that the other three saints canonized in 1622 are absent; it celebrates the Jesuit saints only.

[55] Biblioteca Nacional de España, Madrid, MS 9140, f. 6r, Philip to the city of Jaén, September 27, 1627. For a complete study of the efforts to elevate Teresa to co-patron and its discontents, see, Rowe, *Saint and Nation*.

on a throne, surrounded by side tables full of relics and reliquaries, precious gems, and candles of costly white wax.[56] Visitors who wished to venerate Teresa's image throughout the first seven days were treated to music and singing organized by Pedro Testai, the royal chaplain. On the eighth day, the chapel was filled with the music of instruments and the sound of voices from the Royal Convent of Discalced Franciscans (which had close ties to the Habsburgs) and the Convent of the Incarnation. The liturgy of the first vespers and mass were celebrated by Don Alonso Pérez de Guzmán, a royal almoner and chaplain, who was also the son of the powerful duke of Medina Sidonia and a distant cousin to the count-duke. The service was attended by the representatives to the Castilian parliament and other important courtiers. Similar lavish displays filled the Convent of Santa Ana, the house of the female Discalced Carmelites, although these were on a slightly smaller scale.[57] The king's presence was everywhere in the details of the event; the material objects being displayed, to the people in attendance and performing the celebration, to the sight of the saint on a throne, an unusual way to depict a saint, emphasizing here the function of the new patron as the king's alter ego.[58]

The procession took place on the eighth day of celebrations, the capstone event of the festival. The central figure of the procession was an image of Saint Teresa. Carried on a float by Carmelite monks, the statue was dressed in black and white cloth embroidered with silver thread (*telas de plata*), a cloth so lavish that it required no other decoration.[59] While the

[56] Alejandro Cañeque and Michael Wintroub have both demonstrated the significance of the canopy for royal claims to sacredness, since only the king and the host could walk under it: Cañeque, *The King's Living Image*; and Wintroub, *A Savage Mirror: Power, Identity, and Knowledge in Early Modern France* (Stanford, CA: Stanford University Press, 2006). Presumably, the canopy in the church covered the altar where the host was kept, but placing the image of Teresa underneath might have invoked similar associations.

[57] *Relacion sencilla y fiel de las fiestas que el rey D. Felipe IIII nuestro Señor hizo al Patronato de sus Reinos de España Corona de Castilla que dio a la Gloriosa Virgen Santa Teresa de Iesus, año 1627* (Madrid, 1627), 4^{v}-6^{r}.

[58] It was a rare occurrence but not unique. A painting of St. Rose of Lima displayed during the celebrations for her beatification in Granada (1668) depicted her sitting upon the throne of angels (typically associated solely with the Virgin): Reyes Escalara Pérez describes this image: *La imagen de la sociedad barroca andaluza* (Malaga: Universidad de Málaga, 1994), 310. The enthronement of Spanish saints led to what García Bernal has described as the erasing of boundaries between the courtly and the celestial: *El Fasto Público*, 277.

[59] There are two royal portraits by Velázquez, one depicting Philip IV and one of Queen Isabel, in which the king and queen are both painted wearing dark cloth spun through with silver thread: *Diego Velázquez, Philip IV in Brown and Silver, 1631–32*, London, National Gallery; and Diego Velázquez, *Isabella of Bourbon, First Queen of King Philip*

cloth was otherwise unadorned, the saint wore a large diamond necklace, a small stone for her cloak clasp, a large jeweled broach, and a diadem. The author remarked that some of the pieces were estimated to be worth two hundred thousand ducats, and were possibly lent by the king's sister, the Infanta María, Queen of Hungary. The most striking detail about the image is the necklace, specifically described by the author as designed in the style of the Order of the Golden Fleece.[60] This chivalric order was headed by the Habsburg monarchs beginning with Charles V, and the Order's insignia was a common feature of royal portraiture.[61] The presence of the necklace created a visual and symbolic link between the saint and the Habsburg monarchs, just as the image in the chapel was presented sitting on a throne, like a king. She was to act as the king's viceroy, metaphorically sent out to all the major cities and towns in Castile to consolidate royal authority and promote a specific vision of community. The drama of Teresa's representation at the festival was heightened by the king's own absence here. Unlike the 1622 festivals, the king watched these processions from a balcony in the palace, hidden by curtains, present but invisible.

As the saint's image was carried in the procession, it was led by a standard in silver and crimson carried by Don Nuño Moxica, knight of the Order of Santiago and Avila's representative in the Castilian Cortes. Moxica was followed by "almost all" [of?] the other representatives to the Cortes. The procession began at the Church of Santa María, then moved northward to the royal palace along the Corpus Christi route. At the palace, the king and queen watched the procession from the windows, hidden from public view, as Philip was still too weak from a recent illness to participate. The author of the *relación* remarked here that when the image of the saint passed underneath the palace windows, every one (including the king and queen) made "particular and humble reverences," just as they would have during the Corpus Christi celebrations, which also passed by the palace windows. After passing by the palace, the procession progressed southward to the Puerta de Guadalajara, via San Gil and the Calle de Santiago. The central sites of the event existed on opposite ends

IV, Copenhagen, Statens Museum for Kunst. These paintings are briefly discussed in Laura R. Bass, *The Drama of the Portrait: Theater and Visual Culture in Early Modern Spain* (University Park, PA: Pennsylvania State University Press, 2008), 95–7.

60 *Relacion sencilla*, 6^r^.

61 On a portrait of Charles II that figures two separate images of the Golden Fleece (one on Charles II and one on a bust of Charles V), see Bass, *The Drama of the Portrait*, 63. Bass argues that the Golden Fleece provides one element to the "visual genealogy" provided the viewer in the portrait.

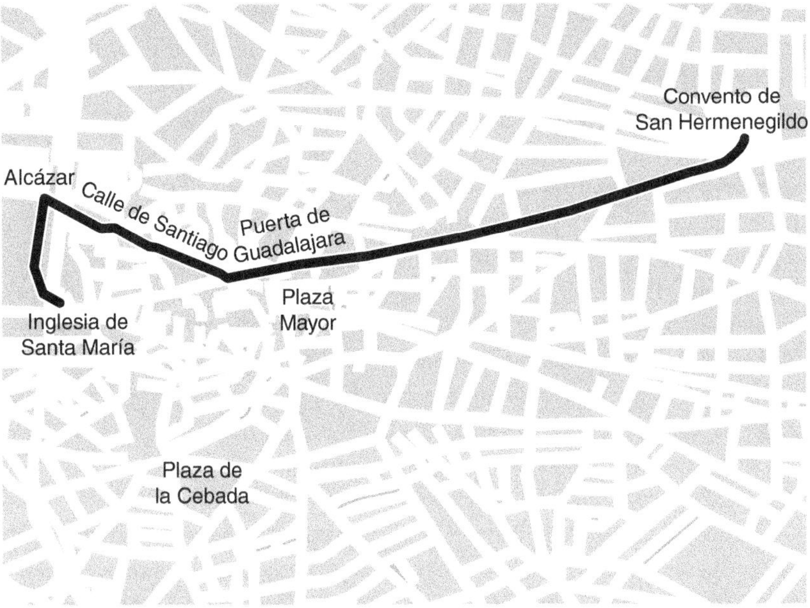

MAP 3.2. Map of procession route, 1627 feast day of Saint Teresa, Madrid

of a long pole, east to west, with the royal palace on one end and the Discalced Carmelite monastery on the other, the entire city stretched out between king and saint. The spatial dynamics of the procession acted as a visual metaphor; if Madrid could be understood as metonymy for Spain (situated in its center), then the king and the saint stood each on one end, marking their dominion over the interior territory.

The author of the *relación* claimed that from San Gil all the way to San Hermenegildo coaches filled with the nobility lined both sides of the street, as well as innumerable (*infinidad*) ordinary people on foot; so many people crowded the streets in fact that they hindered the movement of the procession, which had difficulty squeezing through the crowd. Like other religious festivals, the joyful and sumptuous holiday culminated in extraordinary displays of fireworks. The master of the fireworks was another powerful courtier: the count of Saldaña, Diego Gómez de Sandoval y Rojas, the son of the duke of Lerma, connected by marriage to the influential Mendoza/Infantado clan and married to one of the queen's ladies-in-waiting.[62]

[62] *Relacion sencilla*, 5v-9v. On the count of Saldaña, see Feros, *Kingship and Favoritism*, 98, 235, 240. The anonymous author of the *relación* cites Saldaña's devotion – and the devotion of his deceased father – as motivations for his involvement.

The *relación* is thin on details – unlike Lope's account of the 1622 festivals, the 1627 one does not provide much information on the decorations.[63] We do not know what kind of floats or ephemeral arches were created for the octave, which limits our ability to read the visual rhetoric of the event. In fact, near the end of the *relación*, the anonymous author addressed the absent details head-on. He declared that he made the king the sole focus of the *relación*; that is, everything he described came from the king himself, specifically from his order and regulations.[64] The author added that while other things occurred as part of the festival, he declined to include them in his description, because he did not want to mix the work of His Majesty with the works of other people, with the exception of the fireworks sponsored by the count of Saldaña. Thus, every aspect of the *relación* was deliberately meant to reflect the royal nature of the festivities and the king's direct participation in the elevation of the saint to patron.

We do, however, have details about another important aspect of the event: the sermons. For Teresa's octave, sixteen sermons were preached – that is, two per day – by fifteen different preachers, some of whom held the office of royal preacher.[65] After the octave was over, Philip requested that the preachers have their sermons printed, and granted them licenses to do so. The sermons were integral parts of the celebrations, one of the key events that would have drawn large crowds. Thus, the messages that preachers put forward formed a part of the larger message of the octave.[66] Preachers set out not only to justify the worthiness of the saint

[63] It is possible that the 1627 account was written beforehand, which might explain the paucity of detail. On this possibility, see Watanabe-O'Kelly, "The Early Modern Festival Book," in *Europa Triumphans*, I, 9.

[64] *Relacion sencilla*, 9^{r}: "Todo lo referido hasta aqui, fue orden y disposicion del Rey nuestro señor, y por esso no se han dicho otras cosas que huuo en esta fiesta, por no mezclar con la accion de su Magestad accion de otra persona."

[65] On the politics of the office of royal preacher, see Fernando Negredo del Cerro, "The Royal Chapel as the Setting for Political Struggle: Praise and Attacks on the Royal Favourite during the Time of Philip IV," in *The Royal Chapel in the Time of the Habsburgs: Music and Ceremony in the Early Modern European Court*, eds. Juan José Carreras and Bernardo García García, trans. Yolanda Acker; English edition edited by Tess Knighton (Woodbridge: Boydell & Brewer, 2005), 229–45. On the printing of the sermons, see *Relacion sencilla*, 7^{v}.

[66] We do not know whether or not the listeners were in fact persuaded. But the rhetorical power of some of the sermons is evident even from reading them. And José Maravall reminds us, Baroque preachers were masters in the art of provoking emotional responses from listeners: "The efficacy in affecting, in awakening and moving the affections, was the great motive of the baroque" (74). He points out that persuasion, in contrast to mandate, demands "a greater participation on the side of the guided" (74) which perhaps

to act as patron but also to emphasize the king's role in her election. To this end, nearly all the preachers wove together the themes of Teresa's spiritual virtues, her patron sainthood, Philip's greatness and piety, and the divinely mandated successes that would follow from the relationship between the two. They argued that patron sainthood played an integral role in the defense and expansion of the monarchy, in preserving and augmenting its power and greatness. They elucidated a vision of sacred kingship in which the saints were mobilized on behalf of the king, due to the great debt they owed him as Spaniards. This sent a clear message to the listeners and readers as to the proper loyalty and obedience owed to their sovereign. The theological content of the sermons (i.e. exegesis on a specific scriptural passage and its relationship to the saint) was often superseded by what we can call, for lack of a more nuanced term, political messages.

Teresa's role as the king's advocate required her to maintain her focus as patron saint on securing the longevity and power of his kingdoms.[67] Discalced Carmelite Tomás de San Vicente explained to his listeners: "We have ... great confidence in our Patron, since she interceded with her most loving Spouse on behalf of the health of our Catholic King, that she will secure victory for our armies, rectitude of judges, abundance of crops, reformation of customs, perfection of the religious orders, defense of the Faith and Justice – in short, that she will fulfill all our necessities so that this Monarchy can always be an example to the world."[68] Philip's choice of co-patron saint, therefore, would aid him in fulfilling the duties of good kingship and usher his people into an unprecedented period of prosperity; like an agent of the king, she would ensure that the king's desired policies went into action and were successful. The saint supported the king and facilitated his greatness; the king did not supplicate to the

makes sermons the ideal rhetorical form for the Baroque: José Antonio Maravall, *Culture of the Baroque*, 74–5.

[67] The king's health was of great concern during the autumn of 1627. He had been gravely ill and was just recovering. Simultaneous with Teresa's octave were processions celebrating the king's recovery: Gerónimo Gascón de Torquemada, *Gaçeta y Nuevas de la Corte de España desde el año 1600 en adelante* (Madrid: Real Academia Matritense de Heráldica y Genealogía, 1991), 269–77.

[68] Tomás de San Vicente, *Sermon predicado por el Padre Fray Thomas de San Vicente, Religioso Descalço de la Orden de Nuestro Señora del Carmen, en el Conuento de sus Monjas, el dia sexto de las Octauas, que el Rey Don Felipe IIII. Nuestro Señor celebrò a una en los dos Conuentos de Carmelitas Descalços, y Descalças de Madrid, a la fiesta del Patronato de la gloriosa Virgen Santa Teresa Fundadora desta Reformacion, Patrona de los Reynos de España Corona de Castilla; impresso por mandado de su Magestad* (Madrid, 1627), 15r.

saint. As a subject of the Habsburg dynasty, Teresa owed loyalty to her native land and to her king; her miraculous healing of Philip was characterized in sermons as a spontaneous gift from a loyal and loving subject. If the saints were models for Christian behavior, here Teresa was the model for the proper behavior of a subject as well, mobilized to establish an obedient and unified population.

Mercedarian preacher Francisco Boyl's rhetoric reached the heights of messianism when he reminded his listeners of the history of Spain's expansion from a tiny kingdom in the north to its present status as a global empire. He credits St. James' patron sainthood for the growth of the kingdom from a tiny corner of Asturias to twice its size, then declared: "I profess that, because it is the will of Heaven, inspired by God, the Holy See approved it [Teresa's election] in the time of our glorious, triumphant, Catholic, and magnanimous King Philip the Fourth, a clear sign has been given that there are more kingdoms to add to these. And until it [the Spanish scepter] occupies both ends of the globe, including the holy City, glorious tomb of God and theatre of His unique marvels, God will not sheath His steel in its defense."[69] This prophetic announcement made it clear that Teresa's role was to facilitate the will of God in elevating the Spanish monarchy to world domination. Moreover, in Boyl's rhetoric it was God who defended Philip, rather than the more traditional trope of Spanish king as defender of the faith. While a promise that Philip might one day re-conquer Jerusalem was unrealistic at best, its rhetorical value was immense: only when Spain had reached across the entire globe and encompassed Jerusalem would God's efforts on behalf of the Habsburgs be complete.

* * *

Religious symbols played a key role in the production of royal authority in seventeenth-century Spain. While many scholars claim that the Habsburgs placed themselves in proximity to sacred objects in order to transfer sacred power onto themselves by association, these festivals suggest that placing the person of the king next to holy objects promoted a sense of similitude rather than difference between the king and the sacred. Throughout his reign, Philip IV increasingly relied on the sacred nature of kingship to consolidate his authority in the monarchy and to position himself among and above the saints. The emotion generated by the

[69] Francisco Boyl, *Al Rey Nuestro Señor D. Felipe IIII, el Catolico Invicto, Magnanimo, Fundador zeloso del nueuo Patronazgo de sus Reynos en Santa Teresa Virgen, el año de su salud reparada* (Madrid, 1627), 13^{r}.

holy and careful staging of the city streets (including the king's embodied presence) acted powerfully in concert for the performance of a vision of sacred kingship and royal policies. The events in honor for Teresa were followed by the development of the Corpus Christi festivals in Madrid as key performances of sacred kingship in the 1630s and 1640s.[70] Michael Wintroub has noted a similar trend in early modern France, wherein representations of King Henry under the canopy alongside of the Host served as "a means of effecting a ritual equivalence between Henri and the Eucharist, as the *real* presence of Christ on earth."[71] Jeffrey Schrader has pointed to how Philip IV made repeated efforts throughout the 1640s to institute a monarchy-wide novena to the Virgin for the continued protection of the monarchy (in this example, the Virgin replaced Teresa, since the latter's patron sainthood failed in 1629). Philip did so by requiring all Spanish towns and cities to process with their local cult of the Virgin for the novena, while praying for the monarchy. By doing so, the king harnessed powerful local devotions for the purpose of uniting the monarchy in prayer for a common cause, thereby transforming the monarchy from a patchwork of disparate places to a unified spiritual community.[72] The sacred could become vehicles through which royal power was transmitted and augmented, where unified visions of community could be illustrated and national identity promoted.[73] The king did not need to seek proximity to the saints in order to absorb their holiness; rather, he increasingly sent forth the saints as his representatives, the ultimate symbols of a sacred and glorious monarchy. As their identity as Spaniards gained primacy over their universal holiness, the saints could be subject to the king.

[70] María José del Río, *Madrid, Urbs Regia.*

[71] Wintroub, *A Savage Mirror*, 165. Wintroub sees this as part of a trajectory towards absolutism, in which royal authority essentially replaces ecclesiastical authority, *by means of* the appropriation of the sacred.

[72] Jeffrey Schrader, "Rallying the Cities around Miraculous Images: How a King Handles a Difficult Situation," unpublished conference paper, presented at the Association for Spanish and Portuguese Historical Studies annual meeting, Boston, MA, 2012. Perhaps Philip had learned through the experience of St. Teresa's patron sainthood that imposing one saint on such diverse places would not be accepted.

[73] The sacred, of course, was a multivalent force that resisted fixed and lasting meanings or uses. In this article, I argue for considering one perspective, not that this perspective functioned exclusively to others.

PART II

RELIGION, RACE, AND COMMUNITY

4

A Minority Within a Minority: The New and Old Jewish Converts of Sigüenza, 1492–1570

Sara Tilghman Nalle*

Today the road between Zaragoza and Guadalajara runs through one of the most depopulated regions of Spain, but around 1490, the region was dotted with small towns and villages, many of which fell under the seigneurial rule of such powerful lords as the dukes of Medinaceli and the Infantado, the counts of Tendilla, Monteagudo, and Cifuentes, and the bishop of Sigüenza. Several thousands of Jews, and later, their converted descendants, found refuge and employment in Sigüenza, Cifuentes, Almazán, Medinaceli, Guadalajara, Molina de Aragón, and Soria. As remote as some of these towns were, they did not escape the notice of the inquisitors in Cuenca and Toledo, whose large districts included all of these towns except for Soria. From the late fifteenth century through most of the sixteenth century, the Inquisition of Cuenca systematically prosecuted the *judeoconverso* families of the region and left in its wake a mountain of documents recording the lives of its victims.[1] Until now,

* The research for this article was made possible by the John Simon Guggenheim Memorial Foundation and William Paterson University. The author thanks Ida Altman, Miriam Bodian, Thomas Glick, Richard Kagan, H. P. Salomon, and Alison Weber for their interest, support, and comments.

[1] Inquisitorial jurisdiction over the district alternated between the tribunals of Cuenca and Toledo. Ultimately, in 1522 jurisdiction over the bishopric of Sigüenza was permanently ceded to the tribunal of Cuenca. For the complete details, see Dimas Pérez Ramírez, *Catálogo del Archivo de la Inquisición de Cuenca* (Madrid: Fundación Universitaria Española, 1982), 15–23. The early years of the prosecution of *conversos* in this region have attracted considerable attention from historians who follow the documentary trail in detail until 1530. See the work of F. Cantera Burgos, Carlos Carrente Parondo, Javier Castaño González, and Manuel Diago Hernández. Carrente Parondo published extracts of these early trials in the multi-volume series *Fontes Iudaeorum Regni Castellae*, which has been used extensively by other scholars, notably David Gitlitz, in *Secrecy and*

historians have treated the *converso* minority of Spain as a single group, but in fact, the inquisitorial records from Cuenca reveal that the *converso* community of the bishopric of Sigüenza was deeply split between those who had converted long before 1492 (who ultimately came to be known as the *confesos de los antiguos*) and those who converted in 1492 (known as the *nuevos convertidos de judíos* or *cristianos nuevos*).[2] The divide affected this population's existence to such a degree that not one, but two distinct *converso* minorities lived in the region. In the first half of the sixteenth century, the *confesos de los antiguos* were well on their way towards absorption into the Christian majority, but the *nuevos convertidos* lagged far behind in the process of assimilation.

Until quite recently, the vast majority of scholarly and popular *converso* studies have focused on proving or disproving the minority's loyalty to their Jewish identity by examining their religious beliefs and customs.[3] Since evidence of Judaizing is available primarily through an extremely biased source, the Inquisition's records, further scholarly arguments follow concerning the reliability of the inquisitorial evidence used to prove an individual's "Jewishness."[4] Then again, even as some families did manage to cling to their Jewish faith, as each generation died off, the written texts and communal underpinnings essential to rabbinic Judaism disappeared so that knowledge of what it meant to be Jewish became ever more attenuated. At best, what was retained has been called

Deceit: The Religion of the Crypto-Jews (Philadelphia, PA: Jewish Publication Society, 1996). Javier Castaño González's dissertation, "Las comunidades judías en el obispado de Sigüenza en la Baja Edad Media. Transformación y disgregación del judaísmo en Castilla a fines del medievo" (Universidad Complutense de Madrid 1994, edocument, 2004) utilizes all of the available primary and secondary sources. No work seems to have been carried out on the communities after 1530.

2 In an unpublished paper, H. P. Salomon also notes the change in vocabulary denoting the two categories of *conversos* and studies the Judaizing activities of a group of *nuevos convertidos* from Buitrago. H. P. Salomon, "Las 'prácticas (creencias) judaicas' de los conversos de primera generación en Buitrago: Tres procesos de muestra," paper presented at the Quinto Congreso Internacional sobre los Conversos y los Moriscos, Alcalá de Henares, June 16, 2010. My thanks to the author for sharing his research with me.

3 The literature is vast; two examples in English of heroic crypto-Judaism are Gitlitz, *Secrecy and Deceit* and Renée Levine Melammed, *Heretics or Daughters of Israel? The Crypto-Jewish Women of Castile* (New York: Oxford University Press, 1999). On the other hand, in *Toward the Inquisition: Essays on Jewish and Converso History in Late Medieval Spain* (Ithaca, NY: Cornell University Press, 1997) Benzion Netanyahu has maintained that the *conversos* were sincere converts to Christianity who were victims of anti-Semitism, and discounts the reliability of the inquisitorial records.

4 H. P. Salomon has been particularly vociferous in pointing out evidentiary problems in well-known publications; see his article, "Spanish Marranism Re-examined," *Sefarad* 67 (2007): 111–54; 367–414.

a "community of memory faith," sustained by "familistic values, regendering of social roles, and the transformation of the religious practices into a culture of memory."[5]

More recently, because of these problems, researchers have begun to pay more attention to the *conversos* as an ethnic minority. Focusing on ethnic identity allows the researcher to sidestep the polemical and essentially unanswerable question of how ideologically committed *conversos* were to the tenets and rituals of Judaism. When ethnic identity is the researcher's primary subject, then religious beliefs and customs along with residence, employment, and partner choice all become cultural markers that help identify an individual as a member of a particular ethnic group, regardless of that person's purported religious identification. Since overt religious allegiance to Judaism was banned, social, economic, and political transactions became the means of maintaining the group's boundaries, perpetuating the group's identity, or allowing for its assimilation. Running through all of this, one cannot ignore the role of the family, which social scientists agree is the basic building block of ethnic identity. Through marriage, occupational training, and socialization, the minority family either helps to perpetuate the group identity or hastens its assimilation into the majority population.[6] Finally, there is self-identification: one belongs to particular group because one sees oneself as belonging to it, and others do so as well.[7]

In fact, it was the question of self-identification that first led me to suspect the existence of two separate *converso* groups. Working on a large quantitative project involving ethnic identity and the family in Spain, I noticed the diversity of responses given in the first half of the sixteenth

[5] Hilda Nissimi, "Religious Conversion, Covert Defiance and Social Identity," *Numen* 51 (2004) 383, commenting on the work of Nathan Wachtel, *Le foi de souvenir* (Paris: Seuil, 2001).

[6] See Rogers Brubaker and Frederick Cooper, "Beyond 'Identity,'" *Theory and Society* 29 (2000): 1–47; Robert Bartlett, "Medieval and Modern Concepts of Race and Ethnicity," *Journal of Medieval and Early Modern Studies* 31(1) (2001): 40–56; David Graizbord, "Religion and Ethnicity Among 'Men of the Nation': Toward a Realistic Interpretation," *Jewish Social Studies: History, Culture, Society*, 15 (2008): 32–65; and Thomas F. Glick, "On Converso and Marrano Ethnicity," in *Crisis and Creativity in the Sephardic World, 1391–1648*, ed. Benjamin R. Gampel (New York: Columbia University Press, 1997), 59–76.

[7] Walter Pohl, "Telling the Difference: Signs of Ethnic Identity," in Walter Pohl and Helmut Reimitz, *Strategies of Distinction: The Construction of Ethnic Communities, 300–800* (Leiden: Brill, 1998), 21. See also Yosef Kaplan, "Self-Definition of Sephardic Jews of Western Europe and Their Relation to the Alien and the Strange," in *Crisis and Creativity in the Sephardic World*, ed. Gampel, 121–45; Natalia Muchnik, *De paroles et de gestes. Constructions marranes en terre d'Inquisition* (Paris: Éditions EHESS, 2014).

century by *converso* subjects to the question posed by inquisitors, "*¿qual es su casta y generaçion?*"[8] I also noted that certain towns and occupations appeared to be associated with the type of answer given, "*confeso*" or "*nuevo convertido*." Intrigued, I sorted my subjects and their families according to how they had answered the question on lineage, and discovered that there was statistically a high correlation between the answer given and other aspects of the subjects' identity; in other words, what I was seeing was neither accidental nor random. The result moved me to return to my transcripts, recheck defendants' answers to the question concerning their "*casta y generaçion*" and recalibrate my coding to reflect exactly the diversity of responses given.[9]

One's "*casta y generaçion*," or genealogical memory as David Nirenberg writes, was the by-line of the fifteenth and sixteenth centuries. Lineage brought with it not only membership in a particular social group, but was thought to pass on latent religious as well as biological attributes through the blood. At first the idea was confined to the Sephardim, but gradually extended to the converts, and finally to the Old Christians, resulting in a society obsessed with lineage and "purity of blood."[10] Through the fifteenth and seventeenth centuries, Spanish intellectuals argued over what it meant to have Jewish blood in one's veins – did such blood bring with it real social, cultural, moral, and biological deficiencies, not to mention a propensity for religious backsliding?[11] The inquisitors' genealogical questioning of defendants came out of the contemporary obsession with lineage. As we can see in the responses of ordinary *judeoconversos* from

[8] The question translates roughly as "what is your race and ancestry?"

[9] The project is being carried out using SPSS, Statistical Program for the Social Sciences. As noted above, the coercive nature of inquisitorial interrogations can lead to misinformation. To address this concern, I compared the responses of Castilian *conversos* arrested on charges of Judaizing to questions about their ancestry to the answers given by Portuguese conversos arrested on the same charges a century later. Among the 140 Castilians, 136 admitted to their background even though to do so added to the presumption of guilt. Among the seventy-two Portuguese, the majority (forty-nine out of seventy-two) did just the opposite: in the courtroom they began by denying their *converso* ancestry. In other words, for whatever reason, the Castilian *judeoconversos* were candid about their background until around 1560. (See the article's conclusion for a discussion of changes that came around 1550–60.)

[10] David Nirenberg, "Mass Conversion and Genealogical Mentalities: Jews and Christians in Fifteenth-Century Spain," *Past & Present* 174 (January 2002): 3–41. In his essay, "Was there Race before Modernity? The Example of 'Jewish' Blood in Late Medieval Spain," Nirenberg puts Spanish "racism" into historiographical context. See Miriam Eliav-Feldon, Benjamin H. Isaac, and Joseph Ziegler, eds., *The Origins of Racism in the West* (New York: Cambridge Univeristy Press, 2009), 232–65.

[11] Nirenberg, "Was there Race before Modernity?"

the bishopric of Sigüenza, they were quite aware of their "*casta y generaçion.*" For those real-life *Judeoconversos* who used these terms to identify themselves, words such as "*confeso*" or "*nuevo convertido*" were not abstract concepts, but referred to two communities that differed in measurable ways between themselves and from the majority Old Christian population.

With few exceptions,[12] historians have tended to treat the *conversos* of Castile as a monolithic group, but they were far from one single group with shared characteristics. The most basic distinction depended on when the individual, family, or community converted to Christianity – as early as the 1391 pogroms or one hundred years later, in 1492. In the bishopric of Sigüenza, the oldest *converso* communities were in Molina de Aragón and Cifuentes. Molina's Jewish community was forced to convert en masse in 1391, but in 1411 most of the community in Cifuentes and a good many more individuals around the bishopric converted voluntarily as the result of Fr. Vincent Ferrer's preaching campaign in the region.[13] According to some sources, the Jewish community in Atienza also had largely converted to Christianity before the Catholic Monarchs issued the Expulsion Edict. In addition to the mass conversions early in the fifteenth century, there was a continuous process of individual conversion throughout the century.[14]

Apart from scattered references in the chronicles, another way to track the process of conversion is through genealogical declarations given to the Inquisition by *converso* defendants. The testimony of 164 *converso*s from the region who were born between 1440 and 1537 sheds some general light on what had been the rate of conversion during the fifteenth century. Among the thirty-nine *confesos de los antiguos* in the sample, who were born between 1464 and 1535, not a single one admitted to having had a parent or grandparent who died as a Jew. Among those who

[12] For example, David L. Graizbord, *Souls in Dispute. Converso Identities in Iberia and the Jewish Diaspora, 1580–1700* (Philadelphia, PA: University of Pennsylvania Press, 2004). Also, Carrete Parrondo, whose work inspires Alisa Meyuhas Ginio's article, "Self-Perception and Images of the Judeoconversos in Fifteenth-Century Spain and Portugal," *Tel Aviver Jahrbuch für deutsche Geschichte* 22 (1993): 127–52. Ginio begins to develop the distinction between newer and older converts from Judaism by relying on what witnesses say about each other, but does not follow through with the idea. The subjects of Linda Martz's book *A Network of Converso Families in Early Modern Toledo* (Ann Arbor, MI: University of Michigan Press, 2003) are entirely pre-1492 converts. I have not encountered an article or book which explicitly takes up the question of the differences between the pre- and post-1492 converts.

[13] Castaño González, 300–02.

[14] Castaño González, 152. It is also apparent in 1492 (see below).

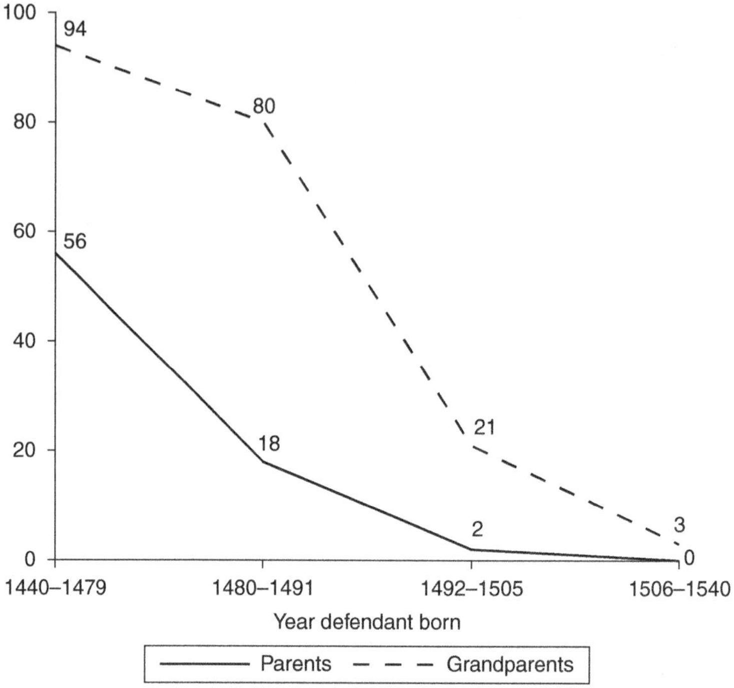

FIG. 4.1. Percentage of new converts' ancestors who died as Jews

identified themselves as new converts, as to be expected, those who were born in the middle of the fithteenth century grew up in their faith among Jewish grandparents and parents who, if they converted at all, did so late in life. The transitional generation was that born between 1480 and 1491, which was brought up as Christian by parents who had been raised within Judaism, as had been most of their own parents.

Despite the ongoing conversions, in the fifteenth century the bishopric of Sigüenza still harbored up to fifty Jewish communities, the largest of which were official ghettos (*aljamas*) sanctioned by royal law. The largest *aljamas* were those of Soria (actually just outside the bishopric) and Almazán, with perhaps as many as 300 families in each. Sigüenza, Medinaceli, and Berlanga were also important Jewish centers right down to 1492.[15] The genealogical evidence from the Inquisition records shows

[15] F. Cantera Burgos, "Las juderías medievales en la provincia de Guadalajara," *Sefarad* 33(1) (1973): 3–51; Máximo Diago Hernando, "Almazán en época de los Reyes Católicos: Estructura social de una pequeña capital de estado señorial," *En la España medieval* 16 (1993): 248; Henry Kamen, "The Mediterranean and the Expulsion of the Spanish Jews in 1492," *Past & Present* 119 (1988): 42. Kamen (33–5) concludes that the

that defendants from these towns had few relatives who converted to Christianity well before 1492. Nonetheless, it is difficult to know exactly what the situation for each community was really like in 1492: how many individuals had already become Christian, how many converted as a result of the Edict of Expulsion, how many chose to leave, and finally, how many decided to return.

What is clear, however, is that the 1492 expulsion edict created an entirely new generation of converts. Instead of experiencing a decades-long process of conversion and adjustment to their new identity, overnight, Jews became Christians, and with the Inquisition now in play, there was tremendous pressure on individuals to conform at once to their new identity. The Inquisition and its victims were quite aware of the difference in condition. The 1492 converts immediately became known as the "*nuevos convertidos de judíos*" (the new Jewish converts) or "*cristianos nuevos*" and identified themselves to the inquisitors as such, as opposed to members of the previous group, who called themselves "*confesos*" or "*conversos*," and whose parents and grandparents had died as Christians.[16] The *confesos*, as will be seen below, refused to mingle with the new converts, while in the bishopric of Sigüenza, the 1492 converts blamed their conversion, the exile of their relatives, and/or general loss of property and status on the *confesos*. And who can blame them? The Edict of Expulsion said as much in its rationale for the decision to expel the Jews, citing their continued presence as a hindrance to the *confesos*' full assimilation.[17] Before leaving, some of the departing Jews deliberately avenged themselves by testifying in the Inquisition's ongoing trials of the *confesos*. Technically, the court could not hear testimony from non-Christians, but it bent the rules in order to gather as much evidence as possible before witnesses departed for parts unknown. [18]

Jewish population of Castile at the time of the expulsion could have been no more than 70,000.

16 All of these terms were in play in the fifteenth century for the converts of that time, but in the Sigüenza/Cuenca region after 1492 they take on distinct meanings. The basis for this terminology is easily confirmed by comparing the two groups' ancestry according to their self-identification as "*conversos/confesos*" or "*cristianos nuevos*." See Brubaker and Cooper's comments on the usefulness of self-identification as opposed to categorization by others (such as modern historians!).

17 "... we were informed that in our realms there were some bad Christians who Judaized and apostatized from our holy Catholic faith, whereof the chief cause was the communication between the Christians and the Jews ..." The document is reproduced, transcribed, and translated in Haim Beinart, *The Expulsion of the Jews from Spain*, trans. Jeffrey M. Green (Oxford: The Littman Library of Jewish Civilization, 2002), 46–54.

18 Ginio citing Carrete Parrondo, 134; Castaño González, 341–3.

In the years to come, the gaping wound in the *judeoconverso* community was not easily healed. The children of the "*nuevos convertidos*" did not merge into the general population of converts. Although their Jewish grandparents were long dead, they continued to identify themselves as "*nuevos convertidos de judíos*" or "*nuevos cristianos.*" Meanwhile, in the 1540s some descendants of the original *conversos* began to underscore their difference from the "new converts" by calling themselves "*confesos de los antiguos*" (the old [Jewish] converts) instead of just saying they were "*confesos.*"[19] Even as late as the 1550s, individuals were careful to identify their lineage (*generaçión*) from either of the two groups, now saying that they were of such-and-such *generaçión* or that they were the "son of a New Christian" or that their grandmother was a "*confesa.*" As we shall see, the two groups were different in more ways than just terminology and hard feelings for one another. The differences begin with the very way in which the inquisitors treated their prisoners.

The Inquisition waged nearly continuous campaigns against the *judeoconverso* population of the bishopric of Sigüenza from 1492 to 1570. During the earliest period, the tribunals' cases involved primarily the *confesos*, since those who had converted in 1492 were given a ten-year grace period to become good Christians.[20] The historian Javier Castaño González counts 270 extant trials dating between 1492 and 1506 in the archives of the tribunals of Toledo and Cuenca; over half of the defendants hailed from the longstanding *confeso* population centers of Cifuentes and Molina de Aragón. The tribunals were merciless: thirty-two persons (12 percent) were burned in person, ninety-seven (36 percent) were burned in effigy, and twenty-eight (10 percent) received the next most severe sentence of reconciliation. (On a more positive note, fifty – 19 percent – were found innocent of the charges.)[21] Between 1522 and 1570, the inquisitors shifted their focus to the *nuevos convertidos* in the bishopric – New Christians were tried at twice the rate as were *confesos de los antiguos*.[22] *Nuevos convertidos* were almost always

[19] This may be appreciated by writing down verbatim individuals' responses and tracking them chronologically. "*Confeso de los antiguos*" does not appear in my sources before 1540. The term was also used by *confesos* in La Mancha, who are not part of this article.

[20] Castaño, 319 n. 9.

[21] Castaño, 319, n. 9. It would be quite interesting to return to these trials and sort them by *confesos* and *nuevos convertidos* to see if the differential treatment meted out by the Inquisition to the two groups began then or later.

[22] To be precise, 138 *cristianos nuevos* versus 26 *confesos de los antiguos*. These numbers include persons of mixed heritage – thirteen *cristianos nuevos* who had married Old Christians, and sixteen *confesos de los antiguos* who had done the same. On this campaign

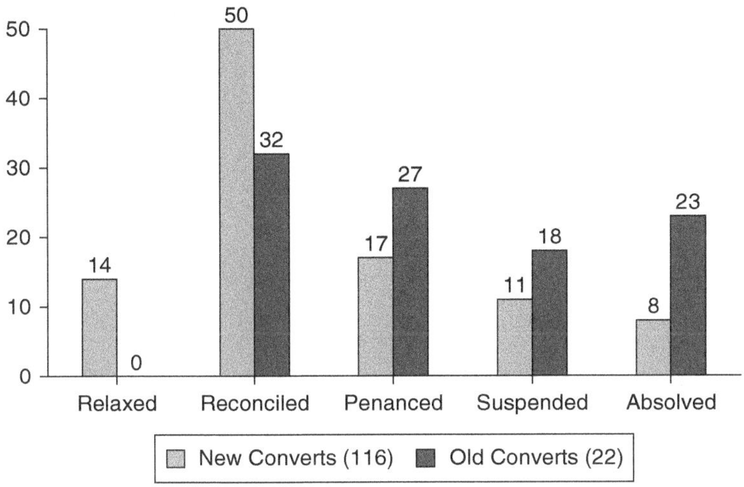

FIG. 4.2. Sentencing for Judaizing (1522–70), by category of convert (in percent)

(82 percent) arrested on charges of Judaizing whereas only 35 percent of the *confesos* were. Instead, *confesos* were most frequently picked up on charges of hindering the Inquisition's activities or for making blasphemous or heretical propositions not related to Judaism.[23] Even when tried for Judaizing, compared to their ancestors and the new converts, *confesos* were treated leniently – none were executed, and 41 percent were absolved or released without sentencing. Based on their sentencing practices, clearly the inquisitors no longer regarded the *confesos de los antiguos*' Judaizing activities, whether real or trumped up, as dangerous. By comparison, the *nuevos convertidos* suffered much harsher punishment, with two-thirds of those arrested receiving the two most severe convictions possible, reconciliation and death.[24]

against the *conversos*, see my chapter, "A Forgotten Campaign Against the Conversos of Sigüenza: Pedro Cortés and the Inquisition of Cuenca," in *The Conversos and Moriscos in Late Medieval Spain and Beyond*, vol. 3, ed. Kevin Ingram (Leiden: Brill, 2016), 6–28.

[23] Leonor Zozaya Montes notes a similar hostility towards official religion on the part of *confesos* from Molina de Aragón. See "Popular Religious Practice and Converso Dissidence in the District of Molina de Aragón," in *The Conversos and Moriscos in Late Medieval Spain and Beyond*, ed. Kevin Ingram (Leiden: Brill, 2009), 161–86. Zozaya's article is based on a small cache of trials from Molina that are archived at the Archivo Histórico Nacional.

[24] Compared to the ferocious sentencing meted out to Judaizers in the 1490s, when nearly half of trials ended with the most severe sentence of relaxation (either in person or absentia), the Inquisition's treatment of the *nuevos convertidos* in the sixteenth century was rather less lethal.

The *confesos de los antiguos* had been replaced by a new enemy of the faith, the *nuevos convertidos*. In the sections that follow, I shall trace the emergence of the new minority within a minority, and then compare the families of *nuevos convertidos* with those of the *confesos*. By doing so, we will see what strategies families in each group adopted in order to retain their Jewish identity or to assimilate into the Old Christian majority.

1492: CREATING A MINORITY WITHIN A MINORITY

Since its promulgation, both contemporary observers and historians have argued over the intent and long-term impact of the 1492 Edict of Expulsion. Until about thirty years ago, the general view has been that the entire Jewish population of Spain, some several hundred thousand strong, left, taking their expertise and entrepreneurial spirit with them. This view no longer is sustainable. For example, after reviewing the evidence, Henry Kamen concluded that about 70,000 Jews lived in Spain in 1492. Of that number, only about half chose to emigrate, and of those, many returned, particularly in the region of Guadalajara, where the Jews had enjoyed the protection of the duke of the Infantado.[25] Kamen's observation is amply borne out by my survey of *nuevos convertidos* from this region who later were tried by the Inquisition, many of whom reported on their experience of exile, conversion, and return.[26]

The hardships, violence, and penury collectively experienced by the exiles have been amply documented.[27] The turmoil of these years is reflected by the fact that so many individuals who were born between 1487 and 1507 claimed that they did not know who their grandparents were and in what religion they had died. Some of this could be a defensive measure, but not entirely so, because those who were born *before and after* the expulsion period had no difficulty remembering their grandparents' religious affiliation, be it Jewish or Christian. Yet, many of those born around the time of the expulsion claimed such absolute ignorance

[25] Kamen, 42 n. 45, 44.

[26] I explore these stories in "Cross and Cross Again: First-Hand Accounts of 1492 Exiles' Return to Castile," in *Religious and Ethnic Minorities in Early Modern Europe*, ed. Yosef Kaplan (forthcoming).

[27] The quincentenary of the Expulsion prompted a surge in publications specifically about the expulsion or the period in general. The most exhaustive history is Beinart's *The Expulsion of the Jews from Spain*.

as to not even know in which religion their ancestors had died.[28] This lack of knowledge could be feigned or real; there is no way of knowing which except to point out that Old Christians' familiarity with their grandparents' identity was quite similar to *conversos*' level of knowledge, except around the Expulsion period (1487–1507), when *conversos*' knowledge dipped but the Old Christians' memory remained constant. Descendants of converts born after 1507 returned to the same level of genealogical awareness as had prevailed before the Expulsion.

Some among the first generation of *nuevos convertidos* continued to practice covertly their faith. The San Clemente family from the small city of Sigüenza is a good example of crypto-Judaizers who were able to escape the Inquisition's notice for some thirty years. After initially emigrating to Portugal, Gaspar de San Clemente Sr. and his wife returned to Sigüenza, where they maintained their faith by cautiously indoctrinating their teenaged children, sending their sons to visit relatives in Portugal, where Judaism was practiced somewhat more freely, preparing kosher foods, and using their home as a center for secret worship meetings known in Spanish as "*conventículos*."[29] The holding of conventicles began soon after the Expulsion and as nearly as can be told from the evidence, involved almost entirely the new converts in the communities of Sigüenza, Berlanga, Medinaceli, and Cifuentes. Out of several dozens of people implicated in the meetings, only two can be identified as *confesos*

[28] The scribe would record whatever information, no matter how skimpy, that the defendant gave concerning relatives who had been baptized. In the case of Jewish relatives, if the defendant recalled that the relative died in Judaism or emigrated, the note-taking stopped there. Collectively defendants had nearly 100 percent recall about their parents' religious identity, but memory of grandparents took a significant dive among those who were born between 1487 and 1507. Among those born prior to 1487 and after 1507, typically defendants could recall the religious identity, and often much more, of over 80 percent of their grandparents. The 1492 generation, however, could supply information for only 60 percent of their grandparents. By way of contrast, Old Christians had similar recall about their grandparents' religious affiliation but during the period 1487–1507, their memory slips from 78 percent to 71 percent.

[29] Archivo Diocesano de Cuenca, Sección Inquisición (henceforth, ADC, Inq.) exp. 1772 (1540), trial of Gaspar de San Clemente. At virtually the same time as some of the *nuevos convertidos* were carrying on their faith in secret meetings, in the same general region of Guadalajara, parallel *conventículos* were being established by *conversos* who were building a new Christian spirituality known as illuminism. Although the religious intent of the meetings was vastly different, the meetings draw on the same group memory of Judaism as practiced in small prayer groups. Structurally they share the same characteristics of "faith communities" as described in García-Arenal and Pereda's article on the *alumbrados* in the present volume. On illuminism as a *converso* religion, see Stefania Pastore, *Una herejía española: conversos, alumbrados e Inquisición (1449–1559)* (Madrid: Marcial Pons, 2010).

de los antiguos. One ongoing conventicle emerged in Sigüenza in the 1490s and met off and on for about ten years in the home of Juan López, a shoemaker. It appears to have begun as a family-only affair. The orphan Magdalena de Soria remembered that after she returned from Portugal (*c.* 1493, at age seven) she first lived in an attic at her uncle's house. In a dark, hidden room, she witnessed her uncle reading prayers from a small book to his wife, mother, and daughter. Magdalena recognized the ritualistic swaying back and forth as *sabadeando* (shuckling or davening) because she had been to the synagogue before the Expulsion.[30] López's home soon became the center for larger gatherings, attended by men and women alike, many of whom were not relatives. Two more witnesses to the López conventicle, one a regular participant and the other an accidental observer, described essentially the same scene. The conventicles, now led mostly by Lorencio del Castillo (who later would be tried and executed as a rabbi),[31] were attended by a core of "regulars" and other, occasional participants. Del Castillo would read from a Hebrew book, and the group would sit, stand, and pray at determined points in the service. In the 1510s, the meeting place in Sigüenza was transferred from the López house to the San Clemente home, with several of the same cast of characters attending. Gaspar de San Clemente senior sometimes would lead the services; sometimes other of the most senior men would.[32]

The other conventicles in Berlanga, Medinaceli, and Cifuentes were reported as meeting around 1512, 1516, 1521, and 1530. The conventicle in Berlanga met off and on from 1516 to 1530 in various homes but was led just by two persons, Dr. Juan López de Uceda and Juan de Buenaventura, Sr. Excluding the López family prayers described above and one exceptional conventicle in Medinaceli, witnesses consistently recalled the names of between nine and twelve men who attended the meetings; in other words, the communities seemed to be trying to gather a minyan (the minimum quorum of ten males) to conduct services. The exceptional and perhaps fictional meeting in Medinaceli took place around 1521, and was described by Juan de la Peña thirty years later in

30 ADC, Inq. exp. 1771 (1536). After her return from Portugal Madgalena was passed around from house to house and married at age thirteen.

31 ADC, Inq. exp. 1630 (1501–34).

32 ADC, Inq. exp. 1572, Hernando de Soria (1532) and exp. 1788, María de Soria (1532). Although the two were siblings, Hernando did not live in Sigüenza and stumbled in on a meeting at his uncle's house when invited in for a meal. He describes quite vividly the suspicion with which the other men treated him and how they pretended to have some other reason for meeting.

1551. Despite the fact he was only eight or nine years old at the time and observing from his grandfather's bed, he claimed to remember that the meeting took place on a Wednesday and it involved a total of some forty men, women, and children. The inquisitor wondered why Juan remembered that the meeting occurred on a Wednesday without remembering what time of year it had been; we can wonder how a boy of eight could remember to perfection thirty years later all the people who attended the meeting as well as their various relationships to one another.[33]

Despite these conventicles which attempted to carry on the practice of Hebrew worship, one thread that runs through the various confessions that appears in trials from 1525 to 1555 is how quickly the second generation of *nuevos convertidos* lost touch with aspects of Judaism. The ability to read and write in Hebrew, which depended on special teachers, schools, and books (all of which were outlawed in 1492), was an early victim. Tomás López, testifying about a conventicle he attended around 1513 in Cifuentes, where most of the Jewish community had converted in the early fifteenth century, remembered that they depended on the rabbi Pedro de Carrión, from Sigüenza, because no one else was capable of reading the prayer service.[34] Carrión reassured them that although they did not know the prayers, if they said them with him that would be enough.[35] Gaspar de San Clemente, Jr., born in 1502 and literate in Latin and Spanish, claimed that he had been unable to identify the Hebrew language as such when he heard it read aloud. Agostín Martínez, born in 1501 and by profession a public notary (*escribano*), attended a conventicle in Berlanga in 1515 or 1516 where he got a good look at the book that Dr. López de Uceda was using. He recalled, "it was a parchment book, in 4°, and two fingers thick, ... I did not recognize the script it was in and I asked Dr. Uceda, 'what script is that?' and he replied that it was Hebrew." Apparently Martínez, about fifteen at the time, had

[33] ADC, Inq. exp. 2197 (1551). Juan by profession was an apothecary. Juan's laundry list implicated one of the two *confesos de los antiguos* accused of attending conventicles. The other, Juan del Monte, accused of attending a conventicle in Cifuentes, was absolved of any guilt and none of the other men accused in that conventicle seem to have been tried (ADC, Inq. exp. 1754).

[34] This loss of knowledge of Hebrew was also reported in Cuenca. After their rabbi "Don Symuel" was executed, Luis de Madrid read the services in Castilian to the crypto-Jews of Cuenca. Pérez, *Catálogo*, 32–3.

[35] ADC, Inq. exp. 1754 (1539) "ya yo sé que vosotros no sabeys estas oraciones del semac ni del modin mas yo os he dicho que con lo que yo digo conplis vosotros diziendolo de la manera que yo lo digo." The Inquisition used the word "rabbi" to refer to any person who dogmatized (Gitlitz, 516). This denunciation may be false: the witness was visiting Cifuentes on business and no one he mentioned was indicted and tried.

never attended such a ceremony before and asked the men what they were doing at each step – what was that language being spoken, why the men were swaying backwards and forwards, and where had Dr. Uceda's special belt come from. Martínez even asked the men where the old synagogue had been and what had happened to the Torah, but he did not know exactly what the Torah was except that he had heard that it was a parchment book that the Jews used for worship.[36] Finally, Licenciate Juan de Buenaventura, also born around 1500 and from Berlanga, remembered that when he came home for summer vacation from the University of Salamanca, he twice attended his father's conventicles. His father and Lázaro Alvarez "took a book and began to read and pray something that I did not understand nor did many of those of us who were there except that after it was read and prayed they ... gave us to understand that the Jewish religion was the good and true one and we would be saved in it."[37]

In the late 1520s and continuing without interruption until the 1550s, the Tribunal of Cuenca systematically gutted the communities of *nuevos convertidos* in the bishopric of Sigüenza.[38] Tens of *nuevos convertidos* were sent to the stake, and hundreds more were penanced or reconciled.

[36] ADC, Inq. Exp. 1790 (1540), confession of 27 January 1540. "Vido como el dicho doctor Uzeda estaba subido de pies ençima de vna silla de madera de Castilla arrimado a vna pared de la dicha pieça y tenia en la mano un palo y las otras dichas personas estavan asentado de dos bancos largos ... tenían cubiertas las cabeças con los cabos de sus capas y el dicho doctor tenía en la cabeça vn sonbrero grande y vestido vn mongil de paño negro largo hasta en pies con vnas mangas ansi mismas largas y el dicho mongil tenya vna capilla larga y la tenia echada ençima del honbro derecho y ençima del dicho mongil ceñyda vna çinta de cuero ancha y tenya en la mano vn libro de pergamino el qual este confesante vio y no conosçio la letra que se hera y demando al dicho doctor Uzeda que, qué letra hera aquealla, y el dho doctor le rrespondio que hera hebrayca y que alli estava la ley de Moysen, (...) y luego dende a vn poco el dicho doctor Uzeda se desçiendio de la dicha sylla y se asento en ella y el començo a leer en el dicho libro y lo que lo que leya lo dezia en ebrayco, y este confesante le pregunto que, qué hera aquello que dezia porque este confesante no lo entendía, y el dicho doctor Uzeda le rrespondio que lo que leya hera todo de la ley de Moysen ..." This went on for about half an hour, then the men began to daven, which one of the men explained to Agostín "que se acostumbraba ansy hazer en tiempo de los judios quando estavan en la sinoga y acababan de leer lo que se leya." The man also told Agostín that Dr. Uzeda's belt had been kept in Lázaro Alvarez's house. The belt seems to have had special significance, perhaps as a substitute for the tefillin.

[37] ADC, Inq. exp. 1825 (1525–40) "tomaron un libro e comenzaron a leer e rezar cosa que yo no entendia ni muchos de los que alli estavamos mas de que despues de leido e rezado dezia e divulgaba que la ley de los judios era la buena e verdadera e que en ella se habian de salvar ..."

[38] From 1540 to 1542, relatives of the victims from Berlanga and Almazán mounted an appeal to the *Suprema*, without success. They claimed that the extremely harsh conditions in the secret prisons had led prisoners to form cabals, lie, and do whatever they could to save themselves. See Nalle, "A Forgotten Campaign."

In some cases, one by one, all the adult members of a family were arrested, tried, and convicted. The harsh persecution succeeded in its purpose, to frighten at least some *nuevos convertidos* away from Judaizing. Around 1530, one of the rabbis who had routinely held services in his home in Sigüenza told his congregation that it was time to give up their activities before they were betrayed and destroyed. More astonishingly, he told them that their belief in redemption was in vain. After leading prayers one final time, Fernán Gómez put down his book and announced,

Friends, you've known for some time that we have been blind and wrapped up in this blindness and folly that came to us with the Exile (although not to me because I stayed here and didn't spend my fortune like you did yours [by leaving and coming back]). No remedy or rescue has come from Him for whom we waited ...[39] Let us gather ourselves together and quit these ceremonies which profit us no longer ... surely we will be found out by some people and if the authorities find out we will all be lost and finished. You well know a passage in the Bible or in one of those Hebrew books that says that nothing, however well hidden it may be, can be hidden forever,[40] and this is my opinion, and each one of you from here on forward should do whatever you like, but I will receive you no longer in my house.[41]

Gómez told his congregants that he would tell the other two rabbis, Lorencio de Castillo and Juan López, about their pact and have them promise to do the same in their communities. According to one witness, after this date (about 1530) there were no more conventicles, and no

[39] A reference to the many rumors that the Messiah had appeared in Spain or Portugal. See my article, "The Millennial Moment: Revolution and Radical Religion in Sixteenth-Century Spain," in *Toward the Millennium*, eds. Mark Cohen and Peter Schäfer (Leiden: Brill, 1998), 153–73.

[40] There are several passages in the New Testament (!) that echo this thought, especially Luke 8:17, "For there is nothing hidden that will not be disclosed, and nothing concealed that will not be known or brought out into the open."

[41] "... se asentó el dicho Fernán Gómez e dexó el libro de sus manos e dixo, 'sentaos, señores, que quiero deziros dos palabras,' e dixo, 'amigos ya sabéys como ha tanto tiempo que estamos çiegos y metidos en esta çeguedad e perdiçión que nos vino en el destierro (aunque no a mí porque yo acá me quedé e no gasté mi hazienda como vosotros las vuestras) y no nos ha venido ningund rremedio ni socorro de quien esperavamos, y ... que tornąsedes a vuestras casas bien sería que diesemos maniaviesa [mano aviesa] al diablo e nos rrecogésemos y nos apartásemos destas çerimonias que no nos aprovechan ya pues que ha tanto tiempo no puede ser menos sino que avemos de ser sentidos o barrantados por veçinos o por algunas personas que lo sientan e sy la justiçia lo siente todos seremos perdidos e acabados, que bien sabéys que dize vn testo en la blyblia o tytulín o en vn libro desos ebraycos que ninguna cosa por yncubierta que se haga puede ser que no sea descubierta, y esto es mi parecer y cada uno de aquí adelante haga lo que quisiere, que yo no os acogeré más en mi casa," ADC, Inq. exp. 1572, Hernando de Soria (1532), recalling his attendance at various conventicles.

mention of them after that date has turned up in the inquisitorial records. From here on the trail becomes ever more obscure as memory of the time when Judaism was openly practiced faded and genuine crypto-Judaizers were driven further underground. At this point, rather than focusing on religious markers of identity it makes more sense to shift our attention to social ones.

THE PATH TO ASSIMILATION

In a provocative article published in 1997, Thomas Glick rebuked those who point to crypto-Judaizing as proof of the essential Jewish identity of the *conversos*.[42] In the crucible of the inquisitorial process, witnesses and defendants' statements about their beliefs, actions, and intentions were nothing if not self-serving. A solution to this problem is to concentrate on, as Glick puts it, "the dynamics of group membership, solidarity, and conflict, and the network of structured relationships both within and between groups."[43] The fact that several different witnesses, arrested at different times in different places, could describe their participation in the same conventicles in several towns in the bishopric of Sigüenza, for example, is better proof of their collective commitment to a Jewish identity than one individual's coerced confession. Of more interest to us, therefore, are the *extra*-religious activities of Sigüenza's *converso* families, because through them we may begin to discover how the minority as a whole adapted to its new circumstances.

Geographic proximity is the life blood of social networks. And, when a group is scattered, how effectively its members maintain contact with one another over long distances is compelling proof of their collective commitment to the group identity. In the first years after 1492 some *nuevo convertido* families went to the extreme of shuttling back and forth between Portugal in order to maintain ties with Judaizing family members.[44] Far easier, however, was to live in the same community. When

[42] Glick, "On Converso and Marrano Ethnicity." Graizbord notes a shift in emphasis so that historians are less concerned today "with the question of whether conversos were or were not Jews in essence," *Souls in Dispute*, 10–12.

[43] Glick, "On Converso and Marrano Ethnicity," 60, 68, 73.

[44] The best example is the San Clemente family, which tried to arrange an across-the-border marriage between one of their sons and a relative in Portugal. An excerpt of Gaspar de San Clemente's trial with translation is available at the "Early Modern Workshop: Jewish History Resources, Volume 8: Egodocuments: Revelation of the Self in the Early Modern Period, 2011, University of Texas at Austin, August 21–23," www.earlymodern.org/workshops/2011/nalle/texto2/original.php?tid=177.

the residences of the *judeoconversos* of Sigüenza are plotted on a map, it immediately becomes apparent the degree to which the *confesos* and the *nuevos convertidos* occupied separate physical worlds: instead of building networks, they built walls between themselves. On the principle that a person does not elect where he is born but can move to where he wants, I plotted the 164 defendants and 914 of their relatives (grandfathers, fathers, uncles, aunts, siblings, and adult children) according to their current residence as reported to the Inquisition by the informant/defendant.[45] Plotted in this manner, the geographic segregation of the *confesos* and *nuevos convertidos* immediately becomes apparent.

The southeastern half of the bishopric and extending into Cuenca's Alcarría was home to the *confesos de los antiguos* while the northwestern half of the region, bleeding over into Soria, was the New Christians' territory. Three-quarters of the *confesos* living in the bishopric of Sigüenza resided in or near the towns of Molina de Aragón and Cifuentes, while just over half of the *nuevos convertidos* and their families lived in or near Almazán, Atienza, and Medinaceli. So pronounced was the segregation that Almazán's *judeoconverso* population was almost entirely (98 percent) made up of *nuevos convertidos*.[46] When relocating within the region, the *nuevos convertidos* shuttled back and forth between familiar haunts. And why not? The towns of Medinaceli, Almazán, Atienza and several smaller ones were close together and functioned more like one large metropolitan area than many separate communities. The *confesos*, on the other hand, were just as likely to move to a community that was not a stronghold of other *confesos* as they were to stay within their group's geographical limits. Their geographic mobility in part was fed by their willingness to marry Old Christians.

The sixteenth century for Spain was a period of expansion, and the *judeoconversos* took full advantage of the phenomenon. About one-fifth of each group resided outside of its core region, but the patterns of geographic segregation carried over to places of residence even further afield. It is well known that individuals tend to immigrate to areas where

[45] As much as is possible, the database has been carefully purged of duplicate and even triplicate entries resulting from close marriages or multiple members of the same family being tried by the Holy Office.

[46] This residence pattern is not the result of the Inquisition's trial activity. For example, we might posit that if for some reason the judges tried only New Christians in Almazán, then it would appear that only New Christians resided there. However, all of the families in the study group married into families from other communities, so if there were intermarriages between the two groups it would show up in the residence patterns.

MAP 4.1. Northern district of the Inquisition of Cuenca

TABLE 4.1. Division of *judeoconverso* communities by subminority

Community (includes districts)	New Converts	Old Converts	Community	New Converts	Old Converts
Molina de Aragón	10%	90%	Medinaceli	87%	13%
Cifuentes	11%	89%	Atienza	95%	5%
Sigüenza	85%	15%	Soria	97%	3%
Almazán	98%	2%	Alcarría (Cuenca, Guadalajara)	58%	42%

one family member already has settled or has connections; this was as true for Spain's colonies in the sixteenth century as it is today in the United States.[47] Just outside the core area defined by the old bishopric of

[47] See Ida Altman's pioneering work on this subject, *Emigrants and Society. Extremadura and Spanish America in the Sixteenth Century* (Berkeley, CA: University of California Press, 1989).

Sigüenza, many *nuevos convertidos* clustered in villages along the northern border in the bishopric of Osma and across the border in Aragon, placing themselves outside the Inquisition of Cuenca's reach. Some *confesos* gravitated towards the *confeso* communities of New Castile. Both groups took advantage of their connections to the two powerful noble houses of the area, led by the dukes of the Infantado (Mendoza family) and Medinaceli (de la Cerda), who over the ages had employed many of the Jews and then the *judeoconversos* in the Guadalajara-Sigüenza region. These noble houses had important interests in Andalusia; the count of Tendilla (Mendoza) was captain-general of Granada, while the duke of Medinaceli was lord of the Andalusian port city, Puerto de Santa María. Thus it is no surprise that, following their connections, some *judeoconversos* from the Guadalajara-Sigüenza region wound up in Andalusia. An example of such parlaying of connections is the de la Fuente family. The New Christian Diego de la Fuente lived with the duke of Medinaceli around 1500; another branch of the de la Fuente family, also from Medinaceli, had one son in the empress's mounted guards, probably thanks to the duke's influence, and two sons settled in the Puerto de Santa María.[48] Martín de las Juberías, born around 1500, belonged to a second New Christian family that served the dukes of Medinaceli and followed their interests to Andalusia. Martín's paternal grandfather, Martín de las Juberías, had served "the old duke" as his head muleteer. As he remembered it, his maternal grandmother had served Doña Ana, widow of Don Domingo de la Cerda, lord of Mandayona (Guadalajara). Martín junior worked his way up through the duke of Medinaceli's private guard, starting as a footman (*hombre de pie*), then a crupper (*mozo de espuelas*), and traveling with the duke wherever he went. Eventually Martín also chose to settle in the Puerto which he had come to know through his service to the duke. After a few years, his wife joined him there and his oldest daughter married in the town.[49] In general, following the patronage of their home-town seigneurs, the *nuevos convertidos* gravitated to western Andalusia (duke of Medinaceli), while the *confesos de los antiguos* wound up in eastern Andalusia, in Granada or Málaga, the major port of the region, where the count of Tendilla (Mendoza family) dominated.

[48] ADC Inq., exp. 1858 (1540) and exp. 2277 (1555).

[49] ADC Inq., exp. 2271 (1555). Juberías quite clearly states "Domingo de la Cerda," but perhaps he meant "Diego Hurtado de Mendoza," the count of Melito and husband of Doña Ana de la Cerda, who owned the señorío of Mayadona in the early part of the sixteenth century. James M. Boyden, *The Courtier and the King; Ruy Gómez de Silva, Philip II, and the Court of Spain* (Berkeley, CA: University of California Press, 1995), 25–6.

Perhaps it was the *nuevos convertidos*' connections with the Puerto de Santa María, a stepping-off point for Atlantic navigation, that inspired some of them to keep going west. Still other members of the extended de la Fuente family slipped around the prohibition against *conversos* emigrating to the New World. First was Rodrigo de Uceda, a glovemaker from Medinaceli and son of yet another Diego de la Fuente. Rodrigo left as a young man, probably in the 1520s. His nephew, Pedro Martínez, a weaver from Brihuega, emigrated around 1540 and perhaps joined (or given the early date, helped to establish) the community of Brihuegan clothmakers in Puebla, Mexico.[50] Another *nuevo convertido*, Lázaro Martínez of Sigüenza (no relation to Pedro), left home at age twenty-two to enlist with Francisco Pizarro, who had returned from Panama in 1528 to recruit volunteers for the conquest of Peru.[51] Several *confesos* also emigrated to the New World during this period.

EMPLOYMENT

Behind the *judeoconversos*' movement around the region and even abroad was often the prospect of employment: either to train for a career, or to take a job. The two groups' geographic segregation carried over into the division of economic space in the region, which was carved up into different areas of specialization. In general, the *confesos* embarked on professions that reflected their longer experience adapting to Christian society and taking advantage of its privileges. The most striking difference was the number of *confesos* who chose a career in the Church – 16 percent as opposed to just 5 percent of new converts. Moreover, some 6 percent of *confesos de los antiguos*, through advantageous matches and successful business activities, had become independently wealthy or sported a noble title such as "don" or "escudero," while just three (0.7 percent) of the *nuevos convertidos* had been able to do so. For their part, the *nuevos convertidos* dominated in the same professions that had been their specialties prior to conversion: tax and rent collection, accounting and estate management, medicine, and certain crafts.

[50] ADC Inq., exp. 2204 (1553). On on the settlement of Puebla by clothmakers from Brihuega, see Ida Altman, *Transatlantic Ties in the Spanish Empire. Brihuega, Spain and Puebla, Mexico, 1560–1620* (Palo Alto, CA: Stanford University Press, 2000).

[51] ADC Inq., exp. 1649. Information provided by Lázaro's father, Pedro de Palazuelos. As of 1536, Lazaro's father had yet to hear from his son. Palazuelos was born in 1471 to Jewish parents and converted in 1492.

TABLE 4.2. Specialization of *judeoconverso* population by general activity and group affiliation*

Occupation	New Converts No.	New Converts Percent	Old Converts No.	Old Converts Percent
Agriculture	20	5.1	12	8.8
Crafts	**196**	**50.5**	**58**	**42.6**
Shopkeepers	4	1.0	0	0
Merchants	32	8.2	12	8.8
Money handlers	**24**	**6.2**	**2**	**1.5**
Medical services	**22**	**5.7**	**2**	**1.5**
Legal, administrative	27	7.0	12	8.8
Religious	**20**	**5.1**	**22**	**16.2**
Wealthy, noble	**3**	**.7**	**8**	**5.8**
Miscellaneous	27	7.0	7	5.1
Absent, lost	**13**	**3.5**	**1**	**.7**
Totals	388	100	136	100

*Excludes spouses and Old Christian relatives; includes women, as in nuns.

The area of manufactures, with its innumerable specialities, was carved up into niche occupations even within a general area of activity. For example, in the area of cloth finishing, the new converts monopolized the tailor's craft (11 percent v. 4 percent) as well as other related professions such as hosier. The *nuevos convertidos* shared the shoemaker's and furrier's trades with the *confesos*, but *nuevos convertidos* completely controlled the manufacture of saddles. Interestingly, vis-à-vis *judeoconverso* participation in the major industry of the region, the wool trade and manufacture of woolen cloth, the *confesos de los antiguos* dominated production at the top, since only *confesos* were cloth merchants and manufacturers (*mercaderes de paños* and *tratantes de fazer paños*), and master wool producers (*maestros de lanas*). The rest of production was divided evenly between the two groups, except for the skill of dyer, which was almost completely dominated by the New Christians, perhaps because many of the dyes and other chemicals necessary for their trade would have been supplied to them by the apothecaries, who were also entirely New Christians.[52]

[52] For the history of this industry, see Carla Rahn Phillips and William D. Phillips, Jr., *Spain's Golden Fleece: Wool Production and the Wool Trade from the Middle Ages to the Nineteenth Century* (Baltimore, MD: Johns Hopkins University Press, 1997). The authors

To know anything about pre-modern society, however, is to know that one's family greatly influenced career paths and marriage choices. Up until this point, I have discussed the economic and geographic stratification of the *confesos* and *nuevos convertidos* by looking at the aggregate information concerning individuals. Such an approach is misleading in that it suggests that individuals had complete autonomy in their decisions. What happens to this information when we consider the families themselves as actors? If individuals had autonomy in their choice of career, we might expect the diversification of jobs within a family. In pre-industrial economies, however, learning a craft was a time-consuming, expensive, and sometimes secretive process, as artisans strove to maintain an advantage over their competitors. Famously, merchants relied on family members to maintain their connections to far-flung markets. Thus, we have the received wisdom that pre-modern families conserved their market advantage by training their children to follow in the family business; if there were no sons, then a son-in-law was acquired. When a minority population is involved, the pressure to specialize increases so that members of the minority can dominate one niche in the economy and thereby prosper despite their small numbers and overall disadvantaged position in the marketplace. According to this model, we should find a high degree of economic specialization among the *judeoconverso* families of Sigüenza.[53]

To test this idea, I looked at the occupational profile of families in which the occupations of at least five individuals (including spouses) is known, and then arranged the families on a continuum from extreme specialization within niche occupations associated with the Jews to integration into the Old Christian economy. I also tracked families according to their ethnic identification. Once this is done, it becomes clear that few *judeoconverso* families of Sigüenza, whether *confeso* or *nuevo convertido*, conformed closely to the model of niche specialization. As to be expected, the *confesos* had traveled furthest down the path of assimilation into the Old Christian economy – six of the nineteen *confeso* families in the

do not specifically address the *conversos'* participation in the trade. Later, Portuguese New Christians, probably descendants of 1492 exiles from this region, handled the trade. See Máximo Diago Hernando, "La irrupción de los conversos portugueses en el comercio de exportación de lanas de la Corona de Castilla en el tránsito del siglo XVI al XVII," *Sefarad* 70 (2010): 399–434.

[53] The literature goes back to the early twentieth century, most notably, Max Weber's *The Protestant Ethic and the Spirit of Capitalism*. See also Thierry Volery, "Ethnic Entrepreneurship: A Theoretical Framework," in *Handbook of Research on Ethnic Minority Entrepreneurship: A Co-evolutionary View on Resource Management*, ed. Léo Paul Dana (Northampton, MA: Edward Elgar, 2007), 30–41.

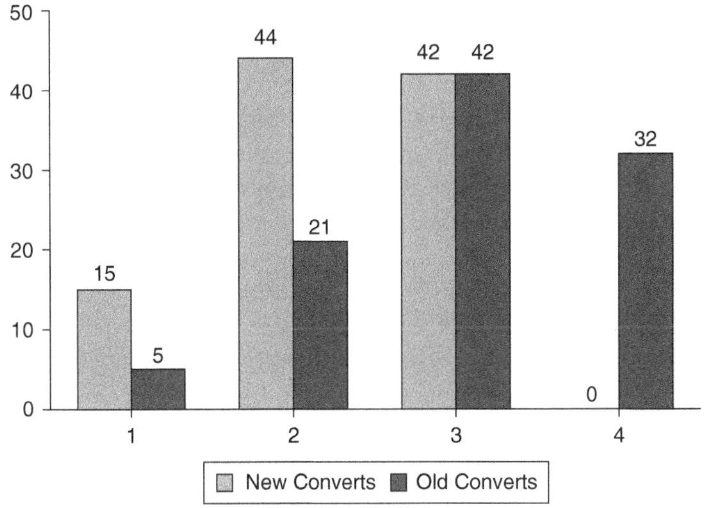

FIG. 4.3. Economic assimilation of seventy-four *judeoconverso* families: 1. Specialized in a one "Jewish" occupation; 2. Variety of "Jewish" trades practiced; 3. Some expansion into "Christian" occupations; 4. Only "Christian" occupations

study had entered entirely into professions identified with Old Christians. Much of the expansion was achieved through the time-honored strategy of intermarriage with Old Christian families (more on this below). An example of such strategizing is the Castillo-Hombrados family, as described by fifty-year-old Diego Malo, a resident of Prados Redondos, near Molina de Aragón. On his maternal grandfather's side, Diego was descended from assimilating *confesos*. His mother's father, Diego del Castillo, had been the *escribano del número* (a royal appointment subject to an examination) of the village of Prados Redondos. Diego del Castillo had married well to Beatriz de Zorita, an hidalga, and their two sons had continued the process of assimilation by becoming a parish priest in Molina de Aragón and an *escribano*, and their daughter had married into the hidalgo Hombrados family. Diego's paternal grandfather, Juan de Hombrados, had been the governor of the Castle of Zafra, an important fortification on the frontier with Aragon. Neither Diego nor his father, Juan de Hombrados, Jr. (who had been arrested on blasphemy charges in 1529) appeared to have any occupation.[54]

[54] ADC Inq., exp. 2220, Diego Malo, *incredulidad*, suspended, 1554. Diego had originally denounced himself to the Inquisition in 1532, but it took the judges twenty-two years

While the *confesos* had the most success in taking on Old Christian occupations, the *nuevos convertidos* also had made gains, with nearly one half of the fifty-five families entering into Old Christian fields. By the same token, only a few of the *nuevos convertido* families had opted to specialize in a single trade. An example might be the López family of Almazán. Living in the town with the highest concentration of *nuevos convertidos*, the López worked in the leather trade as furriers and shoemakers. Both grandparents had been furriers, as had one uncle, the father, and his son, Martín López. The women in the family had married other furriers or shoemakers. Yet, even the López family was not frozen in place. Martín was the first person in his family to marry an Old Christian (the woman's father was a shoemaker), and he told the inquisitors that he planned a career in the priesthood for his son, also a first.[55] The López, however, were more specialized than most New Christian families, who preferred to diversify within a general area of the economy – crafts, commerce, liberal professions, or even to cross over between sectors.

MARRIAGE STRATEGIES

As is clear from the anecdote of the Castillo family's success or even from Martín López's tentative moves, integration into the dominant economy and society could be accomplished through marriage into Old Christian families. In the fifteenth century, the *confesos* proved this so well that they brought down on themselves the wrath of Old Christian society, which counteracted by passing purity of blood statutes meant to exclude *confesos* from public life.[56] By the same token, marriage was also the mechanism by which *judeoconverso* families could combat the centrifugal forces of life – emigration, death, inheritance laws – that could pull them so apart that they could preserve neither their fortune nor their

to get around to his case. Diego's father apparently was given to swearing at the dinner table. In exasperation Diego's brother Garci exclaimed, "For once why can't we eat in peace!" and Diego piled on, blaspheming, "Not even God has the power to give us peace!" (i.e. not even God could shut their father's foul mouth). Denying that God had the power to do a particular thing was considered heretical, even as a common form of blasphemy.

[55] ADC, Inq. exp. 2247, Martín López, Judaizing (1554). Martín was penanced; others of his family were not so lucky – those who had not been arrested had fled. Two family members had been burned and several were reconciled. *Conversos* were prohibited from entering the priesthood, but Martín seems unfazed.

[56] For a recent overview of the topic, see José Javier Ruiz Ibáñez and Juan Hernández Franco, "Conflictividad social en torno a la limpieza de sangre en la España Moderna," *Investigaciones históricas: Época moderna y contemporánea* 23 (2003): 35–56.

identity. As was true of everything else in their lives, the *confesos* and *nuevos convertidos* differed in their marriage strategies. The Inquisition materials make one thing very clear: the first rule for both groups was to never intermarry with one another.

The 164 defendants questioned by the Inquisition submitted a large amount of genealogical information about their family members. In 575 cases I was able to pinpoint the ethnicity of both partners in a marriage, mostly that of the defendant (ego, in anthropological terms) and his or her spouse, parents, and grandparents. In the case of grandparents, the information often reaches far back into the fifteenth century. Many of those marriages were between Jews, although occasionally, a surviving partner was forced to convert to Christianity in 1492, thus giving the impression of a Jewish/Christian match. Legally, that was impossible, so such cases have been counted as Jewish pairings. In some cases, because of a Converso-Old Christian union between the ego's parents, one set of grandparents might both be Old Christian. In Fig. 4.4, the marriage information is presented by generations – that of the grandparents (Generation 1), parents and uncles (Generation 2), and the ego, his siblings, and adult children (Generation 3). The large majority of both *nuevos convertidos* and *confesos* preferred to marry within their own ethnic community, although with time, we see more willingness on both sides to contract matches with Old Christians.[57] What the new and old converts were *not* willing to do, however, was to exchange marriage partners between their two communities. Of 575 matches, only twelve (just 2 percent) were matches between *nuevos convertidos* and persons of *confeso* ancestry, while fifty (9 percent) were with Old Christians. [58] In other words, while most *Judeoconversos* preferred the status-quo of endogamy, when moved to action, their marriage choices showed that the future lay in a match with an Old Christian, not with a member of the other minority group. This was true above all for the *nuevos convertidos*, who were responsible for only two of the twelve *nuevo–antiguo* matches, even though they outnumbered the *confesos* in the sample by four to one.

57 Although moriscos also lived in the Bishopric of Sigüenza, they are so absent from the judeoconversos' lives that one might think that they did not exist.

58 Defendants were so precise about their lineage that I had to create a separate category for persons who were a mixture of *nuevos convertidos* and Old Christians. They tended to marry others of this type, *nuevos convertidos*, or Old Christians, but not *antiguos*. For display purposes, the few *antiguos* who married *nuevos cristianos* are included in the category.

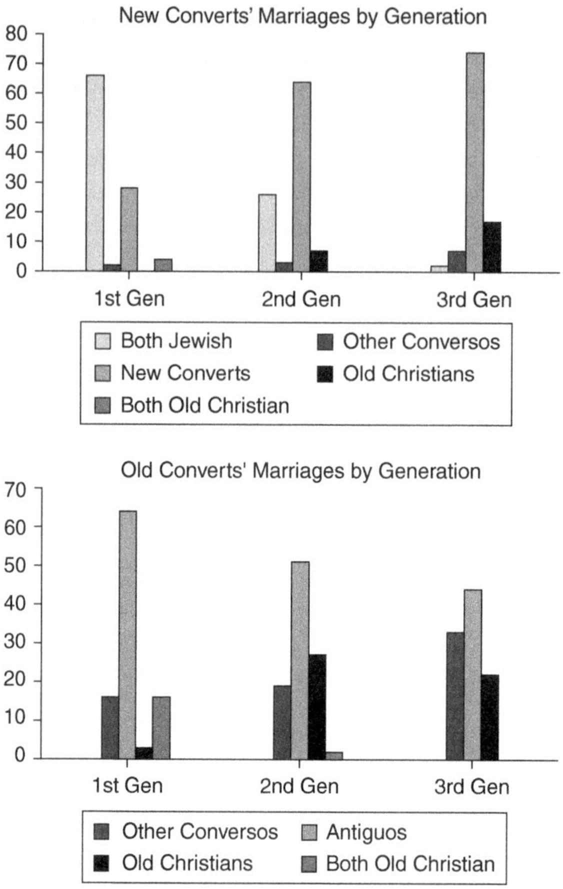

FIG. 4.4. Assimilation of converts by generation and minority identification, fifteenth and sixteenth centuries

On the whole, the preferred match among the *judeoconversos* of Sigüenza was with someone of the same ethnic background. Not only did the matches take place within the same ethnic subgroup, but not infrequently they were consanguineous. Spain's three religions had radically different positions on the question of consanguineous marriage. The Catholic Church's famously strict teaching on marriage forced families into exogamic partnerships, but Judaism and Islam's laws did not. In Table 4.3, regardless of the fact that all of the couples were technically Christians, we can see that ethnicity played a strong part in determining how many consanguineous matches and of what type would be celebrated between families. Among the 164 *judeoconverso* families that

TABLE 4.3. Consanguineous marriages of converts, by ethnicity, before 1580*

Type of marriage	New Converts $n = 125$	Old Converts $n = 39$
Cross cousin (FZD, MBD)	3	
Parallel cousin (FBD, FBS)	2	
Maternal uncle	2	1
Paternal uncle		
Sibling pair	BZBZ (2) BBZZ (3)	BBZZ (2)
Other close relatives/in-laws	3	
Totals (ratios)	15 (ratio 1:8)	3 (ratio 1:13)

*Preliminary results from the rest of my data base suggest that in the sixteenth century, Old Christians' use of close marriage was infrequent (one in thirty-four families surveyed) and was limited to sibling matches. The moriscos were the most endogamous of all, never intermarrying with the other two ethnicities. One in every nine families reported a sibling match or less frequently, a match between parallel cousins. Because of the incomplete nature of the inquisitors' questioning, results from the sixteenth century are only suggestive of trends.

make up the sample under study, we find fifteen instances of consanguineous marriages or sibling matches that will lead to the creation of double cousins.[59]

The *nuevos convertidos* were behaving as if they were still Jewish, among whom close marriages were the norm, rather than the exception. Moreover, with total disregard for canon law, they engaged in the very closest form of consanguineous marriage, a match between a girl and her maternal uncle ("girl," because many were in fact that, thirteen to sixteen years old). The *nuevo convertido* Pedro García el Rubio explained the rationale behind such matches when he reportedly said "the law of the Jews was true, that cousins should marry in order not to disperse property."[60] The practice of cousin marriages is typical of traditional societies in the Middle East, with a match between

[59] This is despite the fact that the inquisitors' genealogical inquiries mostly did not include wives and until 1600, never cousins or nephews. If these individuals' identities were to be known, the number of known consanguineous marriages would be even higher.

[60] Carlos Carrete Parrondo, "Nostalgia Among Castilian Judeoconversos," *Mediterranean Historical Review* 6(2) (1991): 35, citing ADC 749 exp. 15 (1505).

a man and his father's brother's daughter (FBD) being seen as ideal. Such a match between parallel cousins represents a complete refusal of exogamy because no new male line is brought into the family, and no woman (and her dowry) is lost to another lineage.[61] Uncle–niece marriages, needless to say, are completely endogamous and were permitted within Sephardic tradition.[62] To use a knitting analogy, those *nuevos convertidos* who resorted to consanguineous marriages to preserve their identity, wealth, and community were creating a tight-fitting garment indeed, despite Christian condemnation of such practices. A generation or two after 1492, the *nuevo convertido* community of Sigüenza had not changed its basic thinking about who could be a member of the group. On the other hand, if partner choice is any indication of group identity, the *confesos* had.

* * * *

Although Spaniards' obsession with purity of blood would continue for another two centuries, in the 1560s something began to happen to the *Judeoconversos* of Sigüenza – they began to forget who they were. The trend was repeated in other parts of the district of the Inquisition of Cuenca. Whereas once individuals matter-of-factly and precisely had identified themselves as "*confesos*," "*nuevos convertidos*" and everything in between, now, seventy years on from 1492, defendants were having a hard time figuring out how to answer the question, "*¿qual es su casta e generaçion?*" Their forgetting was certainly aided by the widespread adoption of purity of blood statutes in the 1550s, making owning up to Jewish ancestry even more undesirable. When defendants' verbatim responses are lined up chronologically, from the 1550s through the 1560s we see a very natural progression from surety to self-doubt, confusion, and even surprise. Even the word "confeso" loses its precise meaning, and becomes the generic term for a Jewish convert that we know today. And, given a choice, converts would certainly prefer to identify themselves as "confesos" since the term added another one hundred years or so to one's pedigree as a Christian. Agustín de la Peña (b. 1529), from a *nuevo convertido* family in Medinaceli, knew in

[61] Robert F. Murphy and Leonard Kasdan, "The Structure of Parallel Cousin Marriage," *American Anthropologist* (New Series) 61(1) (February 1959): 17–29.

[62] Such marriages still take place among the Sephardic Jews in Israel. Alan Bittles, "When Cousins Marry: A Review of Consanguinity in the Middle East," in *Perspectives in Human Biology, vol. I, Genes, Ethnicity and Ageing*, eds. Lincoln Heinze Schmitt, Leonard Freedman, and Rayma Pervan (Nedlands: University of Western Australia, 1995), 71.

1569 which persons in his family tree had been convicted by the Holy Office, but described himself as a "*confeso.*" Eighteen years earlier, when his uncle, Juan de la Peña (b. 1513), was asked this question, he had answered correctly, "*nuevo convertido.*"[63] The confusion over identity was particularly marked for those of *judeoconverso* descent who were arrested on charges not related to Judaizing, for example, blasphemy or denying that fornication was not a sin. First they would say, "Old Christian," then remember an incident that contradicted that assertion, and then finally admit that they did not know what their "casta" really was, after all. Or, remarkably, they would note all the contradictions, and still conclude that they were Old Christian.[64] After 1570, except for those very few defendants arrested for Judaizing and as "*inhábiles*,"[65] individuals routinely answered that they were Old Christian and the inquisitors accepted their answer at face value.

The inquisitors had moved on, and the *conversos* of Sigüenza no longer were of interest to them. Although the descendants of Judaizers were punished from time to time for violating their condition as "*inhábiles*," and many were picked up for vaguely hostile minor religious "crimes,"[66] no natives of the region were arrested again for Judaizing. Without the Inquisition's active interest and in the absence of notarial records from the region[67] the area's *converso* communities simply disappear from sight. Historians argue over whether there were real crypto-judaizers in La Mancha. What is clear from the evidence presented here,

[63] ADC Inq., exp. 3290, Agustín de la Peña, inhábil (1569) and exp. 2197, Juan de la Peña, judaísmo (1553). There is no doubt that they were 1492 converts.

[64] ADC Inq., exp. 3381, Juan Martínez Mazon, Lutheranism (1570). A self-identified Old Christian from Molina de Aragón, Juan said that his schoolmates would taunt him by calling him a "Jew." Despite his claim to Old Christian status, everything about Juan's family profile suggests that he was descended from *confesos*. Mid-century generally is seen as the time when the *conversos* slipped into the majority population, partly in response to the widespread adoption of purity of blood restrictions at that time. Thereupon sprang up a cottage-industry specializing in whitewashing backgrounds. See Ruth Pike, *Linajudos and Conversos in Seville: Greed and Prejudice in Sixteenth- and Seventeenth-Century Spain* (Bern: Peter Lang, 2000).

[65] *Inhábil* – a person disqualified from public office, riding on horseback, and wearing certain clothes by the conviction of a relative for Judaizing.

[66] In her study of the trials from Molina de Aragón housed at the Archivo Histórico de Madrid, Zozaya also notes that the *conversos* "maintained a subversive and resistant attitude towards the state religion" (p. 161).

[67] A visit to the Archivo Histórico Provincial de Guadalajara proved that almost no records from the sixteenth century survive. An exception is a 1568 notarial register from Molina de Aragón (P-1892.02) that does record a few transactions by some of the families studied in this article.

is that in the first half of the sixteenth century there existed not one, but two *converso ethnic* minorities within the region, each with its own distinctive identity and communal memory. We cannot recover the truth of their religious convictions, but by tracking their actions, we can detail the social and ethnic evolution of the *judeoconverso* community in Castile.

5

On the *Alumbrados*: Confessionalism and Religious Dissidence in the Iberian World

Mercedes García-Arenal and Felipe Pereda

Translated by Consuelo López-Morillas

A longstanding practice in historiography has described the phenomenon that we call *alumbradismo* (Illuminism) as an offshoot of the great Spanish mystical tradition, a sort of wayward branch grafted onto the healthy trunk of sixteenth-century Spain's exuberant spirituality. In the paradox-filled pages of *En torno al casticismo*, Miguel de Unamuno first placed mysticism at the core of the Castilian spirit and then diagnosed its malady (and by extension that of Castile itself) as a case of "moral hypertrophies." Unamuno wrote, "Those of Castilian mysticism were an egotistical quietism that was a plunge into nothingness, and a coarse illuminism that tended towards pleasure and satisfaction of the instincts, resulting in the horrible conjunction of a drowning of the intellect in the void, conceived as a carnal union of the sexes ... in the last extreme of what St. John called spiritual lust and gluttony."[1]

[1] The first version of this article was based on the presentation made by its two authors at the opening of the meeting "Alumbrados y disidencia religiosa en el mundo Ibérico" (Madrid, Consejo Superior de Investigaciones Científicas, June 24–25, 2010). It was published in Spanish in *La Coronica* 41(1) (2012): 109–48. The English version that it is published here has been consistently enlarged and updated in terms of bibliography but also in terms of discussion. The research leading to these results has received funding from the European Research Council under the European Union's Seventh Framework Programme (FP7/2007–2013)/ERC grant agreement n° 323316, project CORPI (*Conversion, Overlapping Religiosities, Polemics and Interaction: Early Modern Iberia and Beyond*). "... hipertrofias morales. Las del misticismo castellano fueron el quietismo egoísta del abismarse en la nada o el alumbradismo brutal dado a la holganza y al hartazgo del instinto, que acababa con el horrible consorcio del anegamiento del intelecto en el vacío conceptualizado con la unión carnal de los sexos ... en el último extremo de lo que llama San Juan lujuria y gula espiritual." Miguel de Unamuno, *En torno al casticismo* (Madrid: Cátedra, 2005), 229.

Unamuno could have known little or nothing about the true contents, practices, beliefs, and attitudes adopted by those whom the Holy Office persecuted as *alumbrados* for more than a century; in fact, only long labor in the archives has gradually uncovered the documents that now allow us to assess this elusive phenomenon. Nevertheless, the dramatic picture of Spanish spirituality whose outlines Unamuno sketched – a spirituality that he saw as caught between the overwhelming force of its search for Meaning (with a capital) and an Inquisition that resembled a "customs-house of purist unitarianism" – impressed a young historian named Marcel Bataillon.

Bataillon not only translated Unamuno's essays into French – incidentally, changing their title to the no less eloquent *L'Essence de l'Espagne*[2] – but devoted the rest of his life to studying the uniqueness of this intellectual and religious tradition.[3] His work was enriched in successive editions by the comments of such enthusiastic yet critical readers as Eugenio Asensio. As a result the French scholar was able to identify, at the very center of Spain's spiritual adventure, a phenomenon that had resisted doctrinal repression just as it would later resist historians' understanding; its significance proved proportional to the difficulty of establishing its precise doctrinal frontiers, its sources, and its exact chronology. Bataillon made a special attempt to respond to and incorporate specific criticisms by Asensio, who had shown that the medieval roots of the sixteenth-century's spiritual movements lay along three separate axes: biblical studies by Jews, *conversos*, and Old Christians; Franciscan spirituality; and the seeds of religious renewal that arrived, by different channels, from Italy.[4] Therefore Erasmianism, which was fundamental in shaping certain spiritual and ideological currents in Spain, in Asensio's opinion neither stood alone as these ideas were forming nor was the only response to concerns about religious renewal.

[2] Marcel Bataillon, *L'essence de l'Espagne. Cinq essais (en torno al casticismo)* (Paris: Librairie Plon, 1923).

[3] His monumental work *Erasmo y España* first appeared in 1937, but attained wide circulation only after it was translated into Spanish in 1950 (México: Fondo de Cultura Económica).

[4] In his famous article, Eugine Asensio, "El erasmismo y las corrientes espirituales afines. Conversos, franciscanos, italianizantes," *Revista de Filología Española* 36 (1952): 31–99; and in "La peculiaridad literaria de los conversos," *Anuario de Estudios Medievales* 4 (1967): 327–51. Discussion and debate between this historian and others, like José C. Nieto and Francisco Márquez Villanueva, and Marcel Bataillon may be found in Bataillon, *Erasmo y el Erasmismo* (Barcelona: Crítica, 1977). Responses to the work of Bataillon continue: see Lu Ann Homza, *Religious Authority in the Spanish Renaissance* (Baltimore, MD: Johns Hopkins University Press, 2000), among others.

As Bataillon refined these arguments, successive editions of *Erasmo y España* focused his analysis of *alumbradismo* on its unquestionably autocthonous nature.[5] Although Bataillon's strategy was to study the spiritual history of the sixteenth century in a European context, thus making it part of the same reformist current that flowed through the rest of Europe, his "Illuminism" – as he baptized the movement – resisted attempts to force it into imported categories. On the contrary, as the French scholar already suspected by the middle of the last century, it was "essential to view it as submerged in the depths of a vaster spiritual movement, one that the Inquisition tried to hold back at the time with a dike: *alumbrados*, *dejados* and *perfectos*."[6]

Bataillon portrayed *Alumbradismo* as an aspiration of reform outside the Church, an "interiorized Christianity" inspired by an "optimistic feeling of grace."[7] This renewed Christianity was repelled by external signs of formalized religiosity, which was imbued with gestures, formulas, and ceremonies that were seen as unnecessary obstacles to one's experience of faith. It rejected the mediations imposed by a Church which, under the pretext of making God more accessible to man, actually made Him more distant, to the point of replacing God with the vicarious and idolatrous cults of the Eucharist, the saints, relics, and images. As Old Christians saw it, the practice of heresy arose from a wish to free oneself from rites and ceremonies, payment of tithes, bulls, and other restrictions; in their eyes such freedom could only lead to sexual license. As Bataillon wrote in 1937, "At the heart of the debate, under the names of Lutheranism and Illuminism, was the great defense of cultivating one's spirit in opposition to ceremonial religion. The official announcer of this defense to

[5] "It has always been evident to me that Erasmian modernism had deep roots in the fifteenth century and earlier ones. The same would have been true of Spanish Illuminism, which found one of its favorite sources in Erasmus. I see it more clearly all the time ... what we historians of the sixteenth century need above all is to understand the fifteenth [century] better." Prologue to the Spanish translation, 1950: xiii. The second Spanish edition in 1966 incorporated Asensio's points of view ("El erasmismo"). See both included in Bataillon, *Erasmo y España*, 1966 and later reprints.

[6] On the difference between *recogimiento* (spiritual withdrawal) and the doctrine of *dejamiento* (abandonment of the self), whose chief exponent was Isabel de la Cruz, see Bataillon, *Erasmo y España*, 166–76. See also Melquíades Andrés, "Alumbrados, erasmistas, luteranos y místicos, y su común denominador: el riesgo de una espiritualidad más 'intimista'," in *Inquisición española y mentalidad inquisitorial* (Barcelona: Ariel, 1984), 373–409; and John E. Longhurst, "La beata Isabel de la Cruz ante la Inquisición," *Cuadernos de Historia de España* 25–26 (1957): 270–303.

[7] Bataillon, *Erasmo y España*, 175.

contemporary Spaniards was Erasmus."[8] This provocative statement gives rise to reflection and discussion.

Alumbradismo entails the double angle of its deep roots in Spanish soil and the ease with which its branches seem to intertwine with those of other European movements of reform and spirituality; as a result, attempts to interpret it have tended to a pendulum-like oscillation. Bataillon himself, in later years, would revise his own stance and suggest new directions of research that he would not have time to explore in person. As he would write at the end of his career, "today I would pay less attention to a common denominator of interiority, and much more to the believer's non-mystical conviction of profound dependence upon the God who grants him grace."[9] Bataillon's new thinking connected both Isabel de la Cruz and Alcaraz closely to Lutheranism, while proposing a more exact analysis of intellectual exchanges and of the contexts in which those ideas had flourished. It drew on contributions by Angela Selke and José Carlos Nieto, who had earlier interpreted Illuminism not as traditional mysticism but as a radical form of Christianity that tied it to Lutheranism.[10]

At the other extreme, Miguel Asín Palacios, from the mid-1940s onward, ardently defended the view that Castilian *Alumbradismo* had its origins in Islamic Sufism transmitted by the Morisco population, Spain's other *conversos*.[11] Only a few other scholars have continued to support this theory up to the present day: it has still not been set on a firm foundation.[12] We shall return to the topic later in this article.

In light of these complicated ingredients – described here briefly and sketchily from both the historical and the historiographical points of view – Illuminism came to be defined less and less as a hypertrophic form of mysticism, a "faulty understanding and experience of true mysticism, something like its shadow," in the words of a well-known scholar of

[8] Bataillon, *Erasmo y España*, 473.

[9] Bataillon, "Acerca de la influencia de Erasmo," in *Erasmo y el Erasmismo*, 245–54.

[10] Angela Selke, "Vida y muerte de Juan López de Celaín, alumbrado alcarreño," *Bulletin Hispanique* 62 (1960): 136–62; José C. Nieto, *Juan de Valdes and the Origin of Spanish and Italian Reformation* (Geneva: Droz, 1970).

[11] He maintained it in several works, principally a series of articles that appeared in *Al-Andalus* (1944–51) and were collected in Miguel Asín Palacios, *Sadilíes y Alumbrados* (Madrid: Hiperión, 1990).

[12] See Luce López-Baralt, *Huellas del Islam en la literatura española. De Juan Ruiz a Luis Goytisolo* (Madrid: Hiperión, 1990); also, *La literatura secreta de los últimos musulmanes de España* (Madrid: Trotta, 2009).

Spanish spiritual literature.[13] Instead, it came to be seen as a symptom of a religiosity that had developed within a historical and human landscape unique and unparalleled in Europe.

HISTORICAL ANTECEDENTS AND HISTORIOGRAPHICAL PRECEDENTS

The phenomenon that the Holy Office baptized with the name of *mal alumbrado* (Illuminist malady) appears on history's stage in 1525. The first received denunciation against the group had been provided by Mari Nuñez in 1519, but no action was taken on her accusations until years later. The inquisitors, unsure at first of the true nature of the problem they faced, soon bent every effort to its eradication. The arrests of two leaders of the group (Isabel de la Cruz and Pedro Ruiz de Alcaraz) occurred in 1524 and their interrogation provided the additional information needed for the inquisitors to provide a definition of what this group supposedly believed. In 1525 the Inquisitor-General Alonso Manrique famously released the first Edict of Faith condemning the propositions of this group that the inquisitors had elected to name *alumbrados*. This was the first official document identifying this group of individuals as a heresy, the first document to outline their beliefs. However, those grouped together as a sect were in reality a fractious group of lay religious individuals competing for followers. They never identified themselves as *alumbrados*, and as far as we know none of them ever composed a formal statement of belief or doctrine. This famous edict placed the locus of this new heresy in the archbishopric of Toledo, but once defined and delimited by the Inquisition its ramifications would soon be discovered all over central Castile, principally in Valladolid.[14] The *alumbrado* heresy would spring up once more in Llerena in the 1570s and in Seville during the second decade of the seventeenth century; it would reappear in Madrid at mid-century and, as Marilena Modica has recently demonstrated,[15] would reach

[13] See Melquíades Andrés Martín, *Los Recogidos: nueva visión de la mística española (1500–1700)*(Madrid: FUE, 1975), 43; also *Nueva visión de los "Alumbrados"* (Madrid: FUE, 1973), and "La espiritualidad española en los siglos XV y XVI. Movimientos y disputas," in *Historia de la Iglesia en España*, 5 vols (Madrid: Biblioteca de Autores Cristianos, 1979), 3: 327–61.

[14] See Vicente Beltrán de Heredia, "El edicto contra los Alumbrados del reino de Toledo," *Revista española de Teología* 38 (1950): 105–30; Milagros Ortega Costa, "Las proposiciones del Edicto de los Alumbrados. Autores y calificadores," *Cuadernos de Investigación Histórica* 1 (1977): 23–36.

[15] Marilena Modica, *Infetta dottrina. Inquisizione e quietismo nel Seiciento* (Rome: Viella, 2009).

even faraway Sicily in the seventeenth and eighteenth. *Alumbradism* took other shapes also in the New World as important work done on colonial Mexico has shown.[16] But those later groups fall out of the present essay, focused on the intellectual and religious origins of the first *alumbrados*.

The recurrent history of Illuminist spirituality illustrates how naturally diverse was the heresy that the inquisitors fought so hard to eliminate. The *alumbrados*, rather than forming a succession of interrelated groups that passed along definite codes of conduct, arose at different historical moments and professed different ideas and customs, but were always united by their search for an immediate experience of grace and their corresponding rejection of ecclesiastical control.

They were united also, and most especially, by the way in which the Holy Office had defined it and identified it as a heresy. Illuminism's spread into Europe reveals, on the one hand, how hard it is to define a specific Illuminist credo, and on the other, how strangely Spanish the phenomenon was. As Sophie Houdard has shown in another seminal book, in Europe it would encounter both an enthusiastic reception and militant resistance.[17]

In depositions made by the accused that survive in extant documents, Illuminism invariably displays two distinct characteristics. First, it shows a combination of practices, beliefs, and even feelings that were shared within small clandestine groups or *conventículos*, united more by emotion than by strictly confessional or ideological ties: "emotional communities," in Barbara Rosenwein's phrase.[18] As Alastair Hamilton has shown in his illuminating study,[19] we find in the earliest trials of *alumbrados* in Toledo (especially that of Pedro Ruiz de Alcaraz, still unpublished) competition among a small number of individuals, preachers, and devout women, who sought to impose their charisma on small, closed circles of converts. In the second place, all of these groups shared a limited set of ideas and behaviors that combined, to a greater or lesser degree, a feeling of purity, the practice of mental prayer and an abundance of ecstatic experiences with a determined anti-ritualistic stance. *Alumbrados* believed themselves especially enlightened by God and advocated for a

[16] Nora E. Jaffary, *False Mystics: Deviant Orthodoxy in Colonial Mexico* (Lincoln, NE: University of Nebraska Press, 2004); Adriana Rodríguez Delgado, *Santos o embusteros: los alumbrados novohispanos del siglo XVII* (Mexico: Gobierno del Estado de Veracruz, 2013).

[17] Sophie Houdard, *Les invasions mystiques. Spiritualités, heterodoxies et censures de l'époque moderne* (Paris: Les Belles Lettres, 2008).

[18] Barbara Rosenwein, *Emotional Communities in the Middle Ages* (London: Cornell University Press, 2006).

[19] Alastair Hamilton, *Heresy and Mysticism in Sixteenth-Century Spain: The Alumbrados* (Cambridge: J. Clarke, 1992).

more interior spirituality divorced from the prescriptions of the Catholic faith. According to the Inquisition, this group encouraged mental rather than vocal prayer, belittled the importance of nearly all the sacraments, denigrated the veneration of saints, the cross, the Virgin Mary, and even representations of Christ. Ultimately, they advocated an abandonment of one's soul to God resulting in a personal connection with the divine that made both the Church and its functionaries unnecessary. These were, broadly speaking, the only common denominators of the various communities: a desire for regeneration or reform founded on a return to an internalized religiosity, and a claim to a space in which to practice it that was at least autonomous, if not frankly anti-institutional. That is how events unfolded first in Medina de Rioseco, with the quixotic adventure patronized by Alonso Enríquez, *Almirante* of Castile; later on in small rural communities in the diocese of Badajoz, and fifty years later in the Sevillian congregation of La Granada, a special and very heterodox religious fraternity united in defense of the Immaculate Conception.[20]

The historical reconstruction of Illuminism has yielded important results in recent years, ever since renewed interest was focused on it in the 1990s. As early as 1958 Román de la Inmaculada published an article suggesting a renewed study of heresy and heterodoxy that would take *alumbradismo* as its point of departure, since it represented a unique contribution to the history of sixteenth-century Spanish spirituality.[21] And a few years later, in 1962 in the same journal, Ignacio Tellechea Idígoras issued a call to revise the prevailing opinion of Illuminism by applying a new hermeneutics to texts that historians had come to see as empty of meaning.[22] If his suggestions had been followed, they would have led to a new historiographic formulation of the concepts of heresy and religious marginality; but they inspired no response among Spanish

[20] On its origins, see Francisco Javier Campese Gallego, especially "Un profeta singular y sus revelaciones desaparecidas: una tradición centenaria en el Siglo de Oro," in *Testigo del Tiempo, Memoria del Universo; cultura escrita y sociedad en el mundo ibérico (siglos XV–XVIII)*, ed. Manuel Fernández, Carlos-Alberto, González-Sánchez, and Natalia Maillard Ávarez (Barcelona: Rubeo, 2009), I, 349–69, and Campase Gallego "Gómez Camacho, in profeta paradójico en el Siglo de Oro," *Investigaciones Históricas* 28 (2008): 11–28; also Michel Boeglin, "Valer, Camacho y los 'cautivos de la Inquisición'. Sevilla 1540–41," *Cuadernos de Historia Moderna* 32 (2007): 113–34; and Antonio González Polvillo, "La Congregación de La Granada, el Inmaculismo sevillano y los retratos realizados por Francisco Pacheco de tres de sus principales protagonistas: Miguel Cid, Bernardo de Toro y Mateo Vázquez de Leca," *Atrio* 15–16 (2009–10): 47–72.

[21] Román de la Inmaculada, "El fenómeno de los alumbrados y su interpretación," *Teresianum, Rivista della Pontificia Facoltà Teologica 9 (1)* 79 (1958): 49–80.

[22] Jose Ignaciao Tellechea Idígoras, "Textos inéditos sobre el fenómeno de los alumbrados," *Teresianum 13 (1–2)* (1962): 768–74.

historians at a time when studies of spirituality, even of religion or the Church, were left chiefly in the hands of Catholic clergy. To this day, the history of spirituality and religion in Spain suffers the consequences of that confessional focus and a reduction of the whole field to social history.[23] But in the meantime, great advances were made in Spain in publishing the richest records of Inquisition trials from the Archivo Histórico Nacional. Some of the most important studies were those by Angela Selke on Alcaraz and Francisco Ortiz; by Milagros Ortega Costa on María de Cazalla; Carlos Puyol's minute reconstruction of the late episode of San Plácido; the monumental *Historia* by Álvaro Huerga; and Vidal Abril Castelló's equally extensive publication of the trial of Francisco de la Cruz, the *alumbrado* prophet of Peru. All of these have left us with a much more detailed knowledge of the social reality of the Inquisition, as well as ready access to the documentary record.[24]

Alongside of or inspired by these publications other attempts have been made to define the *alumbrado* phenomenon on a more intellectual ground. Such was the proposal of Antonio Márquez, whose 1972 monograph[25] identified Illuminism specifically with Neoplatonism, in contradistinction to Hebrew Messianism. Márquez presents Illuminism not as a diffuse "movement," as most of their authors claim, but on the contrary as a clearly delineated "mystical sect" with a concrete ideology and a traceable intellectual genealogy.[26]

Unfortunately, the necessary condition of separating the history of religion from confessional history did not occur in Spain. It was from Italy that a renewal of the field would come, as we shall see below. Among the great historians of Italian religiosity in the age of Church reform were Massimo Firpo, Adriano Prosperi, Carlo Ginzburg, and Michele

[23] Naturally there have been exceptions, in large part through work done by Spaniards outside of Spain: for example, José C. Nieto San Juan, *Juan de Valdes and the Origins of the Spanish and Italian Reformation* (Geneva: Droz, 1970), originally a PhD dissertation at Princeton.

[24] Angela Selke, *El Santo Oficio de la Inquisición: proceso de Fray Francisco Ortiz (1539)* (Madrid: Guadarrama, 1968); Milagros Ortega Costa, *Proceso de la Inquisición contra María de Cazalla* (Madrid: FUE, 1978); Carlos Puyol, *Inquisición y política en el reinado de Felipe IV: los procesos de Jerónimo de Villanueva y las monjas de San Plácido (1628–1660)* (Madrid: CSIC, 1993); Álvaro Huerga, *Historia de los Alumbrados, 1570–1630*, 5 vols. (Madrid: FUE, 1978–94); Vidal Abril Castelló, *Francisco de la Cruz. Inquisición. Actas.* (Madrid: CSIC, 1992–97). Also José Manuel Carrete Parrondo, *Movimiento alumbrado y Renacimiento español: proceso inquisitorial contra Luis de Beteta* (Madrid: FUE, 1980); Alastair Hamilton, *El proceso de Rodrigo de Bivar (1539)* (Madrid: FUE, 1979).

[25] Antonio Márquez, *Los Alumbrados: orígenes y filosofía: 1525–1559* (Madrid: Taurus, 1972).

[26] See an early critique of this theory in José C. Nieto, "En torno al problema de los alumbrados de Toledo," *Revista Española de Teología* 35 (1975): 7–93.

Olivari.[27] The work of these historians has shown that the intense spirituality of the 1520s and 1530s encompassed complex and diverse forms of religious ferment, which conditioned and to an extent diluted both the reception of reformist doctrines and the inevitable theological confrontation with them – at least until the mid-1540s, when the creation of the Roman Inquisition and the first conciliar deliberations had the effect of connecting Illuminism to Nicodemism.

In Firpo's words, the *alumbrados* and *espirituales*, "by the very act of turning inward and hiding in their own secret and unknowable consciences, seem also to point to the only path for those in Italy who could not, or would not, choose to take the bitter and harsh road of exile: that of Nicodemism."[28]

In this way, dissimulation was legitimized: by a conviction that knowledge of God can be obtained only through the light of faith, and that faith can exist only as "secret knowledge" that each person keeps within his own soul as a "secret revelation" that is granted to very few, because it is "hidden and concealed."[29] This state of affairs requires that the historian possess a fine sensibility for reading and interpreting the texts that we will turn to below; it also suggests that further cultural artifacts, such as images, be included in this still loosely defined history of spirituality in the years immediately preceding the Council of Trent.[30] The latter line of

[27] From their extensive bibliographies we note in particular Massimo Firpo, *Tra Alumbrados e "spirituali": studi su Juan de Valdes e il Valdesianesimo nella crisi religiosa del '500 italiano* (Florence: Leo S. Olschki, 1990); Carlo Ginzburg and Adriano Prosperi, "Juan de Valdés e la Riforma en Italia: Proposte di ricerca," in *Doce consideraciones sobre el mundo hispano-itáliano en tiempos de Alfonso y Juan de Valdés* ed. Francisco Ramos Ortega (Rome: Instituto Español de Lengua y Literatura de Roma, 1979), 185–97; and Michele Olivari, *Fra trono e opinione. La vita politica castigliana nel Cinque e Seicento* (Venecia: Marsilio, 2002), and Olivari, *Avisos, pasquines y rumores. Los comienzos de la opinión pública en la España del siglo XVII* (Madrid: Cátedra, 2014).

[28] Firpo, *Tra Alumbrados e 'spirituali'*, 220.

[29] Carlo Ginzburg, *Il nicodemismo. Simulazione e dissimulazione religiosa nell' Europa del '500* (Turin: Einaudi, 1970), 14. See a discussion of Ginzburg's even more radical proposal in Albano Biondi, "La giustificazione della simulazione nel cinquecento," in *Eresia e Riforma nell'Italia del Cinquecento. Miscellanea*, (Florence and Chicago, IL: Sansoni; The Newbury Library, I, 1974), 7–68; and especially Carlos M. Eire, *War Against the Idols: The Reformation of Worship From Erasmus to Calvin* (Cambridge: Cambridge University Press, 1986), ch. 7, "Calvin against the Nicodemists."

[30] Massimo Firpo has explored this possibility in several recent studies: see especially his *Gli affreschi di Pontormo a San Lorenzo. Eresia, politica e cultura nella Firenze di Cosimo I* (Turin: Einaudi, 1997); *Artisti, gioiellieri, eretici. Il mondo di Lorenzo Lotto tra Riforma e Controriforma* (Rome: Laterza, 2001); and Firpo and Fabrizio Biferali, *Battista Franco "pittore viniziano" nella cultura artistica e nella vita religiosa del'500* (Pisa: Scuola Normale Superiore, 2007).

research remains entirely unexplored in Spain. Paradoxically, it is in Italy where Bataillon's proposals and hypotheses have been adopted; he believed that the Erasmists' Irenist and anti-dogmatic tendencies must have been related to Nicodemist attitudes. Bataillon affirmed in the 1970s that "new studies of Spanish heterodoxy [should] shed more light on the roots and varieties of Nicodemism, as well as on the thorough anti-Nicodemism of early Calvinist orthodoxy."[31] He went on to suggest the significant overlap between Nicodemists and *judeoconversos* that inclined the latter towards heterodoxy within Christianity establishing implicitly a correspondence between assimilation and dissimulation. On the other hand, a deep commitment to Nicodemism was also to be part of the roots of secularization. This connection has recently been noted by Marilena Modica who, in her study of "quietism" or "relinquishment," has shown that the issue of the individual's power over his own soul was a strategic element in the politics of Roman Catholicism, which had already been undermined by a continual process of secularization.[32] These proposals have placed the study of *alumbradismo*/Illuminism at the center of a new debate and at the leading edge of a new religiosity opened to new modes of religious skepticism.

If entering upon this path calls for a greater attention to contexts, specifying the nature of Illuminism requires a deconstruction of inquisitorial discourse. As we have just observed, historians continue to see Illuminism as a phenomenon both elusive and resistant to sharp categorization; it is as if it were a fantasy that existed only in the minds of the inquisitors of the Holy Office, who were always prepared to place apparent heterodoxy in the inherited categories of a threatening *alumbrado* "nightmare." It is difficult to find any spiritual figure of the sixteenth and seventeenth centuries – from Juan de Ávila to Fray Luis de Granada, from the Archbishop of Toledo Bartolomé de Carranza to Ignatius of Loyola,[33] Juan de

[31] M. Bataillon, "Juan de Valdés nicodémite?," in *Aspects du libertinismo au XVIème siècle* (Paris: Vrin, 1974), 278–85.

[32] Modica, *Infetta dottrina.*

[33] Luis Fernández was the first to reconstruct the contexts in which Ignatius of Loyola's biography intersected with Illuminist circles. For the first perspective on the Jesuit order in Spain in this regard, see Stefania Pastore, "Tra conversos, gesuiti e inquisizione: Diego de Guzmán e i processi di Ubeda (1549–1552)," in *Le Inquisizioni moderne. percorsi di ricerca*, ed. Giovanna Paolin (Trieste: EUT, 2002), 216–51; and "I primi gesuiti e la Spagna: strategie, compromessi, ambiguità," *Rivista storica italiana* 117 (2005): 158–78. See also Carlos Ortega, "S. Ignacio en el *Libro de los alumbrados*," *Arbor* 107 (1980): 163–74. Marcel Bataillon had made number of suggestions already in the course that he devoted to the subject in 1945–6, published recently in Bataillon, *Los Jesuítas en la España del Siglo XVI* (Valladolid: Junta de Castilla y León, 2010), 232: "Frequently – and the confusion is significant – the Inquisition used the terms *alumbrado* and Lutheran as synonyms."

Yepes,[34] and Teresa of Avila[35] – whose orthodoxy was not measured at some point with the yardstick of *alumbradismo*.[36] From this complementary perspective, Illuminism becomes not a specific phenomenon but an intellectual category, a tool that allows us to consider the singularity of a whole spiritual tradition not in its identity but in its alterity, precisely as a heterology. It is a "discourse that is other and of the other," in the felicitous phrase of Michel de Certeau – like a negative image of the orthodox discourse that was constantly being produced by the institutional Church.[37]

The task of reconstructing this heterological discourse, which stood in opposition to the hegemonic discourse of the Church and its related institutions, requires the historian to set aside, at least for the moment, the history of events, elites, and institutions – concerns that have been at the forefront of scholarship, perhaps to excess, in the last two decades. The scholar must, rather, indulge in a more careful reading of the texts and return to the hermeneutic effort that Tellechea called for – in order, in the words of Lu Ann Homza, "to find new meanings in texts whose import seemed to have been fixed already."[38]

A good number of those meanings have been explored using the new methodologies included in the wave of "the linguistic turn" and the history of "gender," as well as in the more recent "history of the emotions."[39] The large number of women who swelled the ranks of the *alumbrados* has provided prime source material for readings in the "rhetoric of femininity" (reflected in the title of Alison Weber's important *Teresa of Avila*

[34] José Luiz Sánchez Lora, *El diseño de la santidad. La desfiguración de San Juan de la Cruz* (Huelva: Bibliotheca Montaniana, 2004); and *Scripture and Pluralism. Reading the Bible in the Religiously Plural Worlds of the Middle Ages and Renaissance* (Leiden: Brill, 2005).

[35] Enrique Llamas Martínez, "Teresa de Jesús y los alumbrados," in *Congreso Internacional Teresiano*, ed. Teófanes Egido Martínez et al. 2 vols. (Salamanca: El Congreso, 1983), 137–67.

[36] See an interesting example of this prejudice in the writings of Jiménez Patón, in J. Garau, "Ideas religiosas del Maestro Bartolomé Jiménez Patón (1569–1640) en *El virtuoso discreto* [c. 1629–1631]," *Hispania Sacra* 64(129) (2012): 237–58. See also A. Weber, *Teresa of Avila and the Rhetoric of Femininity* (Princeton, NJ: Princeton University Press, 1990).

[37] Michel de Certeau, *Heterologies: Discourse on the Other* (Minneapolis, MN: University of Minnesota Press, 1986).

[38] Lu Ann Homza, "The Merits of Disruption and Tumult: New Scholarship on Religion and Spirituality in Spain during the Sixteenth Century," *Archiv für Reformationsgeschichte/Archive for Reformation History* 100 (2009): 212–28, 227.

[39] See for example, Elena Carrera, "The Emotions in Sixteenth-Century Spanish Spirituality," *Journal of Religious History* 31(3) (2007): 235–52; William Christian, "Llanto religioso provocado en España en la Edad Moderna," in *Accidentes del alma. Las emociones en la Edad Moderna*, eds. James Amelang and María Tausiet (Madrid: Abada, 2009); and Tausiet, "Agua en los ojos: el 'don de lágrimas' en la España Moderna," in *Accidentes del alma.*

and the Rhetoric of Femininity).[40] These studies, chiefly by Anglophone women historians, have coincided with a growing engagement in both Italy and Spain with female sainthood and its prominence in the Catholic Counter-Reformation, resulting in a number of works of great interest: Maria Laura Giordano's on María de Cazalla and Isabel de la Cruz, Ángela Muñoz Fernández's on Castilian female saints and devout women, and those by Ronald Surtz, Stephen Haliczer, Jessica Fowler and Gabriella Zarri, among others.[41]

As many of these authors make clear, a rhetorical analysis of the texts does not imply ignoring their contexts; rather, it problematizes the dependence of text on context. The fact that so many charismatic spiritual figures were first investigated as *alumbrados* and later canonized as saints (especially female ones, as we have noted) shows the futility of trying to explain this phenomenon while drawing a sharp line between orthodoxy and heterodoxy. Unless we are prepared to admit that the former was constructed, perhaps in unique fashion in early modern Spain, in part at least as a permanent reaction against the threats – real or imagined – represented by the latter. The "Unamunian" prejudice that

[40] See also Elena Carrera, *Teresa of Avila's Autobiography: Authority, Power and the Self in Mid-Sixteenth Century Spain* (Oxford: Legenda, 2005).

[41] Maria Laura Giordano, *Maria de Cazalla (1487-?)*(Madrid: Ediciones del Orto, 1998); and "Nel nome di Paolo: umanesimo biblico e risonanze converse in Isabel de la Cruz e María de Cazalla (1512–1534)," in *Donne e Bibbia nella crisi dell'Europa cattolica (secoli XVI–XVII)* (Trapani: Il pozzo di Giacobbe, 2014), 49–70. Ángela Muñoz Fernández, *Beatas y santas neocastellanas, ambivalencias de la religión y políticas correctoras del poder (ss.* XIV–XVI*)* (Madrid: Dirección General de la Mujer-UCM, 1994); and "Madre y maestra, autora de doctrina. Isabel de la Cruz y el alumbradismo toledano del primer tercio del siglo XVI," in *De leer a escribir. La educación de las mujeres ¿libertad o subordinación?* Ed. Christina Segura Graíño (Madrid: Asociación Cultural al-Mudayna, 1996), 99–122. R. E. Surtz, *Writing Women in Late Medieval and Early Modern Spain. The Mothers of Santa Teresa* (Philadelphia, PA: University of Pennsylvania Press, 1995); and *The Guitar of God: Gender, Power and Authority in the Visionary World of Mother Juana de la Cruz (1481–1534)* (Philadelphia, PA: University of Pennsylvania, 1990); Stephen Haliczer, *Between Exaltation and Infamy. Female Mystics in Golden Age Spain* (Oxford: University Press, 2002); Jessica Fowler, "Illuminated Islands: Luisa de los Reyes and the Inquisition in Manila," in *Devout Laywomen in the Early Modern World*, ed. Alison Weber (Abingdon: Routledge, 2016), 152–72; G. Zarri, *Le sante vive. Profezie di corte e devozione femminile tra `400 e`500* (Turin: Rosenberg & Sellier, 1990); see also S. Pastore, "Mujeres, lecturas y alumbradismo radical: Petronila de Lucena y Juan del Castillo," *Historia Social* 57 (2002): 51–73; Andrew Keitt, "The Miraculous Body of Evidence: Visionary Experience, Medical Discourse, and the Inquisition in Seventeenth-Century Spain," *Sixteenth Century Journal* 36(1) (2005): 77–96; and *Inventing the Sacred: Imposture, Inquisition, and the Boundaries of the Sacred in Golden Age Spain* (Leiden: Brill, 2005); Moshe Sluhovsky, *Believe Not Every Spirit: Possession, Mysticism, & Discernment in Early Modern Catholicism* (Chicago, IL: Chicago University Press, 2007), 97–150.

relegated *alumbradismo* to the social or intellectual margins is likewise faulty, and cannot be supported in light of the varied social standing of its members (some of whom belonged to the highest ranks of the nobility), or the fact that prominent printers, humanists, and preachers were among the accused. Nonetheless, that attitude persists to some degree to this day.

ALUMBRADOS AND *JUDEOCONVERSOS*: THE FATE AND LIMITATIONS OF AN INTERPRETIVE PARADIGM

Nonetheless, Illuminism's social or chronological breadth or the blurry outlines of its doctrinal positions do not mean – and this is of prime importance – that its interpretation and analysis should be postponed indefinitely while we await a complete reconstruction of the documentary record. The research of the last few years has only strengthened the hypotheses that link its emergence in the second decade of the sixteenth century to processes of religious integration that had been evolving for at least two centuries.

As Bataillon himself had noted, and as the most recent studies have confirmed, some aspects of Illuminists' social origins connect the movement directly to the history of a multi-confessional Spain, and do so by way of one of its most prominent groups, converts from Judaism or (*judeo*)*conversos*. We have known for some time how often persons of suspected Jewish heritage appear linked to the first groups of *alumbrados*: among them, in its earliest days, were Isabel de la Cruz, María de Cazalla, and Pedro Ruiz de Alcaraz. The connection between *alumbrados* and *conversos*, although it was repressed to some degree, became so rooted in the minds of its persecuters that by the Patriarch Ribera's time it was taken as a given.

Juan de Ribera (canonized as recently as 1960 by Pope John XXIII)[42] denounced, in a sermon devoted to the Holy Office in Badajoz in 1565, the heretical vice of Judaism as a sin inherent to the race; that was why Jesus had been forced to address his preaching to the Gentiles, ignoring the portion of his flock that it would have been simply impossible to persuade. For the bishop of Badajoz that people, whom the Holy Office had undertaken to suppress, was incarnated in "corrupt churchmen

[42] Giovanna Fiume, *La cacciata dei moriscos e la beatificazione di Juan de Ribera* (Rome: Morcelliana, 2014).

and other Illuminist persons."[43] Thus this model bishop of the Counter-Reformation – protector of devout women in Badajoz and scourge of the Moriscos in Valencia – offered a formula by which the sin of heresy was linked to the traumatic history upon which Spain's unity of religion had been constructed through decrees of mass conversions and the banning of Judaism and Islam. At the same time, the inquisitors' anti-Jewish obsession, which expressed early modern Spain's anxiety about its impure racial and cultural past, seems to us an ideal vantage point from which to view our phenomenon. We will examine several recent books that deal with this issue,[44] particularly those of Stefania Pastore and Maria Laura Giordano.[45] The terms of the debate rest on the importance of the *judeo-converso* element in the religiosity and interior spirituality of early modern Spain; they presuppose – as Bataillon had already suggested – that the *converso* phenomenon gave the Spanish case a unique social cast, lending special characteristics to religious and ideological debates that were taking place elsewhere in Europe.

Let us consider for a moment the traumatic process of creating a mono-confessional society (a society of enforced and official Catholicism) that produced the historical and human landscape to which we refer. It is the common basis for the works of the two historians just mentioned, and both of them formulate it in a similar way. In brief, from the late fourteenth century and throughout the fifteenth in the case of Jews, and in the first half of the sixteenth century in the case of Muslims, Spain instituted a policy of mass conversions to Catholicism. Through this policy a society that had been ethnically and religiously plural gave way to a single religion in its widest sense: a single Law but also a single revealed text, a single set of social and cultural behaviors, and a single spirituality. The resulting trauma was immense and long-lasting, not only for the

[43] Juan de Ribera, *Sermones*, ed. Ramón Robres Lluch, 6 vols. (Ediciones Corpus Christi, 1987–2001), V, 267. It was a paradoxical claim in light of the support that Patriarch Ribera gave to circles that were thought of as Illuminist, both while he was bishop of Badajoz and later as bishop of Valencia. See Francisco Pons Fuster, *Místicos, beatas y alumbrados. Ribera y la espiritualidad valenciana del siglo XVII* (Valencia: Edicions Alfons el Magnànim, 1991), 13–48.

[44] For the historiographic background of this connection, see A. Selke, "El iluminismo de los conversos y la Inquisición. Cristianismo interior de los alumbrados: resentimiento y sublimación," in *La Inquisición española: Nueva visión, nuevos horizontes*, ed. Joaquín Pérez Villanueva (Madrid: Siglo XXI, 1980), 617–36.

[45] See S. Pastore, *Un'eresia spagnola. Spiritualità conversa, alumbradismo e inquisizione (1449–1559)* (Florence: Leo S. Olschki, 2004), and M. L. Giordano, *Apologetas de la Fe. Elites conversas entre Inquisición y patronazgo en España (siglos XV y XVI)* (Madrid: FUE, 2004).

converted peoples but also for the society which had to absorb them. That society faced the problems both of integrating and assimilating them and of reconsidering its own sacred texts (or different readings of its sacred texts, in the case of the Bible), its prophets, its lineages, and its languages. This process overlapped with pre-reformist Humanism, which in turn was interrogating the pre-Vulgate Greek and Hebrew versions of the revealed text as part of a new critical and philological study of Scripture. At the same time, a life lived among different religions, in a climate of religious polemic, fierce indoctrination, social suspicion and marginalization, and even open persecution – for conversion to a single faith had given rise to the Spanish Inquisition – led some to take refuge in an inward religiosity, many others to dissemble, and still others to adopt a relativist and even skeptical view of religion. In any event it produced a religious unease in which confusion and ignorance combined with a strong desire for religious reform that was engendered and nourished by fifteenth-century Humanism and was certainly not confined to New Christians alone.

A society that was deeply conformist in supporting a single rigid model, with an attitude both aggressively polemical and defensively entrenched after the enactment of the "purity-of-blood" statutes, also produced dissident and resistant movements, especially among New Christians. The problem is to consider the incidence of conversion on stances of religious dissidence. The plural nature of the *alumbrado* phenomenon, even though it never created an organized doctrine, was a fundamental element of the traumatic process to which we have referred. It certainly presupposed a defense of interior spirituality and faith over a religion that was seen as empty of meaning, or at least no more meaningful than any other. It also sometimes presupposed doubt, either hidden or open; the dominant society was keenly aware of this, feared its corrosive influence, and fought it through its new institutions for controlling belief, particularly the Inquisition. We are not the first to affirm that the religious unease of the *conversos* in the bosom of the Spanish Catholic community contributed to the religious complexity of the time, as scholars from Américo Castro to Bataillon have amply shown through their work.[46] Our task now is to assess the weight and the mechanisms of that contribution. But it is our task also to avoid the risk of compartmentalization. Understanding a society as a sum of social groups prevents us

[46] Some followers of the former, such as Alcalá, have done so as well; Angel Alcalá Galve, *Los judeoconversos en la cultura y sociedad españolas* (Madrid: Trotta, 2011).

from studying the relationships between these groups – that is to say the dialectics of change. This compartmentalization, which is methodological in origin, entails the illusion of different cultures within one common culture, and allows for a thesis like that of acculturation.[47]

The process of imposing a single religion and a single belief produced a series of phenomena that characterized early modern Spain; many of these involved drawing new boundaries within what had previously been, in theory, a homogeneous society but was now seriously alarmed when faced with "infiltration" by social groups that had been clearly segregated for centuries. Eleazar Gutwirth has shown how the early modern construction of an image of the *conversos*, which in the fifteenth century had been varied and diverse, was now responding above all to the anxiety produced by the disappearance of clear distinctions between different groups.[48] David Nirenberg has made clear, in a series of seminal studies, how the dominant hegemonic society defended itself against this blurring of lines by a variety of means: not only the control of conscience undertaken by the Inquisition, but also an exaggerated stress on genealogy whose chief result was the purity-of-blood statutes, and whose aim was to close the door to integration of the New Christians into the Catholic world.[49] It was a complicated process in which two social groups confronted each other: the newly converted who hoped to integrate, and those who placed every possible obstacle in their way. And not only were there Old Christians versus New Christians: most appropiately for our argument, Sara Nalle, in her contribution to this volume, shows the difference and even the enemity among diverse generations of *conversos* depending on the moment in which they had converted. In Castile, the Jews converted at the beginning of the fifteenth century were known as *confesos* and were never to mix, in marriage or otherwise, to the converts of 1492 known as *conversos* or *cristianos nuevos*. The Inquisition of Cuenca, Nalle shows, persecuted fiercely the second group and did not

47 Jean Wirth, *Sainte Anne est une sorcière: et autre essais* (Geneva: Droz, 2003), 196.

48 Among Eleazar Gutwirth's many publications, see especially "Jewish-Converso Relations in Fifteenth-Century Segovia," in *Proceedings of the Eight World Congress of Jewish Studies*, 2 vols (Jerusalem: World Union of Jewish Studies, 1983), II, 49–53; and "Music, Identity and the Inquisition in Fifteenth-century Spain," *Early Music History* 17 (1998): 161–81.

49 D. Nirenberg, "Mass Conversion and Genealogical Mentalities: Jews and Christians in Fifteenth-Century Spain," *Past & Present* 174 (2002): 3–41; "Conversion, Sex and Segregation: Jews and Converts in Medieval Spain," *American Historical Review* 107 (2002): 1065–91; and "Enmity and Assimilation. Jews, Christians and Converts in Medieval Spain," *Common Knowledge* 9(1) (2003): 137–55.

mind the *confesos*. What arose from these different tensions was a set of religious phenomena which, while faced with a constant suspicion of heterodoxy, sought to become the "other" Christian church. But exactly in what sense, and to what extent, can the *judeoconverso* element be an explanatory factor?

We have perhaps not reflected enough on the paradox that the information patiently collected by the Inquisition itself forms the basis for our knowledge of the new converts' dissidence or religiosity. As recent scholarship has shown, one of the main objectives of the Inquisition was didactic and pedagogic: not so much teaching how to be orthodox but how to identify heterodoxy, and did so through "visitas," *autos de fe* (edicts of faith). The Inquisition provides a vocabulary of heresy as well as proof of the existence of that heresy in the numbers of punished heretics, in a self-referential discourse created and disseminated by the Inquisition itself. The work of Jessica Fowler among others is showing how the Inquisition, through its public appearances, led members of the audience to later participate in identifiying heresy, and how this process resulted in *alumbrados* appearing across the entire empire despite the lack of any association between different groups.[50] Likewise the Inquisition's own categories, used in interrogating its victims, have shaped the queries with which we historians have sought the "identities" of minority religious groups. The Inquisition's classifications were always based on the origins of the accused, irrespective of one's practices or beliefs. For example, if an individual refused to worship images or tended toward iconoclasm those attitudes were considered symptoms of Judaism, Islamism, or Lutheranism depending on which of those groups the accused belonged to, or on the fact of being a foreigner. Moreover, those three categories into which the Inquisition assigned heretics merely reflected Old Christian stereotypes: of conspirers with foreign enemies who sought to destroy Spain, or of practitioners of sexual freedom or license who undermined society's basic values and threatened to overwhelm it with their high fecundity. The accusation of "Judaizing," for example, might reflect only a perception of Nicodemism, if the accused were suspected of a discrepancy between religious practice and true sentiment. In any event we should avoid relying

[50] Jessica Fowler, "Illuminating the Empire: The Dissemination of Inquisition and the Heresy of Alumbradismo in the Spanish Empire," PhD dis. (University of California, Davis, forthcoming); and "Illuminated Islands." Thomas Werner, *La represión del protestantismo en España, 1517–1648* (Leuven: Leuven University Press, 2001), and *Los protestantes y la Inquisición en España en tiempos de Reforma y Contrarreforma* (Leuven: Leuven University Press, 2001).

too much on the records of Inquisition trials, not only because the guilty were classified according to their origin but, above all, because we must take account of what was left unsaid, of statements by the accused that an inquisitor did not pursue, of the less obvious clues. Written documents and Inquisition trials should be the object of new readings and the new hermeneutics that Tellechea called for.

In this sense the work of Stefania Pastore is exemplary. This scholar's ability to discover hitherto unexplored keys to reading arises from a hermeneutics of suspicion, an awareness of the lack of trust in the delicate position maintained by many actors in the world of Spanish heterodoxy; that cast of mind was always sensitive to the nuances of the doctrinal formulations offered by the accused. Pastore's readings are able to reveal the wandering paths of a discourse resistant to confessional forced normalization, but she is always prepared to extract from the texts a pervasive discourse that criticized the Inquisition's practices and had been, up to now, almost completely silenced. In her first book[51] – which is really a new history of the early Spanish Inquisition – Pastore has reread the classic texts of Spain's religious history and interpreted them in a new way. They extend from the mid-fifteenth-century polemic between Fray Alonso de Espina and the Hieronymite Oropesa on the conversion of the Jews, through Hernando de Talavera (that key figure in the sacred history of Spain) to José de Sigüenza's *Historia de la Orden de San Jerónimo* (1600–05), and they include texts that document the mortal struggle between the Franciscan and Hieronymite orders at a time when the tribunal of the Holy Office was being invested with unprecedented powers and jurisdiction.[52] Pastore reads these texts with extraordinary sensitivity, and detects in them expressions of dissatisfaction with and opposition to the Inquisition's founding and the paths it was soon to take. At the same time, she makes clear how the Holy Office came to dominate consciences, beliefs, and behaviors in both the public and the private spheres. In her second book,[53] the author has returned to some of the characters and texts that appear in the first, but with the aim of examining how an interior religiosity developed in Spain and how it was related to the *conversos*.

Pastore has drawn attention to certain unique aspects of the Illuminist credo and has observed them from a fresh perspective. Some *alumbrados*, believing that while the light of Faith was essential the authority of the

[51] S. Pastore, *Il Vangelo e la Spada. L'Inquisizione di Castiglia e i suoi critici (1460–1598)* (Rome: Storia e Letteratura, 2003).

[52] On Sigüenza, see also Olivari, *Avisos, pasquines y rumores*, 341 ff.

[53] Pastore, *Un' eresia spagnola.*

Law was expendable, went to the extreme of claiming that there was only one Revelation and that therefore Mohammed was a true prophet. Alcaraz in particular – a lay preacher who was close to Isabel de la Cruz's circle – showed in his Inquisition trial his concern for the relationship among the three monotheistic faiths, even though he minimized the role of Scripture as a source of liberation and truth. Alcaraz's reading of the Epistle to the Romans included the Qur'an among the "Scriptures" which, according to St. Paul, offered hope to the believer (Rom 15:4).[54] Juan del Castillo would articulate the same vision of a common revelation: "for the law of Our Lord was *ut vivemos in pace*, and not that we should be saved by that law; for both those who lived by it and those who did not have been saved."[55] The typical Illuminist rejection of rules, rituals, and ceremonies was based on a radicalism that in the end invalidated any sign of acknowledgment or belonging. One of the most original and meaningful chapters of Pastore's book is her analysis of this radical and subversive aspect of the doctrines of many *alumbrados*, notably of Castillo, whose eclecticism she also demonstrates.

In the 1520s and 1530s the Inquisition pursued first the *alumbrados* and then the so-called Erasmians, as well as an impressive collection of Illuminists, Messianic visionaries, prophets, and others. Many of these were *conversos*, but many others were not. The common denominator of this peculiarly Spanish heresy was not, Pastore tells us, crypto-Judaism, but rather "Paulinism" (a concept already employed by Bataillon): a special reading and interpretation of the writings of St. Paul. By stressing certain passages from Paul – on the notion of faith as an illumination by whose light every human distinction becomes meaningless – it offered the newly converted a chance to re-create themselves entirely, as well as to face down the exclusionary stance of most Old Christians. Many of these converts in Castile reclaimed the Pauline dialectic that set the freedom of faith against the coercive power of the law (the triumph of the spirit over the letter);[56] in the fifteenth century

[54] "Exponebat illud dictum Apostoli ad Romanos 15, quaecumque enim scripta sunt ad nostram doctrinam scripta sunt ut per patientiam et consolationem Scripturarum habeamus, dicebat reus quod haec autoritas intelligenda est de ómnibus Scripturis, etiam de Alcorano ipsius Machometi et aliis boniis ac malis Scripturis." Pastore, *Una herejía española*, 218.

[55] This thesis has been elaborated in Schwartz's valuable book, although divorced from the immediate contexts that could lend it greater meaning. Stuart B. Schwartz, *All Can Be Saved. Religious Tolerance and Salvation in the Iberian Atlantic World* (New Haven, CT: Yale University Press, 2008).

[56] This was the same point at which reform movements in the Catholic Church, such as the Benefice of Christ (1543), intersected with the Reformation's "justification through

and the early decades of the sixteenth they revived a dialectic, inherent to Christianity, whose synthesis would heal the fissures of the immediate past to form a new horizon of reconciliation. This proposal was socially inclusive and inevitably presupposed a reform of the Church – which, of course, would have to enter into dialogue with the varied assortment of reforms that had arisen in Europe during those years. Yet paradoxically, its encounter with reality condemned many of its believers to interior exile.

The remarkable personality of Talavera, his own theology – thoroughly imbued with Paulinism, as were the works of his teachers Cartagena and Oropesa – were the most significant fruits of a Spain in which reading and discussing St. Paul had flourished. Cartagena, Oropesa, Talavera, the great converted Biblists of the fifteenth century all read and used St. Paul to defend a peaceful, harmonious, and just reconciliation between Old and New Christians, and that associated external ceremony and ritual with a belief in the superiority of the New Testament.[57] Only the individual who was illuminated by faith through grace could truly know the new law of the New Testament, which was imprinted on the heart and not on tablets or stones. The pages of St. Paul validated the absolute equality of all Christian believers and the exultant experience of the individual illumined by faith, which was presented as an ineffable gift from God. Gospel verses about divine illumination formed the basis for condemning any discrimination between New and Old Christians, and the foundation for rejecting the Inquisition.

Maria Laura Giordano also deals with "Paulinism" and the role of *conversos* in her book[58] that appeared at the same time as Pastore's original Italian edition and considers many of the same issues. Giordano seeks to trace the outlines of the Christian identity of individuals of Jewish origin, and the theological support that they claimed; to this end she studies spiritual literature from the late Middle Ages as well as Inquisition trials like that of María de Cazalla. She concludes, with Pastore, that the Pauline concept of faith left a lasting imprint on every experience of *converso* spirituality in texts from the mid-fifteenth century onward. According to Giordano, Erasmus triumphed in Spain because he could depend upon a previously existing Paulinism. If we compare Erasmian to pre-Erasmian

faith." See C. Ginzburg and A. Prosperi, *Giochi di pazienza. Un seminario sul "Beneficio de Cristo"* (Turin: Einaudi, 1975), 89.

57 On Talavera and his predecessors, teachers and teachings, see the work of another Italian scholar, Isabella Iannuzzi, *El poder de la palabra en el siglo XV: Fray Hernando de Talavera* (Salamanca: Junta de Castilla y León, 2009).

58 Giordano, *Apologetas de la fe*.

Spain, and Spain to other Christian countries of the period, the problem then becomes one of understanding why Pauline language was privileged and why it came to be a sign of Erasmianism in Spain but not elsewhere. Giordano finds a definitive answer in the *converso* factor. Studying spiritual texts by *judeoconversos* of the fifteenth century, she carries out a subtle reading of works by two intellectuals and humanists from the first converted generations, Alonso de Cartagena and his niece Teresa de Cartagena (especially the latter's *Arboleda de los enfermos*, a text that has attracted much attention in recent years). In both cases Giordano can show significant and notable affinities with the Toledo group of Illuminists, forming a current that, a century later, would flow into Juan de Ávila. Thus for over a hundred years "Paulinism" became the spiritual language, even the code, shared by writers, readers, and disciples – a Paulinism that was exclusive to Spanish culture and independent of its alternative form, which was born of Lutheranism and would later come to dominate. And even in the manifestations of Spanish Lutheranism, according Giordano, the *judeoconversos* played an essential role. She believes that the history of the Reformation in Spain is indissolubly linked to that of the *conversos*, who acted as the engine of the history of Spanish spirituality.

Both Pastore and Giordano point to this "Paulinism" as the crucial experience – if not ideologically, at least hermeneutically – within that history. These two historians place its development in a time that preceded the introduction of Erasmus' ideas, and show how it facilitated the reception of those ideas and increased their effects. Giordano makes clear that an inclusive reading of St. Paul's epistles to the Gentiles was a practice deeply rooted in the religious experience of those who had been forced to convert to Christianity since the late Middle Ages, whether through necessity, conviction, coercion, or fear. This Pauline element would become the nexus that united in a single vision such distinct and profoundly personal religious experiences as those of Juan del Castillo and Francisco Ortiz.

Nonetheless, Giordano, unlike Pastore, insists on a strong connection between Illuminists and *judeoconversos*. Although she (and others) tries to define that connection as situational or "positional" vis-à-vis the dominant society (i.e. its exclusive causes would have been marginalization and fear of persecution), her work suggests the existence of a "*judeoconverso* identity."[59] This final step is, in our opinion, questionable at best,

[59] Giordano, *Apologetas de la fe*, 25. For a similar effort to identify a *converso* theology in the fifteenth century, see Rosenstock's book and its review by Nirenberg: Barbara Rosenstock, *New Men: "Conversos," Christian Theology, and Society in Fifteenth*

because its identification of Illuminism with *judeoconversos* follows a sort of "inquisitorial" logic that assumes, and even internalizes, a necessary link between religious heterodoxy and racial prejudice.[60] As we have already seen, however, the Holy Office's criteria for classifying the effervescent religiosity of early modern Spanish society reflect the diversity that it was witnessing at least as much as the ideological roots of its own prejudices. But by the 1590s even the Inquisition had started to demonstrate a certain skepticism about the *converso-alumbrado* nexus. Thus it is impossible, and certainly undesirable, to create a framework in which the barrier between *converso* practice and "heterodox" *converso* spirituality preceded the Inquisition trials, either intellectually or socially.

The Inquisition made assumptions about presumed membership in a community, and about lineage, that the historian should approach with the deepest caution. Can the *judeoconverso* element, after all the changes in context that took place over more than a century, really be an explanatory factor, a key that opens all doors to our interpretation and understanding of Spanish spirituality in this period? What did the "Judaizing" of the *conversos* consist of? Or, to put it another way, what was – and continued to be for decades – the relationship of *conversos* to Judaism? Once again, was it a matter of lineage? Was it a racial prejudice, or an essentially cultural one?[61] What relationship exists between the questions about Spanish culture/religiosity that we modern historians ask, and those that the Inquisition and others asked of the converts of the fifteenth and sixteenth centuries? And finally, are we not perhaps following the trails blazed by Archbishop Juan de Ribera and his contemporaries? In speaking of *judeoconversos*, are we not ignoring the effects of time on agents and on cultural phenomena? In short, we must shift our focus from genealogy to culture in order to study the processes by which Old Christians of the sixteenth century created their own concept of race, transforming social class, cultural difference, and even religious

Century Castile (London: Queen Mary and Westfield College, University of London, 2002); D. Nirenberg, book review in *Speculum* 80(1) (2005): 315–17.

[60] Mercedes García-Arenal, "Creating Conversos: Genealogy and Identity as Historiographical Problems (After a Recent Book by Ángel Alcalá)," *Bulletin for Spanish and Portuguese Historical Studies* 38(1) (2013): 1–19.

[61] See Nirenberg's forceful proposal for the second option in "Figures of Thought and Figures of Flesh: 'Jews' and 'Judaism' in Late-Medieval Spanish Poetry," *Speculum* 81 (2006): 398–426.

belief into immutable products of nature. This is an issue of paramount importance.

This strategy would allow us to rethink not only the nature of the *alumbrado* phenomenon within the Spanish context but also its overlap with the cultural and religious history of early modern Europe. Of all the questions formulated above, the one that has, without a doubt, received the greatest attention and has cast the strongest light on their complexity and breadth is that which interrogates the *alumbrados*' relationship to modernity. The intellectual historian Richard Popkin, in his studies of skepticism and Epicureanism published in the early sixties,[62] was among the first to suggest that the *marranos* stood at the beginning and the base of European modernity.[63] Other more recent works have pushed his ideas further, making them extend not only to the origins of skepticism but also to hybridization, dissimulation, and secularization.[64] Although this strategy seeks to defend a *judeoconverso* identity that is seen as "positive," finding roots of skepticism and secularism indissolubly tied to the very flesh of descendants of converted Jews at the end of the Spanish Middle Ages, the fact is that it offers us once more the same logic of genetic heritage that the contemporaries of those Jews already employed. These issues are certainly very difficult to face, and the pitfalls of essentialist *judeoconverso* identity are hard to avoid, but we must point out their dangers. Firpo and his abovementioned colleagues did not need the *converso* factor to explain the drift between Illuminism and Nicodemism, or how the latter tended towards heterodoxy and skepticism. Perhaps the time has come to return to the debate between Eugenio Asensio and Marcel Bataillon that we cited at the beginning of this article, and to propose with them that the *judeoconverso* element, like Erasmianism, was of singular significance in Spain's religious and social history; that it acted as a guide for certain religious concerns at particular moments, or even catalyzed some of them; but that it was not the only agent, nor the only social grouping, that experienced the impulse to reform.

[62] Richard Popkin, *The History of Scepticism from Erasmus to Descartes* (Assen: Van Gorcum, 1960).

[63] On Francisco Sánchez (1550/51–1623). See also Israel Revah, "Les marranes," *Revue des Etudes Juives* 118 (1959–60).

[64] Fair presents the most radical formulation of this stance that identifies *marranismo* with modernity; José Faur, *In the Shadow of History: Jews and Conversos at the Dawn of Modernity* (Albany, NY: State University of New York Press, 1992). See, more recently, Yirmiyahu Yovel, *The Other Within. The Marranos: Split Identity and Emerging Modernity* (Princeton, NJ: Princeton University Press, 2007).

JEWS BEFORE *JUDEOCONVERSOS*

We may need to go back to a time prior to the mass conversions that began in the late fourteenth century to find other answers or to broaden and refine some of those we have. The last two decades have seen a remarkable development of studies of Jewish communities in fifteenth-century Spain, a period that Bataillon urged examining more closely if we wished to understand the processes that were set in motion in the sixteenth.[65] We cannot comment here on all of these contributions; however, if we place Illuminism within the traumatic process of confessionalization that we have been tracing in this article, we must reconstruct a chronological scenario that has often been obscured by the frontiers that traditionally exist between disciplines.

It has become clear through the most recent research how profoundly Jewish culture in late medieval Spain interacted with the Christian culture that surrounded it, and in particular to what extent it participated in the climate of reform and innovation that characterized Spain in the last third of the fifteenth century and the first third of the sixteenth. Many studies are showing the existence of a shared religious preoccupation, a true "co-production" (in David Nirenberg's term) of religious attitudes and problems.[66] In the last few decades, studies of peninsular Judaism have revealed how creative the Hispano-Jewish communities were in a variety of fields (poetry, philosophy, homiletics, exegesis, Kabbala, jurisprudence), not only in the composition of texts but in their diffusion.[67]

Recent work on translations made by Jews in the fifteenth century from Hebrew into Romance, and from Latin and Arabic into Hebrew, have opened new perspectives for understanding the cultural history of the period.[68] We can add the Romance translations of Hebrew works of authors of the stature of Maimonides: between 1419 and 1432 Pedro de Toledo translated *Moreh Nevukhim*, based on the Hebrew version of al-Harizi, into Castilian. Though this translation[69] was never published,

[65] See note 5.

[66] D. Nirenberg, "Introduction," in *Neighboring Faiths. Christianity, Islam, and Judaism in the Middle Ages and Today* (Chicago, IL: Chicago University Press, 2014).

[67] See, for example, Ma Meyerson, *A Jewish Renaissance in Fifteenth Century Spain* (Princeton, NJ: Princeton University Press, 2004).

[68] Jean-Pierre Rothschild, "Pour évaluer la place de traductions dans la littérature hébraïque du moyen âge Occidental," in *We-zot le-Angel: Raccolta di studi giudaici in memoria di Angelo Vivian* (Bologna: Associazione Italiana per lo Studio del Giudaismo, 1993), 435–60.

[69] Biblioteca Nacional de España, MS KK-9.

Alfonso de la Torre's famous *Vysión deleitable* was, and scholars of the latter work have shown that it incorporated Maimonides' work in both form and content.[70] From the Jewish side there was cultural interchange in fields like medicine and astronomy,[71] logic,[72] and above all Christian moral philosophy and scholastics in general. All these instances of collaboration were marked by a burst of activity in translations: notable examples are the *Nicomachean Ethics* by Rabbi Meir, the various versions by Eli Habillo,[73] and Thomas Aquinas' commentary on Aristotle's *Metaphysics* (Ocaña, 1490).[74] This intense project of translation was accompanied by a remarkable development of the Hebrew book and of printing that was unparalleled at the time in any other center of Jewish culture. The invention of printing constituted a revolution that was to have enormous consequences in the life of Jewish communities: not only did it make the Bible more popular than ever before but it even brought about a reordering of its books.[75]

[70] See Alfonso de la Torre, *Visión deleytable*, ed. Jorge García López (Salamanca: Universidad, 1991); and Luiz Girón-Negrón, *Alfonso de la Torre's Visión deleytable: Philosophical Rationalism and the Religious Imagination in 15th Century Spain* (Leiden: Brill, 2001). Pedro de Toledo's translation awaits edition and publication.

[71] Gad Freudenthal, "Arav and Edom as Cultural Resources for Medieval Judaism: Contrasting Attitudes toward Arabic and Latin Learning in the Midi and in Italy," in *Late Medieval Jewish Identities: Iberia and Beyond*, ed. Esperanza Alfonso and Carmen Caballero Navas (New York: Palgrave Macmillan, 2009), 123–56.

[72] Angel Sáenz Badillos and Arturo Prats, "Shelomo Bonafed y la lógica cristiana del siglo XV," *Revista Española de Filosofía Medieval* 10 (2003): 15–27.

[73] Mauro Zonta, *Hebrew Scholasticism in the Fifteenth Century: A History and Source Book* (Dordrecht: Springer, 2006), 165–201.

[74] On Christian moral philosophy and scholasticism, see in particular to E. Gutwirth's excellent articles: "Fragmentos de Siddurim españoles de la Genizá," *Sefarad* 40 (1980): 389–401; "Ideario de los traductores hispano-judíos del latín," in *Actas del Segundo Encuentro de las Tres Culturas* (Toledo: Ayuntamiento de Toledo, 1985), 189–96; "Religión, historia y las Biblias romanceadas," *Revista catalana de Teología* 13 (1988): 115–34; and "A Medieval Spanish Translation of Avot: Genizah Fragments," *Annali del Istituto Orientale di Napoli* 49 (1989): 289–300. In 1405 Meir ben Salomon Alguadex, chief rabbi of Castile, translated the *Nicomachean Ethics* from Latin into Hebrew: see E. Gutwirth, "The Stranger's Wisdom: Translation and Otherness in Fifteenth Century Iberia," *Portuguese Studies* 13 (1997): 130–42.

[75] Printing with movable type reached Spain around 1472–3, and Jews immediately adopted the new technology and made it their own. From the 1470s onward there were Hebrew presses in Zamora, Guadalajara, Híjar, Puebla de Montalbán, and elsewhere. Colophons and prologues of Hebrew books (which were not all biblical) expressed astonishment at the almost miraculous character of the new invention, which was called a gift from God; see Menahem Schmelzer, "Hebrew Manuscripts and Printed Books among the Sephardim Before and After the Expulsion," in *Crisis and Creativity in the Sephardic World*, ed. Benjamin R. Gampel (New York: Columbia University Press, 1997), 257–66. On the production of Hebrew Bibles in the fifteenth century and the interest in their texts, see Mark Meyerson, *A Jewish Renaissance*; see also Abraham Gross, "Centers of Study and yeshivot in Spain," in *Moreshet Sepharad: The Spanish Legacy* 2 vols (Jerusalem: Magnes Press, 1992), I, 339–410.

This fascination with the Bible – a privileged area for collaboration and interchange, as the Biblia de Alba project alone demonstrates – overlapped with pre-Reformation humanism, at the same time that Christians, with their new critical and philological interest in Scripture, were interrogating the pre-Vulgate Greek and Hebrew versions of the revealed text.[76]

Of all the possible settings for interchange, one was especially significant for the issue that concerns us. In a seminal article, Eleazar Gutwirth has shown that in the late Middle Ages, Spanish Jews shared with Christians a fundamental anticlericalism that was increasing all over Christian Europe in the pre-Reformation years.[77] There was defiance of clerical (in this case, rabbinical) authority, aided by direct reading of the Scriptures. This religious crisis afforded a new perspective on a number of texts, and underlay some cases of voluntary and individual conversions of Jews to Christianity.[78] These observations complicate, but also enrich to a great degree, our vision of Spain in the years prior to 1492.

THE "OTHER" *CONVERSOS*

An alternative hypothesis, which we may call the "Morisco" hypothesis, has not pursued the same genealogical argument. Its roots, however, run deep. The renowned seventeenth-century Orientalist Barthélemy d'Herbelot, author of the great *Bibliothèque orientale*,[79] includes the following comment under the entry "din" [religion]:

> This sentiment is that of the sect of the Illuminated Ones, which had its origin in the Orient, from which it passed with the Arabs into Spain under the

[76] For the Biblia de Alba, see both Fellous and Kupfer. It is important to consider the possible link between this Jewish zeal for printing and circulating the Bible and work by contemporary Christians Biblists and, in general, the trends set in motion by the pre-Reformation in Europe. Sonia Fellous, *La Biblia de Alba. Toledo 1422–1433. De cómo rabí Mosé Arragel interpreta la Biblia para el gran maestre de Calatrava* (Paris: Somogy éditions d'art, 2001); Marcia Kupfer, "Abraham Circumcises Himself: A Scene at the Endgame of Jewish Utility to Christian Art," in *Judaism and Christian Art: Aesthetic Anxieties from the Catacombs to Colonialism* (Philadelphia, PA: University of Pennsylvania Press, 2010), 143–82.

[77] E. Gutwirth, "Conversions to Christianity among Fifteenth-Century Spanish Jews: an Alternative Explanation," in *Shlomo Simonsohn Jubilee Volume; Studies on the History of the Jews in the Middle Ages and Renaissance Period* (Tel-Aviv: Graph-Chen Press, 1993), 97–121.

[78] Piero Capelli, "Jewish Converts in Jewish-Christian Intellectual Polemics," in *Intricate Interfaith Networks: Quotidian Jewish-Christian Contacts in the Middle Ages* (Turnhout: Brepols, 2011), 1–35.

[79] Barthélemy d'Herbelot, *Bibliothèque orientale* (Paris: Compagnie des Libraires, 1697), 296b.

name of Alumbrados, and which has been renewed in our day by the Spaniard Dr. Molinos. The majority of Muslim contemplatives who assume the name of Sufi and Ahl al-tariq(at) [people of the Way] belong to this sect.

It is unusual to find this suggestion of a genetic relationship between Islamic mysticism and Illuminist spirituality, extended even to the quietism of Miguel de Molinos. It almost seems that d'Herbelot was anticipating here, by more than two centuries, the thesis that would be defended by Miguel Asín Palacios.

Nevertheless, research into the complex world of the Moriscos not only continues to deny us any proof of this interchange, but our knowledge of their social reality is making it ever less probable. Spanish Jews of the late fifteenth century (at least, their elite classes) were close to court and aristocratic circles, centers of power and creation of learning, and continued to be so after their conversion. Using these same indicators we can now reconsider the possibility that Sufi ideas could have been transmitted to Christians.

Muslims had been losing their elites, who throughout the Christian conquest had preferred to exile themselves to nearby territories that were under Muslim control. By the end of the fifteenth century their communities were notably isolated from Christian society, and limited by the lack of an intellectual elite and barriers to the circulation of knowledge; they also had no access to printing.

The Moriscos, unlike the Jews, made no effort to translate Latin works into Arabic, nor did either Old or New Christians show any interest in works from the Arab-Islamic tradition; the only exception was the Qur'an, translated into Castilian by Yça de Gebir, *alfaquí* (religious leader) of the Muslims of Segovia, in 1456.[80] The only Islamic works that were known or used were the ones needed to mount religious polemics against Islam. Muslims did not display the same concern as Jews for translating from Latin or Castilian, although we know of cases of individual Moriscos who were influenced by the *devotio moderna*. The famous Mancebo de Arévalo, author of a well-known work titled *Tafsira* that includes chapters steeped in mysticism, made use of the work of Thomas à Kempis.[81]

From Inquisition trials, and works of anti-Islamic polemic and disputation, we know that Moriscos entered into religious brotherhoods that may well have been related to Sufi movements. The Moriscos must have read

[80] Gerard A. Wiegers, *Yça of Segovia (fl. 1450). His Antecedents and Successors: A Historical Study of Islamic Literature in Spanish and Aljamiado* (Leiden: Brill, 1994).

[81] María Teresa Narváez Córdova, *La Tafsira del Mancebo de Arévalo. Transcripción y estudio del texto* (Madrid: Trotta, 2003), and "Writing without Borders: Textual Hybridity in the Works of the Mancebo de Arévalo," *Medieval Encounters* 12(3) (2006): 487–97.

in a mystical light works such as the *Libro de las luces* (Book of Lights), devoted to Mohammed's genealogy as a prophet, which circulated both in Arabic (*Kitab al-anwar*) and in Aljamiado in Morisco communities.[82] Through the Inquisition we also know that Moriscos from Valencia were able to obtain books through relatives who had fled to North Africa. In 1608 Gaspar Rahech was condemned as an *alfaquí* because he had tried to buy and bring from Algiers a copy of *Hizb al-bahir al-bahr* by al-Shadhili, one of the most renowned mystical thinkers of Islam, who gave his name to one of the most important mystical brotherhoods in the Islamic West – the very figure whom Miguel Asín Palacios proposed as a possible influence on the Castilian *alumbrados*.[83] But it is almost impossible to prove a connection between these texts, and the religious spirit they imply, and any transmission to the surrounding Christian society. The meager Morisco literature in either Arabic or Aljamiado (Spanish written in Arabic letters) was almost wholly unknown to the Christians, while Muslims in general and Moriscos in particular lived in isolation from any contaminating contact with Christians. They showed less interest than the *judeoconversos* in partaking of the elites of Christian civil society or of the Church. The fact that these connections are improbable does not, however, imply total independence: Hispanic Muslims, in manifesting their spirituality through brotherhoods and Sufi masters, were also rejecting mediation by the clerical class in their contacts with God. They, like the so-called *alumbrados*, seem to have belonged to *ayuntamientos* and *conventículos* – small, intimate groupings – that reinforced their sense of belonging and served to create emotional ties. Once the community's elites had been lost, these circles might contain charismatic individuals who had been touched by grace.[84]

AN OPEN ITINERARY

A final group of publications, based on meticulous research into archival data, provides not only new material for approaching the hypotheses we

[82] María Luisa Lugo de Acevedo, ed., *El Libro de las luces. Leyenda aljamiada sobre la genealogía de Mahoma. Estudio y edición crítica* (Madrid: Sial, 2008); M. García-Arenal, "*Shurafa* in the last years of al-Andalus and in the Morisco period: *laylat al-mawlid* and genealogies of the Prophet Muhammad," in *Sharifs and Sayyids in Muslim Societies: The Living Links to the Prophet*, ed. Kazuo Morimoto (New York: Routledge, 2012), 161–84.

[83] Carmen Barceló and Ana Labarta, *Archivos moriscos. Textos árabes de la minoría islámica valenciana (1401–1608)* (Barcelona: Universitat de València, 2009), 63.

[84] M. García-Arenal, "A Catholic Muslim Prophet. Agustín de Ribera, 'The Boy Who Saw Angels'," *Common Knowledge* 18(2) (2012): 267–91.

have been reviewing here but also possible itineraries of research for the near future. John Edwards, who has reviewed the cases of the Inquisition tribunal in Soria (1450–1500), has shown that accusations of skepticism, "religious materialism," and more specific sins like rejection of images and other religious mediators fell arbitrarly on Old and New Christians alike, some of them even members of the clergy. His findings force us to rethink not only the geographical nature of the phenomenon but also its supposed intellectual roots.[85]

Sara Nalle's book *Mad for God* likewise raises the issue of a split between Illuminism and the *judeoconversos*.[86] It tells the story of Bartolomé Sánchez, a farmer and wool-carder, of unimpeachable Old Christian ancestry, who underwent a long and well-documented trial before the Inquisition in Cuenca between 1553 and 1560. Nalle's account is more than the reconstruction of a life; it is a tale of a man's descent into madness, framed by a debate about the nature of inspiration and religious enlightenment, insanity, and responsibility for one's own confessions. As the author demonstrates, several of Sánchez's ideas coincide almost exactly with the Edict of Faith of 1525 by which the *alumbrados* were condemned: he refused to kneel before images, rejected verbal confession, and did not believe in the Eucharist. The issues raised by the Sánchez case and his Inquisition trial were both real and present in sixteenth-century Spain: who decides what constitutes an authentic message from God? What exceeds the bounds of orthodoxy and is therefore a crime? What goes beyond the limits of rational behavior? All of these questions were of urgent significance at a time when mystics, seers, prophets, and different types of the "illuminated" exerted an unprecedented influence on Spanish society. But there are other issues that we feel are particularly relevant to the debate about the *alumbrados*: for instance, the hierarchical nature of teaching and learning in the world in which they lived, and the need that some of their members felt to rebel against that hierarchy and even to insist that their ignorance made their religious understanding even keener. (Many of the latter were women, who were scarcely touched by the spread of literacy and lacked access to cheap printed books.)[87] We are faced with a debate about the legitimacy of religious authority.

[85] John Edwards, "Religious Faith and Doubt in Late Medieval Spain: Soria *circa* 1450–1500," *Past & Present* 120 (1988): 3–25.

[86] Sara T. Nalle, *Loco por Dios. Bartolomé Sánchez, el Mesías loco de Cardenete* (Cuenca: Fundación de Cultura Ciudad de Cuenca-Ayto Cardenete, 2009).

[87] A. Gordon Kinder, "'Ydiota y sin letras': Evidence of Literacy Among the Alumbrados of Toledo," *Journal of the Institute of Romance Studies* 4 (1996): 37–49.

Note, for example, how Sánchez took offense when the inquisitor asked where his heretical ideas had come from: "I have learned it from no one but God and His Divine Majesty. How did I learn it from God? How did the Apostles, evangelists and prophets learn what they declared? They did not go to learn it from the universities of Paris and Salamanca!"[88] The mutual lack of comprehension between Bartolomé Sánchez and his inquisitors could teach a valuable lesson to those who seek to interpret today the spirituality of the sixteenth century.

CONCLUSIONS

While the purpose of this article is not to offer a single answer to these questions, we can draw a series of provisional conclusions from the preceding pages. Let us recall that *alumbrados* were individuals who believed themselves especially enlightened by God who advocated for a more interior spirituality divorced from the prescriptions of the Catholic faith. According to the Inquisition, this group encouraged mental rather than vocal prayer, belittled the importance of nearly all the sacraments, denigrated the veneration of saints, the cross, the Virgin Mary, and even representations of Christ. Ultimately they advocated an abandonment of one's soul to God resulting in a personal connection with the divine that made both the Church and its functionaries unnecessary.

All the data we have examined in this essay suggest that every behavior consistent with this definition and indulged in by members of sixteenth-century Spanish society – the climate of reform, spiritual questioning and seeking along paths other than the institutional Church, discontent with or lack of confidence in ecclesiastical hierarchies, intellectual skepticism, and interior exile – were shared in common, irrespective of whether the individual was a convert or an Old Christian. We suggest that the documentation refering to *alumbrados* must be re-read trying to perceive, even to reconstruct, a hetereological discourse, like a negative image of the orthodox discourse that was constantly being produced. It might be that certain groups or individual *conversos* were in the sixteenth century more sensitive or more prone to empathize with this discourse, but it was not a general rule nor even a general tendency. We do not think it is pertinent to introduce a *judeoconverso identity* as an ingredient for the understanding of *alumbradismo*. Together with new work carried out by scholars we

[88] Nalle, *Loco por Dios*, 206.

believe that *conversos*, as Nalle shows in this same volume, were far from being a homogeneous group, not even from a positional point of view.[89]

Nevertheless the development of *converso* ferment on soil fertilized – and later renourished – by the process of confessionalization did not mean that it was irrelevant to the phenomenon that the Holy Office called *alumbrado*. We believe, however, that the role of converts in the movements of religious dissidence was conditioned less by the subterranean survival of crypto-Jewish beliefs than by a required religious Nicodemism (often in the form of withdrawal into oneself) that was directly connected to the social concealment to which an individual's questionable origin sometimes forced him, and again, because of the tendency to produce a negative image to the official discourse, always keeping in mind that dissimulation was paradoxically connected with assimilation.

Second, if we already have been sensitive for some time to the need to re/read anew Inquisition trials and submit them to a close reading, we are now conscious that all other Inquisition material needs to be reconsidered anew taking into account Inquisition strategies and objectives, mainly its didactic and pedagogical role, not only in teaching what orthodoxy was but most specially in defining heterodoxy. The Inquisition was an active agent in defining the contours of heterodoxy, teaching how to identify it and maybe by this same teaching disseminating heterodox ideas.[90] It was also attaching labels and unifying as a sect or a movement different, and not always connected, religious attitudes and beliefs. The Inquisition was not the only actor trying to establish clear, essential differentiations, to define neat categories and ascriptions, and thus to separate, reject, and stigmatize individuals and groups. Many of those were, nevertheless, involved in high levels of local adaptation and negotiation but also reacting to those neat categories and ascriptions.

Third, it is equally certain that the process of conversion undergone by Castilian and Aragonese societies from the late fourteenth century onward conditioned, in a decisive manner, their religious life and created a wish for reform, within the Christian tradition but outside normal institutional channels. It was, in our view, a nutritious soil for this hetereology. A remarkable symptom of this attitude was what we have called "Paulinism," which was destined to prepare the ground for the

[89] García-Arenal, "Creating Conversos."

[90] Natalia Muchnik, *De paroles et de gestes. Constructions marranes en terre d'Inquisition* (Paris: EHESS, 2014).

reception of Erasmian and Lutheran ideas.[91] We should not understand (as has happened and continues to happen) the reading and translation of this Pauline legacy in the interconfessional context of early modern Spain (Alcaraz, Juan del Castillo) as straightforwardly unorthodox. At the least, we should first consider that this phenomenon formed part of a broader process – speculative, dialogic, and above all dynamic – in which the Church was creating its own reasoning, revising history in order to produce a new sense of Tradition, and putting this under the control of a recently founded institution, the Inquisition. The influx of Reformist currents and its convergence or not with the religious positions of converts, coming whether from Islam or from Judaism, is still in need of reevaluation.

These conclusions, perhaps meager for some tastes, hope to remove some obstacles, help find new ventures for research, and yet they return us to the same perplexity with which historians have viewed this fascinating chapter in Spanish religious and intellectual history. Research is now faced with the challenge of reevaluating the perspective from which we have been reading the texts of the *alumbrado* world, so as to disentangle the weave of an episode which, while crossing social, geographic, and cultural barriers, seems to have sent down deep and disturbing roots (as Unamuno said) into Castile during the Catholic Reformation and Counter-Reformation.

[91] For the repeated value of the Pauline Epistles for the idea of reform, see Gerhart B. Ladner, *The Idea of Reform. Its Impact on the Christian Thought and Action in the Age of the Fathers* (New York: Harper & Row, 1967), 49–62.

6

The Spanish Encounter with Islam

Benjamin Ehlers

In December 1566, Philip II issued a royal order naming Don Pedro Luis Galcerán de Borja as governor of Orán and Mazalquivir, Spanish outposts on the North African coast. The letters patent empowered him to rule in the king's name, "in all matters of justice and war," overseeing officials in the presidios and directing troops on land and at sea. Pedro Luis Galcerán traveled to Madrid to receive his commission in a solemn ceremony, swearing once, twice, and three times to serve the Crown faithfully in war and in peace. On the face of it, this appointment conformed to the vision of a monarchy at war with Islam – a vision that would be reiterated once more in the coming months, with the repression of Muslim dress, language, and customs among the Moriscos of Granada. As the son of the duke of Gandía, Pedro Luis Galcerán conformed to the ideal of a nobleman dedicated to military service. Moreover, at the age of sixteen he had been elected Master of the Order of Montesa, an Aragonese military order created in the fourteenth century following the suppression of the Templars. When he traveled from Cartagena to Orán in summer 1567 he brought with him a host of his knights, including his brother Don Felipe Manuel de Borja. Upon their arrival in Orán, Pedro Luis Galcerán received the keys to the city and proceeded to survey its walls, towers, and churches. In the *iglesia mayor*, a converted mosque near the gates to the sea, he observed a red cardinal's hat hanging from the lofty roof before the principal altar. This hat linked the knights of Montesa to Francisco Jiménez de Cisneros, the archbishop who exhorted Ferdinand to conquer Orán in 1508 and subsequently both financed and commanded the assault, riding among the soldiers on a mule. Jiménez de Cisneros urged the men to fight to the death in the name of their faith, and then celebrated their victory in Orán by consecrating the fortress's churches and monasteries in person.[1] Pedro's investiture as governor of this same presidio, in which he swore "loyalty in royal service, and in the defense and expansion of the holy Catholic faith," recalled King Ferdinand's assessment of Jiménez de Cisneros six decades before: "it can be truly said of the Cardinal that he has a very strong desire to wage war against the infidel."[2]

Viewed from other angles, however, the appointment of Pedro Luis Galcerán de Borja to this North African post fails to conform to the Crusading ideal of the age of Reconquest. As this essay will seek to demonstrate, Philip II and his ministers readily invoked the language of religious warfare when it suited their purposes, but these claims served to mask efforts to define and control a diverse Morisco population in Spain, and to aggrandize a token presence of criminals and exiles in the presidios of the Maghreb. To cite but two examples, the repeated laws against Morisco settlements near the Mediterranean coast summoned the specter of collaboration with a Muslim armada, as did the 1563 prohibition of Morisco arms, offensive and defensive.[3] The king's actual policies in Morisco Spain and the Mediterranean, by contrast, varied over time and gravitated towards pragmatism and minimization of losses. In addressing the Moriscos, baptized descendants of the Muslim diaspora in Spain, Philip called for their true conversion, *de corazón*, by means of evangelization and gentle persuasion. Even under pressure from his Council of State, he rejected the proposed wholesale expulsion of the Moriscos from Iberia. Yet the king also authorized the continued segregation of the Moriscos within Spanish society, through restrictions on their professions, their weaponry, their movements, and their contact with the Mediterranean world. From the royal perspective, the outpost at Orán represented not a base of operations for territorial acquisition but a means of gathering intelligence; limiting corsair activity; and impeding Ottoman unification and control of the Maghreb.[4] Pedro

[1] Diego Suárez Montañés, *Historia del Maestre último que fue de Montesa y de su hermano don Felipe de Borja*, eds. Miguel de Bunes Ibarra and Beatriz Alonso Acero (Valencia: Institució Alfons el Magnànim, 2004), 96.

[2] "... lo haría y volvería a hacer de nuevo en fidelidad del real servicio, en defensa y ampliación de la santa fe católica." Suárez, *Historia del Maestre*, 211; Henry Kamen, *Empire: How Spain Became a World Power, 1492–1763* (New York: HarperCollins, 2003), 31.

[3] *Crides, pragmàtiques, edictes, cartes i ordres per a l'administració i govern de la Ciutat i Regne de València en el segle XVI*, eds. Josep Lluís Canet and Diego Romero (València: Universitat de València, 2002), vol. II, 53, 91.

[4] Miguel de Bunes Ibarra, "Prólogo," in Beatriz Alonso Acero, *Orán-Mazalquivir, 1589–1639: una sociedad española en la frontera de Berbería* (Madrid: CSIC, 2000), xvi.

Luis Galcerán's tenure as governor encompassed the famous Battle of Lepanto of 1571, the signal victory for Philip and his Italian allies over the Turkish fleet. This triumph proved ephemeral, and in subsequent years the king's ambitions in the Mediterranean waned as he directed his resources north to the Low Countries and west to the Atlantic world. The king never did realize his dream of capturing the corsair capital of Algiers, and the presidios he maintained – perpetually undermanned, and starved for commodities – stood as a testament to the limits of his reach in the Mediterranean.

The concept of a Spanish Crusade in North Africa fares similarly poorly when assessed through the prism of Pedro Luis Galcerán de Borja's experiences in Spain and Orán. His family had expanded its holdings considerably in the decades since Rodrigo de Borja ascended to the papacy as Alexander VI in 1492. Their lands in the south of Valencia, populated mainly by Moriscos, yielded olives, grapes, honey and above all else sugar, by far their most profitable crop. Pedro Luis Galcerán thus had extensive knowledge of Valencia's baptized Muslims before he ever set sail for the Maghreb. The Moriscos of Gandía lived in comparatively homogenous communities, speaking a dialect of Arabic and practicing Islam to the extent possible in an inquisitorial age. The court records document many instances of Morisco violence in the duke of Gandía's holdings, but not in patterns that would sustain a thesis of Christian-Muslim religious warfare: Moriscos often fought on behalf of their old Christian lords, or against Muslim bandits from abroad. Pedro's election as Master of Montesa, moreover, placed him in a position of authority over a very different Morisco population. The Order's diverse lands, situated primarily in the north of Valencia, included just three Morisco settlements, a small percentage of Montesa's vassals. The Moriscos of Alcalá de Xivert grew to resemble their Old Christian neighbors over time, in diet, dress, and customs, to the extent that they sought exemptions from the expulsion decree of 1609. Recent research on the Moriscos of Campo de Calatrava and the Jiloca valley of Aragon, among many other communities, has demonstrated the many cultural and economic ties that bound Old Christians and New in defiance of laws that sought to segregate them.[5] The Morisco towns of Gandía and Montesa, Pedro's homeland and adopted home, likewise reflect the great diversity that characterized

[5] Trevor J. Dadson, *Tolerance and Coexistence in Early Modern Spain: Old Christians and Moriscos in the Campo de Calatrava* (London: Tamesis Books, 2014); Mary Halavais, *Like Wheat to the Miller: Community, Convivencia, and the Construction of Morisco Identity in Sixteenth-Century Aragon* (New York: Columbia University Press, 2005).

the final century of the Muslim diaspora in Spain. For all their differences, both communities also belied the concept of a Christian-Muslim frontier within Valencia; the Moriscos of Gandía shared a culture of violence with the region's Old Christians, while those of Alcalá de Xivert assimilated into Christian society more quietly.

Pedro Luis Galcerán de Borja's governorship in Orán reveals similar alliances and affinities that crossed religious lines, in contrast to Jiménez de Cisneros' aspirations for the presidio. Pedro Luis Galcerán's biographer, the soldier-chronicler Diego Suárez Montañés, set out to prove the importance of Orán and Mazalquivir to Spanish foreign policy, and the diligence of *el Maestre de Montesa* in pursuing the royal interest. Despite its apologetic depiction of the governor and its reiteration of common tropes with respect to Turkish cruelty, however, the account provides scant evidence for any idealistic interpretation of Spanish activity in the Maghreb. Three themes in particular emerge from the wealth of detail: the armed struggle against *moros de guerra*; the perennial challenge of manning and provisioning the presidio; and the effort to ransom Christian captives in Algiers and other Muslim ports. The battles against Muslims whom the Spanish identified as "warlike" represented not Crusades but rather raids, with a carefully choreographed system of espionage, assault, and plunder. The objective of these repeated *cabalgadas* – which occupy the majority of Suárez's 500-page chronicle – was not conquest or even lasting victory, but rather resources in the form of seized weapons, livestock, and captives. Pedro Luis Galcerán had similar goals via different means in his dealings with the *moros de paz*, or Muslims who signed contracts to supply the presidio in exchange for protection. The shipments of grain and other goods from Spain never came close to meeting demand, and so the governor found himself beholden to the local population for the outpost's very survival. Finally, even in an area in which the Christian-Muslim divide might be expected to prevail, the ransoming of captives, the governor's experience revealed commonalities as well as conflicts. Redemptive orders such as the Trinitarians and Mercedarians have traditionally received the most scholarly attention, but many others took part in the Mediterranean trade in captives, often with less pious motives. Merchants based in Valencia offered the ransom of loved ones for a fee, a business they conducted alongside legal and extralegal trade with North Africa. Pedro Luis Galcerán himself entered into this economy via his sorties from Orán, trading his captives for Christians in Muslim hands. The chronicler Suárez dutifully defended the Master of Montesa for his prudence, just as friars appealed to the piety of their listeners in seeking alms, and merchants cited the purity of the faith in applying for licenses to trade in Algiers. But this rhetoric conformed little to the day-to-day negotiations, alliances, and compromises that collectively defined the Spanish encounter with Islam.

***

Pedro Luis Galcerán de Borja's career sheds light upon two diasporas in the sixteenth century: the diaspora of Muslims in Spain, consisting of Moriscos as well as some slaves and bandits; and the diaspora of Spanish soldiers, officials, merchants, and captives in the Mediterranean world. With respect to the former, Pedro's experiences in his native Gandía presented a stark contrast to his administration of the Order of Montesa. The sprawling territories of the Borja clan encompassed roughly half of southern Valencia, from the coastal city of Gandía into the valley of the River Júcar. These holdings included more than two dozen Morisco settlements, clustered around the majority Old Christian cities on the coast or extending west into the interior. For centuries past, Muslims had provided the bulk of the labor in this region, paying taxes to their Old Christian seigneurs in exchange for a degree of religious and legal autonomy. The end of the era of Mudejars, or Muslims living under Christian rule, came about shortly before Pedro's birth, through a battle – but not a battle between Muslims and their Christian lords. In 1519, groups of artisans in Valencia began forming into militias, ostensibly to defend the coast. Charles V's initial support for these brotherhoods, or *germanías*, dissipated when the men began to voice political objectives such as greater representation in municipal government. The lands of the duke of Gandía remained loyal, but found themselves caught up in the violence nonetheless after the *germanías* seized the fortress of the nearby city of Játiva in summer 1521. The viceroy urged Juan de Borja, Pedro's father, to await the rebel armies in the castle of Gandía. Fearing the destruction of his lands and the flight of his Muslim vassals, the duke insisted on marching forth to meet the *germanías* in open combat.[6] The ensuing rout of the viceregal forces on 25 July 1521 resulted in the sack of Gandía and, even more problematically for the duke, assaults upon the Mudejars. The victorious troops killed many Muslims, but forced hundreds more to receive baptism, as a symbol of divine favor and to disrupt the economic engine of their noble adversaries. According to one witness, "the soldiers

[6] Vicent J. Vallés Borràs, *La Germanía* (Valencia: Institució Alfons el Magnànim, 2000), 114.

took the *moros* and stripped them, and robbed them, and carried them with great fury to the church to be baptized."[7]

The *germanías*, like the German peasants' rebellion to the north, ended in defeat and summary executions, after the nobility closed ranks to defend the society of orders. And yet, by the time of Pedro Luis Galcerán's birth in 1528, Charles V had declared the baptisms of 1521 valid, and his agents carried out (in theory) the baptism of all Muslims in Spain. This decision imposed from above, based on theological subtlety and political expediency more so than on local conditions, fostered the volatile situation in which Pedro Luis Galcerán came of age. Neither his father the duke nor the Muslims had initiated these baptisms, and in their aftermath both conspired to preserve the late medieval status quo to the extent possible. Thus in 1528 a group of Morisco leaders contracted a treaty with the Valencian Inquisition, which now had jurisdiction over them. The Moriscos agreed to pay a substantial fee in exchange for leniency in cases involving minor offenses against the faith. Christian proponents of evangelization may have intended for this arrangement to create a space free of persecution in which the Church could instruct the Moriscos in their new faith, conceding that this evangelization had not preceded the sacrament of baptism. The New Christians, as subsequent events would demonstrate, interpreted this treaty rather as a license to continue practicing Islam as before. The duke of Gandía reinforced this view by perpetuating taxes and services that historically had applied to Mudejars but not Christians, such as the *laudemio*, a tax upon land usage, and the obligation to provide labor for the seigneur's fortifications.[8] A decade after Pedro's birth, the forced baptisms of the Mudejars of Gandía had changed everything in theory, and very little in practice.

This situation changed somewhat with the death of Juan de Borja and the succession of his son Francisco as 4th duke of Gandía in 1543. In the eight years before he renounced his titles and joined the Society of Jesus, Francisco made some concessions to the Moriscos in financial matters, while constructing the monastery of Santa Cruz to provide Dominican preachers to instruct them.[9] The brevity of his tenure, however, combined

[7] "Vido este testigo como los soldados tomavan los moros y los despujavan y robavan y los levavan con gran furia para que se fuessen a babtizar a la yglia." University of Pennsylvania, Henry Charles Lea MS 20, "Informacio super conversione sarracenorum," fol. 21r.

[8] Elia Gozálbez Esteve and Antonio Constán Nava, *Los vasallos moriscos del Marquesado de Llombai en época de Don Juan de Borja y San Francisco de Borja: entre 1533 y 1539* (Valencia: Institució Alfons el Magnànim, 2012), 49–53.

[9] Gozálbez and Constán, *Los vasallos moriscos*, 40.

with the lack of support from other Christian authorities, limited what Francisco was able to accomplish as duke of Gandía. The evidence from inquisitorial records and criminal trials indicates that with few exceptions the Moriscos of Gandía resisted assimilation into the Christian faith, taking advantage of geographical isolation and seigneurial benign neglect to maintain their belief in Islam. A remarkable series of events from mid-century hints at the recalcitrance that so vexed the Valencian inquisitors, who processed hundreds of Moriscos for observing Islamic rituals or failing to conform to Christian ceremonies. In summer 1554, a Turkish corsair outfitted a galley with a crew of sixty men in Algiers and sailed to the Valencian coast, with the goal of capturing Christians for ransom. Near the town of Calpe, under attack from a group of Spaniards in a merchant vessel, the Muslims ran up the white flag and offered to exchange ten Christians already in their possession for fresh water and permission to depart. The men of Calpe agreed, "out of love for the captives," but the exchange went poorly and the corsairs ended up fleeing, leaving one of their men on shore in the process.[10] Two men of Calpe caught him hiding in a forest and then sold him to a carpenter in Valencia, claiming they owed no taxes because they had seized the Muslim in war (*de bona guerra*). A decade later, royal tax collectors arrested a native of Algiers in Gandía for soliciting donations towards his ransom from the same carpenter, whose name he had taken as his own. Upon investigating, however, the officials determined that a group of Moriscos from Gandía had purchased the former corsair some years before, but rather than employing him in their fields they sent him out among Morisco communities to collect alms.[11] By misrepresenting their slave's status the Moriscos revealed their confidence that the Morisco community would be more likely to provide charity to a Muslim castaway in order to ransom him from a Christian master.

As late as 1583, more than fifty years after the forced baptisms, an elderly Morisco from Gandía appeared before the Inquisition for fasting during Ramadan. A witness reported that he had said, "The reason we're having bad weather is that, unlike the Turk, who recognizes the three laws of Moors, Jews, and Christians, the king of Spain does not allow each one to live by his own law."[12] This statement both recalled

[10] "... per amor dels catius que anaven en dita galeota dexaren pendre als turchs los dits dotze barrils de aygua." Archivo del Reino de Valencia [ARV], Bailía, Procesos letra P, sig. 2318.

[11] ARV, Bailía, Procesos letra P, sig. 409.

[12] Archivo Histórico Nacional, Inquisición, libro 936, fol. 358v.

the legal communities of Muslims in the medieval era, and compared Philip II unfavorably to the Ottoman sultan on the question of religious minorities. The nobility by and large tolerated Morisco apostasy, taking a much greater interest in tax collection and controlling – co-opting, even – Morisco violence. Post-baptisms, the Moriscos remained the engine of the Gandian economy, particularly the 20–25 percent of the duke's lands dedicated to sugar cultivation.[13] Carlos de Borja, 5th duke of Gandía and nephew to Pedro Luis Galcerán, sued a former tenant of Llombay who had moved to the Morisco quarter in the city of Valencia and then claimed that he owed allegiance only to the king.[14] This case underscored the collaboration and competition among Christian authorities to extract revenue from the Moriscos. The surviving records indicate a constant effort to catalog and tax a vibrant Morisco economy that otherwise operated in an extralegal manner. The Moriscos of Llombay, for example, kept an Arabic-language account book in which they recorded goods and sales that did not appear in the Romance-language ledger they shared with the duke of Gandía.[15] This pattern also prevailed in cases before the *batlle general*, the official in charge of collecting royal taxes. In 1564, the collectors ordered a Morisco from Gandía to pay the king's fee on two large shipments of sugar, syrup, and candied lemon peels to Castile the previous year.[16] A decade later, another Morisco from Gandía, one Jeroni Cassim, found himself before the royal judge when he was arrested on the road to Sagunto carrying 450 gold crowns, a princely sum. The subsequent investigation revealed him to be a trusted servant of an Old Christian merchant in Sagunto, who had sent him and a partner to sell silk in Gandía, purchase sugar, and return with the remaining funds.[17] The Moriscos of Gandía remained largely a people apart in matters of religion, but intersected with Old Christian networks of production and trade.

Jeroni Cassim's trial illustrates his disadvantaged position as a Morisco in Valencia: the guard who arrested him in 1574 had no particular reason to suspect him, other than the fact that he didn't recognize Cassim. The guard found the leather pouch with the coins only after searching Cassim, and his suspicions grew after he found a license to transport money from

[13] Eugenio Ciscar Pallarés, *Tierra y señorío en el pais Valenciano* (Valencia: Del Cenia al Segura, 1977), 98.

[14] ARV, Bailía, Procesos letra P, sig. 2422.

[15] Gozálbez and Constán, *Los vasallos moriscos*, 70.

[16] ARV, Bailía, Procesos letra P, sig. 433.

[17] ARV, Bailía, Procesos letra P, sig. 2512.

Gandía that bore another man's name. Cassim ultimately demonstrated that this license referred to his partner, and was accused only of defrauding the royal coffers. But the expectation that a Morisco carrying money must have stolen it reflected a common association between Moriscos and crime. The repeated laws against Moriscos bearing arms, migrating towards the coast, or coming into contact with "Granadans, Arabs, Aragonese Moriscos, Moros, or Moriscos from abroad," spoke to the fear that these elements would disrupt the "well-being, peace, and calm of the realm."[18] Though these laws distinguished between Old Christians and New, the often detailed testimony in criminal trials reveals a much less clear-cut situation on the ground. In 1565, on the eve of Pedro Luis Galcerán's departure for Orán, his nephew the duke of Gandía sent his constable and a notary to Beniopa to investigate reports of a Morisco attack upon Christian travelers on the royal road. The survivor they discovered, however, was a Morisco of the Badahui family, and not entirely the victim he claimed to be. The wounded Morisco testified that on returning from the market in Gandía, where he had sold his rice, fifteen Moriscos beat him and stabbed him, "even though he had never harmed any of them." Several other witnesses, however, attested to the longstanding enmity between the Badahui and Faraig clans, recounting murders and assaults going back at least three years. More ominously, the wounded Morisco referred to the desire of his Old Christian lord, the seigneur of Villalonga, to "put an end to these homicides."[19] These events mirrored numerous other documented cases in Valencia, in which violence crossed religious lines and vassals acted at the behest of their masters. The Moriscos of Gandía earned a reputation for apostasy in religion, but in the contest to acquire resources and defend their interests they bore much in common with their Old Christian neighbors.

Though Pedro Luis Galcerán de Borja hailed from a family based in Gandía, as a young man he assumed authority over a very different community with his election as Master of Montesa in 1545. The lands of the order had been divided in two earlier in the century: the Maestrazgo in the north, centered on the castle town of Montesa, was populated almost exclusively by Old Christians, whereas the order's thirteen scattered *encomiendas* included Morisco enclaves in northern coastal towns such as Benicarló and Alcalá de Xivert, as well as in the town of Montroy

[18] "E no res menys ha mostrat que acullir, y receptar Granadins, Alarbs, y Tagarins, o Moros, y Moriscos de allende, o alias, causa gran dany, y mal al dit Regne, y gran inquietut, y perturbacio als poblats en aquell." ARV, Real Cancillería, sig. 699, fol. 68r.

[19] ARV, Real Audiencia, procesos criminales, sig. 190.

to the south. Moriscos represented a small percentage of the more than 5,000 vassals who occupied Pedro Luis Galcerán's new jurisdiction, and those in the north and the south followed divergent paths. The order's southern town of Montroy, a case in point, lay near the duke of Gandía's holdings in Llombay. The Valencian Inquisition reconciled a Morisco from Montroy in 1572 for fasting in Ramadan, performing Islamic rituals, and scaling the wall of the chapel in Dos Aguas "to throw rocks at the cross and cast the sacred altar and instruments of worship on to the floor."[20] The Moriscos of Montroy also resembled their neighbors in disputes over pasturage, murdering rivals and even suing the duke of Gandía in 1589.[21] The Moriscos in the northern towns followed a distinct trajectory even though their forced baptisms also came about through violence in the summer of 1521, when the *germanías* sacked the Muslim quarter of Alcalá de Xivert, burned the mosque, and compelled the panicked Mudejars to accept baptism.[22] The Master of Montesa and his knights led the counterattack, securing victory in the north a few weeks later. Over the course of the sixteenth century, the Master Pedro Luis Galcerán fostered a process many regarded as impossible, the assimilation of his Morisco vassals.

By 1600, the Moriscos of Alcalá de Xivert had adopted the dietary practices, language, dress, and even religious beliefs of the Old Christians who surrounded them. Geographic isolation from other Moriscos and close economic ties with the other vassals of Montesa clearly played a role, but in several key respects Pedro Luis Galcerán also influenced these developments. In 1576, the Order sold the lands of the former mosque in the town and dedicated the funds to the construction of a new church. Four years later, the Master conceded a privilege promoting Alcalá de Xivert to the status of *villa*, a change which granted the Morisco justices "the use and exercise of all jurisdiction, civil and criminal," including over outsiders who committed crimes in their territory.[23] This combination of persuasion and empowerment ultimately led the town council of Alcalá de Xivert to proclaim the adherence of their people to the Christian faith. No evidence survives, furthermore, to suggest that Pedro Luis Galcerán used his northern Moriscos in proxy wars against his rivals. In 1563, the viceroy of Valencia ordered the disarmament of the Moriscos, to be

[20] AHN, Inquisición, libro 936, fol. 77r.
[21] ARV, Real Audiencia, procesos criminales, sig. 494; Bailía, Procesos letra P, sig. 1005.
[22] Vallés, *La germanía*, 108.
[23] ARV, Clero, legajo 938; rep. in Javier Hernández Ruano, *Historia de los moriscos valencianos de Xivert* (Alcalá de Xivert: Centre d'Estudis del Maestrat, 2003), 89, 135.

carried out by their noble lords "with diligence and care, and with all possible secrecy."[24] In the Morisco towns of Gandía, the execution of this order amounted to a house-to-house search, resulting in the confiscation of thousands of swords, firearms, and crossbows. In Alcalá de Xivert, the collection yielded only a few weapons that could be considered dangerous, along with a small pile of "useless" swords and daggers.[25] In further defiance of the concept of a Morisco Fifth Column, the residents had little contact with the Mediterranean, except as fellow victims of corsair raids. The first chronicler of a Turkish attack on the town in 1547 made no particular mention of Moriscos; only years later, amid a dispute over land, did interested parties invoke rumors of local aid to the invaders.[26] Right up until the end, the call to arms against the infidel remained an available weapon to be employed against Moriscos, even among an assimilated community such as Alcalá de Xivert.

Did Philip II then appoint Pedro Luis Galcerán de Borja as governor of Orán on the basis of his deft leadership as Master of Montesa? On the contrary, his service amounted to an honorable exile on account of his involvement in the factional strife that plagued early modern Valencia. In 1553, a drama worthy of a Renaissance play unfolded when the rivalry between two families of the minor nobility drew in their powerful patrons, the house of Aragón and the house of Borja. A knife-fight between the Pardo de la Casta and Figuerola clans in the shadow of the Valencian cathedral led the duke of Segorbe, Alonso de Aragón, to order his clients the Pardo to relocate to Segorbe for their own safety. Pedro Luis Galcerán located them by means of his spies, and in December 1553 he and the Figuerola sent two men to assassinate the Pardo leaders. The attempt failed, leaving one man wounded, and the duke of Segorbe's men captured the perpetrators. Under torture, they confessed that "they had acted on the orders of the Master of Montesa, who had not promised or given them any payment."[27] Confident that the judge in Segorbe would not act severely against men who had simply wounded their target, Pedro Luis Galcerán was enraged to learn that they had been executed, one by hanging, and one by dismemberment. Consumed with "hatred, rancor, wrath and ill will" towards the duke of Segorbe and the officials in his

24 ARV, Real Cancillería, sig. 562, fol. 1v.

25 ARV, Real Cancillería, sig. 562–4; Hernández, *Historia de los moriscos*, 79–80.

26 ARV, Real Audiencia, Letra S, no. 1271; Hernández, *Historia de los moriscos*, 83.

27 "... que había actuado por mandato del Maestre de Montesa y que no se le había prometidoni entregado nada a cambio." Sergio Urzainqui Sánchez, *El asesinato de D. Diego de Aragón* (Segorbe: Fundacion Mutua Segorbina, 2007), 344.

pocket, in January 1554 he sent his brother Diego de Borja with a small army of men from Montesa to the capital.[28] After tracking their target's movements, they trapped him on a bridge near the towers of Serranos, shot him in the chest with an arquebus, and fled the scene. The victim who died several days later was none other than Diego de Aragón, son of the duke of Segorbe.

Recognizing that he had escalated the violence to the highest levels of the Valencian nobility, Pedro Luis Galcerán elected to leave Montesa by night, accompanied by his brothers Diego and Felipe Manuel de Borja and a band of armed men. He understood that his flight placed his followers in a difficult position, and many of them were interrogated subsequently; one reported the Master's plan to challenge the duke of Segorbe to a duel, and his threat, "the Borja who doesn't back me, I will reduce him to ashes." He made the decision to split up his group, and rather than heading to Gandía, he and Diego de Borja went separately to hide in the sierra. The Borja allies in Perpuchent, an *encomienda* of the Order of Montesa, assisted their flight by posting Morisco guards at the gates to watch for "any outsiders, on foot or horseback."[29] The viceroy organized a massive manhunt, sending bailiffs into the sierra, seizing Montesa's assets, and ordering harsh punishments for anyone who harbored the fugitives. These came to nothing, and in the end Pedro Luis Galcerán voluntarily went to Castile to plead his case before Prince Philip. He was not able to save his brother; the royal courts completed the promise the infanta Juana (Philip's sister) made to Alonso de Aragón, "to make the show of force required by such a case," executing Diego de Borja in 1562.[30] Most of the accomplices' sentences were reduced to fines and sequestration of goods, with the exception of the Master and his brother Felipe Manuel. In the interim between the assassination of the duke's son and the trials, Felipe Manuel had attempted to kill a rival nobleman by throwing him from a high window. The king commuted his sentence of death to "exile in the fortress of Orán," an indication of the esteem that service in North African presidios commanded.[31] The trial of Pedro Luis Galcerán proved more complex on account of his status as Master of Montesa, and after some time in

[28] Urzainqui, *El asesinato*, 270.

[29] Urzainqui, *El asesinato*, 287, 238.

[30] Urzainqui, *El asesinato*, 69, 75.

[31] 9 May 1565. ARV, Real Cancillería, sig. 443, fol. 29. In the sixteenth century, Spanish courts reserved banishment to North Africa for nobles and the wealthy, sentencing poor commoners to row in the galleys. Ruth Pike, *Penal Servitude in Early Modern Spain* (Madison, WI: Wisconsin University Press, 1983), 41–3.

FIG. 6.1. "Galleys in the eastern Mediterranean for the fight against the Algerian corsairs" ("Galeras en el Mediterraneo oriental para la lucha contra los corsarios de Argel")
Source: España. Ministerio de Educación, Cultura y Deporte. Archivo General de Simancas, MPD, 67, 024.

seclusion and a good deal of legal wrangling the Council of Orders simply dismissed the case. Even in the absence of a formal sentence, however, it remains clear that Philip II sent Pedro Luis Galcerán to Orán as a means of containing factional violence in Valencia and punishing this most tempestuous member of the Borja clan.[32]

The exile of the Master arguably delivered him from a volatile environment in Valencia; some of the knights of Montesa had sided with the Aragón clan, and at one point Pedro Luis Galcerán debated putting out a contract on one of his own commanders. Joined by his brother Felipe Manuel and a group of loyal soldiers, he set sail from Cartagena, arriving in Orán at the end of June 1567. A map from 1621 captures both the parallelism of the Valencian and North African coasts, as well

[32] In this respect Pedro Luis Galcerán's politically motivated sentence differed from the experience of the Fernández de Córdoba clan, who served honorably under Charles V in both Navarre and Orán, "because of the particular expertise they possessed on frontiers." Yuen-Gen Liang, *Family and Empire: The Fernández de Córdoba and the Spanish Realm* (Philadelphia, PA: Pennsylvania University Press, 2011), 156.

as the precarious position of Orán in the shadow of the corsair capital of Algiers. As Pedro Luis Galcerán passed through the gates of Castanel, he entered a world every bit as violent and even more stark than his native land. Born just as the last Mudejars received baptism, and a generation after the expulsion of the Jews, he had only ever known a realm in which all were officially Catholic subjects of the Habsburg king. In Orán he governed a population that included Christians and Jews within the *plazas* or city walls, while maintaining alliances with "peaceful" Muslims in the surrounding area of Tremecén. The 800 residents of Orán included soldiers, government officials, merchants, and families, some of them second generation by the mid-sixteenth century. Their homes and churches lay below three recently expanded fortresses, which protected the city "as a hen watches over her chicks," in the phrase of the soldier-chronicler Diego Suárez Montañés.[33] In principle, Pedro Luis Galcerán also had authority over the neighboring Christian town of Mazalquivir (Mers-el-Kebir), though in practice the mayor of Mazalquivir had clashed with the previous governor of Orán over provisioning.[34] The Spanish authorities agreed that married soldiers were more stable and less likely to mutiny, but the struggle to feed these families plagued and even defined the governance of the presidio. Given the inadequacy of shipments from Spain, the governor necessarily looked to the Muslims who farmed the fertile river valley of Tremecén. Politically, this region was decentralized by comparison to either Spain or Turkey, and the tribal leaders played the Spanish off the Ottomans and the corsair state of Algiers to the east.[35] While bearing some resemblance to the *convivencia* of medieval Spain, Orán was no kingdom of heaven: for their very survival the Spanish depended upon their ability to commandeer or plunder resources from the Muslim population.

Out of necessity Pedro Luis Galcerán devoted a great deal of his time to what Fernand Braudel called "the favourite pastime" of North African governors, the *cabalgada* or planned raid of nearby settlements.[36] Suárez, who arrived in Orán five years after the Master's departure, nevertheless narrated these sorties in vivid detail. Certain patterns emerge from these ten raids, including the Spaniards' reliance on Muslim or baptized Muslim

[33] Suárez, *Historia del Maestre*, 100.

[34] Suárez, *Historia del Maestre*, 194.

[35] Mercedes García-Arenal and Miguel de Bunes Ibarra, *Los españoles y el norte de África. Siglos XV–XVIII* (Madrid: Editorial Mapfre, 1992), 16.

[36] Fernand Braudel, *The Mediterranean and the Mediterranean World in the Age of Philip II* (New York: HarperCollins, 1992), 623.

spies for information. In June 1568, the governor received word of a band of *moros de guerra*, hostile "enemies of His Majesty and of the name of Christ."[37] Upon deciding that their quarry was near enough and likely to have valuable goods, the governor left some soldiers behind to protect the presidio and led his men to the reported location. Marching through the night and staying off the main roads, the soldiers communicated via birdcalls to maintain the element of surprise. Shouting "Santiago!" to distinguish friend from foe, the men descended on the Muslims at dawn, their swords shooting sparks off the adversaries' scimitars.[38] For all their martial fervor, their goal was to capture, not kill, and on this occasion they seized 193 captives along with 7,000 livestock. Abandoning stealth, the Spaniards returned to Orán in all haste, to avoid ambushes from their victims' allies; the portable altarpiece with which they had heard Mass prior to the battle remained in their saddlebags. They arrived at the presidio having lost just two men, seized after the battle. In dividing the spoils, the governor followed a precise system laid out by Philip II in response to previous disputes. Pedro Luis Galcerán took inventory and then claimed his *joya* or prize, one slave to be sold at auction. After paying off the spy and the interpreter, he assigned loot to his captains and soldiers based on rank and salary. In a nonviolent inversion of the raid, the Spanish then auctioned off a large part of the captives and livestock, either to local Muslim merchants or to the very men who had fought against them just days before.[39] The governor dedicated the remaining funds to the ransom of the two captured Spaniards, and to pious works, including the celebration of the festival of Corpus Christi. In its trappings – the invocation of Santiago, the predawn Mass, the charitable donations – the raid recalled the Reconquest, and Suárez did his best to paint the warlike Muslims as enemies of Christ and king. But the *cabalgadas* did not lead to lasting territorial gains; rather, they served as a means of resource extraction for the perpetually impoverished outpost.

Given the violence that punctuated his interactions with the Muslims, Pedro Luis Galcerán made a concerted effort to distinguish between *moros de guerra* and *moros de paz*. The latter "peaceful" group signed contracts with the governor each year, ensuring their safety in exchange for specified amounts of grain. Suárez praises Pedro Luis Galcerán for not only renewing the existing guarantees, but looking to extend new

[37] "... enemigos de su Majestad y del nombre cristiano." Suárez, *Historia del Maestre*, 242.
[38] Suárez, *Historia del Maestre*, 146, 151.
[39] Suárez, *Historia del Maestre*, 157, 245.

contracts to previously belligerent Muslims, "to secure necessary provisions, and to weaken the enemy Turk."[40] By protecting those under contract, allowing them to live near the city and trade within its walls, the Spanish deprived the Ottomans of tax revenue. The governor honored these arrangements scrupulously, as demonstrated by a failed raid amid darkness and rain in January 1568. Following bad advice from a spy and mistaken directions from a guide, the Spanish soldiers descended upon a tent encampment of *moros de paz*. Seeing his men hesitate upon reaching a ditch filled with rubble, Pedro Luis Galcerán got down from his horse, put his hand on his sword, and shouted at them, "What is this, good soldiers of Orán? Santiago! Santiago! To the *moros*, they are escaping us!"[41] Seeing their own leader scramble through the ditch, the soldiers rallied around him and took thirty-five captives. The Muslims protested that they were under contract, at which point the governor ordered that none of their livestock be taken. Outside the city walls of Orán, he sent a soldier to retrieve the registers, which confirmed the Muslims' claim. Pedro Luis Galcerán freed them, returned their goods, and paid them restitution from his own funds. This case of mistaken identity nevertheless hints at the close ties between the governor's raids and the Muslims under contract. Many of the goods the Spanish traded with their Muslim allies originated in raids on other Muslim groups, in which the *moros de paz* were instrumental as spies, guides, and translators. In the early 1600s, when Philip III attempted to ban the Spanish raids, the governor objected – not only on account of the loss of resources, but also to keep the *moros de paz* from growing arrogant and raising their prices.[42] The threat of violence underlay even outwardly amicable interactions with the Muslims of Tremecén.

By raiding and trading Pedro Luis Galcerán sought both to provision the double presidio and to serve the royal interest by drawing resources away from the Ottomans and the corsairs of Algiers. The Borja brothers' African sojourn coincided with Philip II's greatest victory in the Mediterranean, the Battle of Lepanto in 1571, in which a combined Spanish and Italian naval force defeated the fleet of Ali Pasha. But within just a few years, following the defeat in Tunis in 1573–4, the king adopted a more defensive stance, marking a shift from imperial wars to proxy

[40] "... que en esto estriba el estar abastecidas de lo necesario, y para enflaquecer al enemigo turco." Suárez, *Historia del Maestre*, 217.

[41] "Qué es esto, señores soldados de Orán?, ¡'Santiago', 'Santiago' en los moros, que se nos van todos!," Suárez, *Historia del Maestre*, 235–6.

[42] Alonso, *Orán-Mazalquivir, 1589–1639*, 268–70.

battles via corsairs.[43] Though Philip II never succeeded in orchestrating an assault on Algiers, he did order his Mediterranean commanders to limit Spanish losses through espionage, defense of Spanish interests, and attacks on corsair ships. Pedro Luis Galcerán strove to rise above mere survival in order to pursue Habsburg objectives in the region. In 1570 the governor received an offer of alliance from an unlikely quarter: the Muslims of Bent Arax, who not long before had actively recruited aid from the Turks and Algiers in the destruction of Orán. Following a decision to shift their loyalties to the Spanish, the leaders approached Pedro Luis Galcerán, who surprised some of his counselors by agreeing to send forty-six Spanish soldiers to assist their erstwhile adversaries with the planned attack on Algiers. Suárez refers to these troops as "hostages," and to the governor's subsequent decision to recall them to Orán as "barter," in exchange for the men from Bent Arax who had been given to him as a form of collateral.[44] This series of events is indicative of the governor's interactions with Algiers in two senses: as an inconclusive military engagement, and as an example of the trade in captives. Pedro Luis Galcerán never won a major engagement against the corsair state, but he did play a role in the ongoing trade in commodities, including prisoners.

As the viceroy of Valencia reiterated in the 1550s, Spaniards were officially prohibited from "going to or trading with the city of Algiers," under pain of fines, confiscation of goods, and destruction by fire of the offending ship.[45] And yet the Spanish did business in the corsair capital as a matter of routine. In the best-studied example, redemptive orders such as the Trinitarians and Mercedarians solicited alms, petitioned the viceroy and the *real audiencia* for licenses, and outfitted expeditions to the Maghreb to redeem Christian captives. Royal officials conducted investigations to ensure that all profits from goods exported went into ransom, and even to confirm the upstanding morals of the captain and crew. Recent research has illuminated the key role of lay merchants in the trade with Algiers, which made up a significant percentage of Spain's overseas trade.[46] A 1554 case before the tax collector in Valencia revealed the complexity of this commerce beyond simple redemption. A group

[43] García-Arenal and Bunes Ibarra, *Los españoles y el norte de África*, 109.

[44] Suárez, *Historia del Maestre*, 356.

[45] ARV, Bailía, Procesos letra P, sig. 517.

[46] ARV, Real Cancillería, sig. 620; Rafael Benítez, "La tramitación del pago de rescates a través del reino de Valencia: el último plazo del rescate de Cervantes," and Juan Francisco Pardo, "Mercaders, frailes, corsarios y cautivos," in *Le commerce des captifs*, ed. Wolfgang Kaiser (Rome: Ecole Francaise de Rome, 2008), 165–217.

of investors from Spain redeemed one Anthoni Ximénez, a native of Benicarló and vassal of the Order of Montesa. The merchants initially claimed that the cloths and other goods on board were for trade in Bugía, at that time a Spanish-held port. But further testimony revealed that the investors had purchased the boat from the ruler of Algiers via a Spanish agent who resided in the city, along with an Algerian license to sell their goods in Valencia and Mallorca and return with the boat full of Spanish merchandise.[47] The ransom of Ximénez represented but one aspect of an operation designed to skirt the law and tap into the rich vein of resources in the corsair state – in this case a boat and merchandise the corsairs had assuredly seized from Europeans.

By proximity and by design, Pedro Luis Galcerán both participated in and profited from this trade in captives, even if Suárez attempted to depict the governor's interactions with Algiers in a more heroic light. The chronicler boasts that the corsairs never attacked Orán on the Master's watch, out of fear of Pedro Luis Galcerán and his allies in Valencia, "who would come to his aid even if they had to swim."[48] Suárez tacitly concedes that the governor fared no better in his attacks on Algiers, as when the Muslims captured a boat he sent out to prowl the shipping lanes. Christian captives were as likely to make their own way to Orán by sea; the soldiers marveled at the flimsiness of a cowhide boat in which fourteen men arrived on the beach in 1570.[49] Pedro Luis Galcerán proved much more effective at opposing the Algerians on land, via his raids on their clients in Tremecén. In 1571 he seized a caravan transporting the cargo of a Venetian ship held in Algiers, along with more than 300 prisoners.[50] In the distribution of this bounty, the governor specified a number of these captives to be shipped to Spain for sale at auction, to provide supplies for the presidio. These men thus made a journey denied to Pedro Luis Galcerán, condemned to exile in Orán, and his Spanish soldiers, who petitioned without success to escape service in this impoverished outpost. The Master of Montesa profited from the human traffic in the Mediterranean in order to feed a presidio in which he and his men suffered captivity under a different name.

When Philip II ultimately recalled Pedro Luis Galcerán to Spain in 1572, this was hardly a reprieve. Following the arrest of a young courtier

[47] ARV, Bailía, Procesos letra P, sig. 2324.

[48] "Y encorajinados, viendo a su Maestre sitiado, su caballería y no caballería pasarían, aunque fuere a nado, a socerrerle." Suárez, *Historia del Maestre*, 302.

[49] Suárez, *Historia del Maestre*, 307–8.

[50] Suárez, *Historia del Maestre*, 406–7.

in Madrid for sodomy, one of the knights of Montesa made the mistake of telling a political rival about the Master's "impassioned" response to this news.[51] As the testimony mounted, the Inquisition demanded the detention of Pedro Luis Galcerán and several of his pages. Vowing to return, he ceded his responsibilities to his brother Felipe Manuel, advising him to favor the *moros de paz*, keep up the presidio's fortifications, and pray for the victory of the Holy League over the Turks.[52] Although he escaped the flames that consumed his alleged lover, Pedro Luis Galcerán was stripped of his title as governor of Orán and confined to house arrest in the convent of Montesa. Philip II took advantage of the situation to secure the Order of Montesa for the Crown upon the Master's death. By agreeing to this Pedro Luis Galcerán attained his release and even an appointment as viceroy of Catalonia in his final year of life (1591–2).[53] This decision however compromised his legacy, and doomed his biographer's efforts to publish his manuscript. Upon his return from Orán in 1604, Diego Suárez Montañés embarked on a star-crossed journey through Madrid, Gandía, and Valencia, unsuccessfully courting the patronage of the duke of Gandía, other members of the Borja family, and the town council.[54] Suárez's depictions of Pedro Luis Galcerán as a brave warrior, a leader of men, a fair judge, and a loyal servant drew no interest. In an early chapter, Suárez related the visit of Don Juan de Austria to Orán. When a soldier gave him a letter defaming the governor, the general replied that his half-brother Philip II was satisfied with the presidio, and ordered the soldier hanged with the paper strapped to his chest. Pedro Luis Galcerán begged mercy for his libeler and had him shipped quietly back to Spain.[55] Suárez could not do the same with the charges against his biographical subject that emerged after 1572, and the Borja clan had little desire to commemorate a reconciled sodomite who bargained away the only military order outside of royal control.

[51] Rafael Carrasco, *Inquisición y represión sexual en Valencia. Historia de los sodomitas (1565–1785)* (Barcelona: Laertes, 1986), 197.

[52] Suárez, *Historia del Maestre*, 419.

[53] Fernando Andrés Robres, "De la borrascosa vida de don Pedro Luis Galcerán de Borja y su historiografía: nuevas noticias," in *Francisco de Borja y su tiempo*, eds. Enrique García Hernán and María del Pilar Ryan (Valencia: Albatros, 2011), 711–32.

[54] Suárez, *Historia del Maestre*, 59–68. Unlike the Fernández de Córdoba, whose "family history of empire" emerged from generations of service across Iberia and North Africa, the Borja did not sustain their ties to Orán after Felipe Manuel's return. Liang, *Family and Empire*, 182.

[55] Suárez, *Historia del Maestre*, 258–60.

In a larger sense, Suárez's vision of the Spanish presence in the Mediterranean world no longer resonated with the Valencian nobility or the reading public. Suárez explicitly contrasts his "verified military history" to "convoluted, absurd stories" such as *Don Quijote de la Mancha*, but his noble governor was arguably as far behind the times as Cervantes' knight-errant.[56] In the days of Pedro Luis Galcerán, Spaniards routinely invoked the war against the Ottoman Empire: for veterans and widows seeking pensions, for coastal towns in need of money for fortifications, for merchants soliciting licenses to redeem captives, the threat of the "Armada Turquesca" bolstered their claims.[57] In addition to his chronicle, Suárez also wrote an urgent letter to Philip III arguing for the importance of Orán: "within all the realm of Spain there is no presidio today more worthy of your watchful eye and support," to secure the coast of Spain and to oppose the enemy Turk.[58] By the early 1600s, however, while Philip III continued to trumpet his repeated attacks on Algiers as Christian Crusades, these raids yielded few tangible results in an age now dominated by corsairs, not empires. This disjunction between royal rhetoric and the experiences of Spaniards on the ground culminated in the expulsion of the Moriscos from Spain in 1609–14. Philip III's declaration of their "heresy, apostasy and actions of divine and human treason" elided both the evident diversity of the Moriscos and the Habsburg regime's failure to convert and assimilate the majority of them.[59] Pedro Luis Galcerán did not live to see the tragedy of the Moriscos of Valencia, including former residents of both Gandía and the Order of Montesa, arriving at the port of Mazalquivir. Having received unclear instructions, the governor of Orán attempted to recruit *moros de paz* to guide the Moriscos into Tremecén, but chaos erupted on account of the overwhelming numbers and lack of food. Most of the Moriscos tried to make their way to Algiers, weathering attacks from Spanish soldiers and neighboring Muslims alike. Betrayed by the Valencian lords who had effectively tolerated their practice of Islam for a century, some of these exiles joined the corsairs preying upon Spanish ships and ports. Other

[56] Suárez, *Historia del Maestre*, 77.

[57] ARV, Real Cancillería, sig. 447, fol. 1r.

[58] "… de que no tiene oy dia plaça en todo el ambito de España, que mas le convenga zelar, y guardar." Diego Suárez Montañes, "Avisos importantes para la Magestad del Rey," rep. in Miguel de Bunes Ibarra, "Diego Suárez Montañés, cronista y testigo de la historia de Orán-Mazaquivir," in *Orán: Historia de la Corte Chica*, eds. Miguel de Bunes Ibarra and Beatriz Alonso Acero (Madrid: Ediciones Polifemo, 2001), 349.

[59] "Decree of Expulsion of the Moriscos," in *Early Modern Spain: A Documentary History*, ed. Jon Cowans (Philadelphia, PA: Pennsylvania University Press, 2003), 146.

Moriscos, in a response reminiscent of the assimilated communities of Alcalá de Xivert, secured entry into the city of Orán, and eventually gained the acceptance into society denied to them in Spain.[60] Pedro Luis Galcerán de Borja strove for clarity in defining Moriscos and Muslims, but his policies recognized variations within these communities. While corralling and containing the comparatively unruly Moriscos of Gandía, he empowered those of Alcalá de Xivert; as governor of Orán he courted and rewarded the *moros de paz* upon whom he depended, selling them captives and goods obtained in sorties against the Muslims who rejected his authority. The expulsion of 1609–14 swept aside these distinctions, ending the Morisco century and consigning 300,000 exiles to their fate in the Maghreb and beyond. The Muslim diaspora in Spain lived on only in those Moriscos who eluded the decree or returned quietly. The Spanish presidios in North Africa, a low priority by comparison to the Atlantic theater, persisted as a deterrent to piracy and a vestige of Jiménez de Cisneros' dreams of Crusade. In this new era, Diego Suárez Montañés found himself condemned to wander the land in search of a patron who wanted to remember the last Master of Montesa.

[60] Alonso, *Orán-Mazalquivir, 1589–1639*, 282–8, 311.

7

From Peasants to Slave Owners: Race, Class, and Gender in the Spanish Empire

Allyson M. Poska

In 1780, a young Astorgan couple, Pascuala García and her husband Lázaro Fernández, responded to an edict promulgated by the Spanish Crown calling for poor farming families to volunteer to go as colonists to the Río de la Plata. Although the edict made no mention of it, authorities actually planned to settle the peasants in new colonies along the sparsely inhabited Patagonian coast. In formulating the project, the Crown hoped to move some of the poorest inhabitants of northern Spain (primarily Galicia, Asturias, and northern Castile) where land was tight and poverty intense to a place where they could, in Enlightenment terms, be more "useful to the state."[1] Indeed, many northern peasants saw great opportunity in the Crown's offer. Between 1778 and 1783, Pascuala, Lázaro, and more than 1,900 others made their way to the port at La Coruña to make the transatlantic journey, lured by a contract with the Crown that promised land, cattle, homes, farming tools, seeds, and a daily allowance until they were able to subsist on their own. However, even before the last of the thirteen ships arrived in Montevideo, the project was scuttled due to an array of problems, including an outbreak of scurvy at the sites, problems supplying the outposts, and the costs of the uprisings in the Andes.

With the sudden abandonment of the Patagonia project, authorities decided to settle (or resettle) most of the peasant colonists in the rural Río de la Plata. They sent some families to a line of newly constructed forts outside of Buenos Aires to help secure the territory against Indian attacks, but the majority were transported to new towns in the Banda

[1] Archivo General de Indias (AGI), Buenos Aires 326, fol. 88. A version of this chapter appeared as a part of chapter eight of my recent book, *Gendered Crossings: Women and Migration in the Spanish Empire* (Albuquerque, NM: University of New Mexico Press, 2016).

Oriental outside of Montevideo, where they would reinforce the Spanish presence in lands only recently taken from the Portuguese.[2]

In their new homes, the peasant colonists experienced few changes as dramatic as the transition to living in a multiracial, slave-owning society. Although tens of thousands of peninsular emigrants underwent a similar experience, this process has gone largely unexplored by scholars. The historiography of slavery in the Río de la Plata is both rich and vast. We know much about the slave trade and the use of slave labor in the region, as well as the economic and social lives of urban slaves, yet we know almost nothing about the experience of becoming a slave owner.[3] Indeed, the scholarship tends to take for granted that Spaniards inherently understood the expectations of slave ownership. However, the poor but ambitious men and women who made up the majority of peninsular immigrants to the Americas arrived with no previous experience with slavery. This essay explores the first encounter of these peasant colonists with slavery and slave owning in order to better understand both the complicated relationship between race and class in the region and how readily peninsular Spaniards acclimated to the realities of race and labor in Spanish America.

The peasant colonists came to the Americas with at least a basic familiarity with racial difference. Certainly, from the fifteenth century forward, the legal and cultural emphasis on *limpieza de sangre*, or purity of blood, made Spaniards particularly conscious of such distinctions.[4]

[2] Dauril Alden, "The Undeclared War of 1773–1777: Climax of Luso-Spanish Platine Rivalry," *The Hispanic American Historical Review* 41(1) (February 1961): 55–74.

[3] Among others, Elena Studer, *La trata de negros en el Río de la Plata durante el siglo XVIII* (Buenos Aires: Universidad de Buenos Aires, 1958); George Reid Andrews, *The Afro-Argentines of Buenos Aires, 1800–1900* (Madison, WI: University of Wisconsin Press, 1980); the works of Carlos A. Mayo, including "Iglesia y esclavitud, la estancia colonial en el Río de la Plata," *Revista de Historia de América* 102 (July–December 1986): 91–102; Arturo A. Bentancur and Fernando Aparicio, *Amos y esclavos en el Río de la Plata* (Montevideo: Planeta, 2006); Miguel A. Rosal, *Africanos y afrodescendientes en el Río de la Plata, siglos XVIII–XIX* (Buenos Aires: Editorial Dunken, 2009); Lyman L. Johnson, *Workshop of Revolution: Plebian Buenos Aires and the Atlantic World, 1776–1810* (Durham, NC: Duke University Press, 2011); recent works by Alex Borucki, including "The Slave Trade to the Río de la Plata, 1777–1812: Trans-Imperial Networks and Atlantic Warfare," *Colonial Latin American Review* 20(1) (2011): 81–107; and the many projects connected to the UNESCO Ruta de esclavo, including www.unesco.org/new/es/office-in-montevideo/cultura/proyectos-especiales-ruta-del-esclavo/ and http://rutadelesclavocba.wordpress.com/.

[4] On *limpieza de sangre* laws in Spain, see María Elena Martínez, *Genealogical Fictions: Limpieza de Sangre, Religion and Gender in Colonial Mexico* (Stanford, CA: Stanford University Press, 2008), ch. 1.

Moreover, by the eighteenth century, Africans had lived in Spain for centuries, and since the early years of the conquest, Indians and mixed-race peoples had made the transatlantic journey to Spain from time to time. As a result, both Africans and Indians had both a physical place in early modern Spain and a place in the mentalities of early modern Spaniards.[5] However, unlike their counterparts in southern Spain who lived among small populations of both free and enslaved Africans, sightings of non-white people would have been a rarity in the north of the peninsula. From time to time, they might have encountered an African in the marketplaces of larger towns and almost definitely at northern Spain's major ports; yet, it is more likely that they either met or heard about Africans as a result of intrapeninsular migration. For instance, large numbers of Galegos migrated to Cadiz, Seville, and to neighboring Portugal, all of which had significant populations of Africans. In those diverse cities, their work as muleteers and porters would have brought them into regular contact with the cities' non-white populations.[6]

However, it is unlikely that any of the colonists had any direct experience with slavery prior to their arrival in Montevideo.[7] Rather, from the perspective of Enlightenment thinkers on the peninsula, these peasants were the peninsula's slaves, the workhorses of the rural economy, living in the worst of conditions, often barely able to survive. Of the farmers of Galicia, Asturias, and the mountains of León, Padre Benito Feijóo noted that "there are no people hungrier nor more naked."[8] Englishman James

[5] On slavery on the peninsula, see William D. Phillips, *Historia de la Esclavitud en España* (Madrid: Editorial Playor, 1990) and more recently, Rocío Periáñez Gómez, "La investigación sobre la esclavitud en España en la edad moderna," *Norba. Revista de Historia* 21 (2008): 275–82. On slavery in eighteenth-century Spain, see for instance, Pedro Parrilla Ortíz, *La esclavitud en Cádiz durante el siglo XVIII* (Cadiz: Diputación de Cádiz, 2001).

[6] On Galegos in Cadiz, María José de la Pascua Sánchez, "Los Gallegos en el Cádiz de la Carrera de Indias. Balance secular de un proceso migratorio (1682–1778)," in *Migraciones internas y médium-distance en la península Ibérica, 1500–1900*, eds. Antonio Eiras Roel and Ofelia Rey Castelao, vol. II (Santiago de Compostela: Xunta de Galicia, 1994), 851. On emigration to Portugal and Andalusia, see Antonio Meijide Pardo, "La Emigración gallega intrapeninsular en el siglo XVIII," in *Estudios de Historia Social de España tomo* IV, vol. II (Madrid: CSIC, 1960), ch. V.

[7] There is evidence that some noble families in Galicia owned slaves in the sixteenth and seventeenth centuries: Gonzalo Francisco Fernández Suárez, "De nación turca, de mediano cuerpo peli negra: La esclavitud entre la nobleza gallega durante los siglos XVI y XVII," in *Perspectivas sobre oriente y occidente. Actas del II Curso de Primavera*, eds. María Jesús Vázquez Lobeiras and Alexandre Veiga Rodríguez (Santiago de Compostela: Universidade de Santiago de Compostela, 2008), 55–66.

[8] Benito Jerónimo Feijóo y Montenegro, "Honra y provecho de agricultura," in *Teatro crítico universal* (1726) available at: www.cervantesvirtual.com.

Bruce (1730–94) called Galegos the "slaves of the rest of the provinces of Spain and Portugal"[9] and Fray Martín Sarmiento (1695–1771) referred to them as "the unhappy poor who are slaves of all and of the land."[10] These men were talking about a figurative kind of economic slavery with which the peasant colonists were all too familiar. In their new homes, the meaning of "slave" would change dramatically.

Whatever their previous experience, immediately upon arrival in Montevideo, the racial differences between the peninsula and the Americas would have been evident. Montevideo was a relatively small frontier town with fewer than 5,000 inhabitants; however, according to the 1780 census, people of color made up more than one quarter of the population. Moreover, the non-white population was divided nearly evenly between enslaved and free people – 70 Indians, 199 free Mulattos, 327 free blacks, and 625 slaves.[11] In the larger context, Montevideo was less racially diverse and less reliant on slaves than many other Spanish American cities. Having traveled extensively, Francisco Millau y Miraval noted that Montevideo had "a very small number of Blacks and Mulatos which were only found in some families."[12] Indeed, in 1780, fewer than half of the white households in the city owned a slave.

Nevertheless, the colonists would have interacted with Africans almost immediately upon arrival. African slaves and free black wage laborers may have helped unload the ships as they disembarked and moved their possessions into their temporary housing.[13] Along the banks of the Río de la Plata, the female colonists would have washed their clothes alongside the large numbers of women of color.[14] They would have purchased fruits

[9] Jacobo García Blanco-Cicerón, *Viajeros angloparlantes por la Galicia de la segunda mitad del siglo XVIII* (A Coruña: Fundacion Pedro Barrie de La Maza, 2006), 206.

[10] Meijide Pardo, "La emigración gallega intrapeninsular,"463–606, esp. 481.

[11] *Documentos para la historia Argentina, tomo XII, Territorio y población* (Buenos Aires: Compañía Sud-Americana de Billetes de Banco, 1919), 389.

[12] Francisco Millau, *Descripción de la provincia del Rio de la Plata (1772)* (Buenos Aires: Espasa-Calpe, 1947), 109.

[13] Unfortunately, it is difficult to know how common black labor was at the port. Arturo Ariel Bentancur focuses solely on the white population associated with the port in *El Puerto colonial de Montevideo (I). Guerras y apertura comercial: tres lustros de crecimiento económico (1791–1806)* (Montevideo: Universidad de la República, 1997).

[14] See for instance, the account of John Constance Davie, *Letters from Paraguay: Describing the settlements of Montevideo and Buenos Aires; the presidencies of Rioja Minor, Nombre de Dios, St. Mary and St. John & with the Manners, Customs, Religious Ceremonies & c. of the inhabitants Written during a Residence of Seventeen Months in that Country* (London: G. Robinson, Paternoster Row, 1805), 156. On slaves as laundresses, Marta B. Goldberg and Silvia C. Mallo, "La población Africana en Buenos Aires y su campaña. Formas de vida y de subsistencia (1750–1850)," *Temas de África y Asia* 2 (1993): 38.

and vegetables in the main plaza, presumably from women of all races, as was the case in city markets across Spanish America.[15]

From the beginning, these interactions with the non-white population exposed the complicated relationship between race, class, and labor in the late colonial Río de la Plata. Officially, Spanish law created a segregated society, prohibiting non-whites from engaging in an array of public activities, including holding public offices, receiving higher education, participating in many guilds and in some civic societies; however, the enforcement of those laws and their application in daily life was often uneven.[16] Moreover, among the lower classes, men of different races and statuses worked, drank, and even lived together with few differences in their material circumstances.[17] In Buenos Aires and Montevideo, both skilled and unskilled slaves had considerable physical freedom and often worked for wages. The racial and socioeconomic hierarchy could be so complicated that peninsular immigrants could be found working for creole and casta (mixed-race) masters and among artisans, and it was not unusual for black and casta master artisans to supervise white apprentices.[18]

In the countryside, a general scarcity of workers forced landowners to employ whoever was available and willing to work, regardless of race or status. Thus, the workforce of the region's *estancias* included gauchos (rural poor men, often white), peons (day laborers, usually single migrant men, often mixed race), and slaves.[19] When wages were high, gauchos took on more *estancia* work, but when wages were low, landowners had to rely more heavily on slaves. Indeed, because of their permanence and

[15] On the main plaza as Montevideo's central marketplace, see Emilio José Luque Azcona, *Ciudad y poder: La construcción material y simbólica del Montevideo colonial (1723–1810)* (Seville: CSIC, 2007), 122.

[16] The critical work on the complicated nature of race relations in colonial society is R. Douglas Cope, *The Limits of Racial Domination: Plebian Society in Colonial Mexico City, 1660–1720* (Madison, WI: University of Wisconsin Press, 1994). On Buenos Aires, see Johnson, *Workshop of Revolution*, 105.

[17] Johnson, *Workshop of Revolution*, 18; Arturo Bentancur, "Familia, esclavos y libertos. Algunas pautas sobre el relacionamiento de la colonia gallega con los afromontevideanos," *Anuario del Centro de Estudios Gallegos* (2003), 83.

[18] Johnson, *Workshop of Revolution*, 5, 39 and 73.

[19] On gauchos, see Richard W. Slatta, *Gauchos and the Vanishing Frontier* (Lincoln, NE: University of Nebraska Press, 1992). On the rural workforce, see Juan Carlos Garavaglia, *Pastores y labradores de Buenos Aires: una historia agraria de la campaña bonaerense 1700–1830* (Buenos Aires: Ediciones de la Flor, 1999), 89; Ricardo Salvatore and Jonathan C. Brown, "Trade and Proletarianization in Late Colonial Banda Oriental: Evidence from the Estancia de las Vacas, 1791–1805," *The Hispanic American Historical Review* 67(3) (August 1987): 431–59.

knowledge of the work, black slaves sometimes supervised mixed-race and white peons.[20]

In both the city and the countryside, women of different races and statuses also mingled daily. Poor white women would have done more or less the same work as their mixed-race and black (free and enslaved) counterparts, like cooking, laundering, and spinning. They also bartered or marketed the products of their household labor, including food and textiles.[21]

The terms of the colonists' contracts further complicated their position vis-à-vis the non-white population. Although as free whites their official status was higher than that of non-whites, they faced many of the same constraints as slaves. They could not marry without permission from royal authorities. They could not purchase property, as they had no official "fixed destination," and they had to move whenever and wherever the authorities decided. In fact, in terms of mobility, many slaves in the Río de la Plata had more freedom than the colonists and it would take more than twenty years for many of them to gain the same rights that free people of color took for granted.[22]

Once settled in their new homes in the Americas, the peasants' interactions with the region's black population diminished considerably. According to late colonial censuses, in the countryside outside of Buenos Aires and Montevideo, the population was overwhelmingly white. At the forts on the Buenos Aires frontier, there is only sparse evidence of black inhabitants, either free or enslaved. As late as 1815, Chascomús was 88 percent white and only 6 percent black.[23] Historian Carlos Mayo found that of the 66 *estancias* on the Pampa, 41 had a total of 164 slaves (90 men and 74 women), a median of only four slaves per *estancia*.[24] The situation was more or less the same around the fort and city of Luján.[25] The parish registers reveal only a few black residents. At Luján, on 15 April 1781, Dominga Meneses, a black slave, married Salvador Correa,

20 Salvatore, "Trade and Proletarianization," 440–1.

21 Susan Migden Socolow, "Women and Crime: Buenos Aires, 1757–97," *Journal of Latin American Studies* 12(1) (May 1980): 41.

22 The last of the colonists' contracts were liquidated in 1804.

23 Guillermo Banzato, *La expansión de la frontera bonaerense: posesión y propiedad de la tierra en Chascomús, Ranchos y Monte, 1780–1880* (Buenos Aires: Universidad Nacional de Quilmes Editorial, 2005), 91.

24 Carlos A. Mayo, *Estancia y sociedad en la pampa, (1740–1820)* (Buenos Aires: Editorial Biblos, 2004), 41; Goldberg and Mallo, "La población Africana en Buenos Aires y su campaña," 45–6.

25 Patricia A. Fogelman, "Población de color en una villa en la frontera bonaerense: Luján, 1771–1815," *Signos Históricos* 1(2) (December 1999): 14.

a free black.[26] In 1794, the priest at the fort at Ranchos buried the natural daughter of Rita Martínez, the slave of Manuel Martínez, a soldier.[27] With the exception of these few inhabitants, the absence of free or enslaved blacks and mulatos in the parish documentation is striking.

Most of the peasants were resettled in the Banda Oriental, in towns including Maldonado, Concepción de Minas, San José de Mayo, and San Juan Bautista. In these rural areas, the non-white population was larger than in the forts outside of Buenos Aires, but still miniscule in comparison with Montevideo or other parts of Spanish America. In 1781, Maldonado's population of 1,005 whites owned only seventy-eight slaves, twenty of whom were under ten years old. In town, in addition to the three Indians – two women and an elderly man – lived four free blacks, three men and one woman, and six mulatos. Just outside of Maldonado, there were twenty Indian households, but only six free black households, with a total of twenty-four inhabitants.[28] Even a decade later, Concepción de Minas, with a white population of 400, had only seventeen slaves.[29]

However, over the course of the 1780s and 1790s, the racial makeup of the Río de la Plata began to change as the Crown's desire to develop the region's economy led to a rapid expansion in the number of slaves. Before 1778, merchants imported only a few hundred slaves annually,[30] but after the expulsion of the Portuguese from Colonia de Sacramento in 1778, Montevideo emerged as the region's slave entrepôt,[31] and by 1782–3, that number had grown to more than 6,000 slaves per year. Over the next decade, changes in colonial economic policy, including allowing Spanish colonies to trade with foreign colonies, expanded the legal slave trade with Brazil, and led to the importation of another 13,500 slaves between 1792 and 1799. As a result, between 1778 and 1810, the slave population of Buenos Aires grew by more than 100 percent and the growth in Montevideo was even more intense: between 1791 and 1810, the slave population of the frontier town increased 486 percent.[32] Even

26 Nuestra Señora de Luján, *Libro de matrimonios*, fol. 93–4.

27 Nuestra Señora de Pilar, General Paz (Ranchos), *Libro de defunciones*, fol. 14.

28 Archivo General de la Nación Argentina (AGNA) sala IX, Maldonado 3.7.4, fol. 261ff.

29 Archivo General de la Nación Uruguay (AGNU), Padrón de Minas, 1791, libro 287ª.

30 Johnson, *Workshop of Revolution*, 36.

31 Borucki, "The Slave Trade to the Río de la Plata," 83. Portuguese trading networks were critical to this expansion, see Fabrício Prado, "In the Shadows of Empires: Trans-imperial Networks and Colonial Identity in Bourbon Rio de la Plata (c. 1750–c. 1813)," PhD dissertation (Emory University, 2009).

32 Borucki, "The Slave Trade," 84, 88, 85.

more slaves entered the region illegally from both Brazil and the interior of the viceroyalty.

At first, the colonists' ability to own slaves was limited by the fact that they were tied by contract to the Crown and lived "without a fixed destination." Moreover, at least at the beginning, they were some of the poorest people in the region, almost entirely reliant on the government for their subsistence. Thus, not surprisingly, in Maldonado's 1782 census, none of the colonists owned slaves, although a few of their non-colonist neighbors did.[33] In this way, they were much like other recently arrived immigrants, who were less likely to own slaves than longer-term residents.[34] However, once they were settled, even though the authorities did not provide many of them with their daily allowance for most of the 1780s and 1790s, the fact that the government gave them houses, land, seeds, and farm implements allowed some of them to accumulate enough savings to enter the world of slave ownership.

For poor people who had only recently come into their own land in the Americas, the decision to become slave owners indicates a rapid assimilation into colonial ideas about labor and status. In northern Spain, land was scarce and labor was abundant. In the Río de la Plata, the situation was the opposite – for the colonists, the land was free, but the labor required some capital investment either by paying wage laborers or by purchasing slaves. Slaves were a means to accumulate wealth and slave owning was a sign of upward mobility. In fact, in Buenos Aires, most slaves were owned by people "of modest means" who lived almost exclusively off the income produced by their slaves.[35]

Those peasants who achieved some early success based on their entrepreneurial or artisanal skills were the first to become slave owners, often within a decade of their arrival. Francisco Martínez, a farmer from Astorga, settled in Maldonado in 1781 along with his wife and three children.[36] In their new home, he farmed and worked as a carter. Only seven years later, he purchased a twenty-one-year-old slave named Ventura.[37] Francisco Osorio, the apothecary in San Carlos, also purchased at least

33 AGNA, Sala IX, Maldonado 3.7.4, fol. 443–61.

34 Susan Migden Socolow, *The Merchants of Buenos Aires, 1778–1810: Family and Commerce* (Cambridge: Cambridge University Press, 1978), 77.

35 Johnson, *Workshop of Revolution*, 39 and 203.

36 Juan Alejandro Apolant, *Operativo Patagonia: Historia de la mayor aportación demográfica masiva a la Banda Oriental* (Montevideo: Ediciones El Galeón, 1970), 311.

37 María A. Díaz de Guerra, *Documentación relativa a esclavos en el departamento de Maldonado, siglos XVIII y XIX* (Montevideo: Imprenta Cooperativa Gaboto, 1983), 20.

one female slave in the late 1780s.[38] By 1803, he owned eight slaves, three men and five women.[39] Market forces worked in favor of the peasants becoming slave owners. Just as many of the colonists were gaining some economic traction, the number of slaves in the region increased.[40] As the supply increased, prices declined, making slaves become affordable to more colonists with each passing year. In 1800, Francisco Aparicio purchased a sixteen-year-old female slave named Theresa for 250 *pesos*. That same year, Juan Antonio de la Fuente bought a male slave in his late twenties for slightly more money, 283 *pesos corrientes* and José Rodríguez traded a middle-aged slave named Manuel for 300 head of cattle.[41]

By the 1790s, a number of farming colonists could also afford slaves, including Matías Prieto, Cayetano Ignacio Huertas, and Francisco Pérez Brañas.[42] Isabel Herrera (also Ferrera) and her husband Matías Alonso were very successful farmers, eventually coming to own a large *estancia* which they worked with the help of two slaves.[43] Pascuala García, the peasant woman with whom I began this essay, was widowed in 1788 and in 1790 married a much younger man from outside of the colonist cohort. They owned four slaves in 1803.[44]

Many of the colonists who settled close to Montevideo were artisans and, presumably, their slaves were at least semi-skilled laborers who aided the men with their work. Vicenta Fortuna and her husband, Domingo Nuñez, who had left their homes in Galicia where he was a butcher and farmer, settled in San Carlos in 1781. When Domingo died in 1811, he was a carpenter who owned four slaves. As Domingo suffered from some illness that curtailed his ability to work, his slaves may have been a major source of income. In 1799, despite his physical problems, the royal official who supervised the colonists reported that the family was doing fine.[45]

[38] Her death is recorded in the death register of San Fernando de Maldonado, *Libro de defunciones*, 1789.

[39] A spreadsheet version of the 1803 padrón used to be available (but currently is not) at the website *Pueblos y números del Río de la Plata, 1760–1860*, http://pueblosynumeros.fcs.edu.uy/.

[40] Johnson, *Workshop of Revolution*, 40.

[41] Díaz de Guerra, *Documentación relativa a esclavos*, 21 and 25.

[42] For Prieto, Brenda Pagola, "... *Y ellos le dieron vida" San Carlos: fundadores y vecinos* (Montevideo: Arca, 2001), 534; for Huertas and Pérez Brañas, Apolant, *Operativo Patagonia*, 245 and 253. The slaves also appear in San Carlos Borromeo, *Libro de defunciones, 1765–1823*.

[43] Pagola, "... *Y ellos le dieron vida*," 53.

[44] From the 1803 padrón at *Pueblos y números del Río de la Plata, 1760–1860*.

[45] Brenda Pagola, "... *Y ellos le dieron vida*", 426. Apolant, *Operativo Patagonia*, 243.

Slave owning was not a solely masculine endeavor. In 1790, the parish registers note the death of the child of two black slaves owned by Angela García.[46] Angela and her first husband, Antonio Cantera from Palencia, had enlisted along with their three young children. A farmer in Spain, he also took up farming in Maldonado until he died in 1788. It is unclear whether they already owned the slaves and she inherited them or whether Angela purchased them to help work the farm after his death. Similarly, Antonia Menéndez arrived in the Río de la Plata along with her parents at age twelve. She married twice, but by 1836, she was widowed and owned one male and four female slaves.[47]

Not surprisingly, slave owning became even more common among the younger generation of colonists and was definitely a sign of their commercial success and integration into the colonial economy. According to the 1803 census, Francisco Riestra, who had come to the Río de la Plata with his parents as a colonist at the age of thirteen, later lived in Montevideo with his wife, son, and one slave. At the same time, Cayetano Otero, who had arrived at age fifteen, was living in Montevideo with his wife and children. He owned a *pulpería* (a corner store) and three slaves, and he had three peons living with and working for him.[48]

Many of the peasant colonists would have understood their relationships with their slaves in terms of their own experiences as poor workers. Before coming to the Río de la Plata, some had worked as temporary wage laborers, traveling long distances to engage in hard labor during the harvest season. Thus, it is likely that, at least initially, these poor settlers utilized slave labor to augment their own work on their *ranchos* with little differentiation between the work completed by themselves and their families and that of their slaves. The employment of short-term laborers for a variety of tasks was as common on the peninsula as it was in the Americas, and they would have used their slaves just as farmers on the peninsula had employed peasants like themselves and their neighbors. Those who could not afford to purchase slaves of their own hired slaves owned by others, as well as peons and other temporary workers of all races, when they could afford them.

Once they decided to purchase or hire slave labor, the colonists had to learn how to treat their slaves. Scholars have engaged in considerable

[46] San Fernando de Maldonado, *Libro de defunciones*. Apolant, *Operativo Patagonia*, 271.

[47] Apolant, *Operativo Patagonia*, 353, and Luis Fernando Loriente Brian and Luis Alberto Pintos Anza, eds., *Libro primero de matrimonios de la iglesia de San Juan Bautista 1782–1845* (Montevideo: Instituto de Estudios Genealógicos del Uruguay, 2003), 32.

[48] From the 1803 padrón at *Pueblos y números del Río de la Plata, 1760–1860*.

debate over treatment of slaves in the Río de la Plata. Some have asserted that unlike plantation slavery, the more extensive mobility of slaves and the fact that masters and slaves often lived together in similar conditions led to a more benign form of slavery in the region; based on archival evidence about maltreatment of slaves, others have argued that violence was inherent in all slave regimes.[49] Without a doubt, the full spectrum of slave owners, from the least to the most violent, existed in the late colonial Río de la Plata. The colonists received information about the expectations of slave treatment from a variety of sources, including the examples of slave-owning neighbors and authorities and Crown policy. Moreover, their personal sense of duty and piety also influenced their actions as slave owners.

As they interacted with their new neighbors, the colonists had ample opportunity to learn about the expectations of slavery and slave ownership. As was typical, some of the shopkeepers in Maldonado owned slaves, including twenty-three-year-old Domingo Almeda who owned a nineteen-year-old male and a thirty-six-year-old female slave.[50] The butcher, Pedro José Pérez, owned a female slave and two children. Wholesale merchant Don Miguel Olavarrieta owned five slaves. In these semi-urban settings, male slaves were often used to haul goods or trained in the owner's trade. Others were hired out as day laborers. Female slaves usually engaged in domestic work, including washing, cooking, ironing, and sewing. As they mingled with these artisans and shopkeepers, the colonists would have witnessed a variety of interactions between masters and slaves and become familiar with the community's expectations about those relationships.

The colonists may have also looked to the prominent men in their communities for indications of how to interact with their slaves. Around Maldonado, both Rafael Pérez del Puerto, the royal official in charge of the colonists, and the parish priest of San Carlos, Manuel Amenedo y Montenegro, owned slaves. Although the extent of his slave ownership is unclear, Pérez del Puerto was actively involved in the importation of slaves into the area and some of his slaves appear in parish records. He allowed one slave couple to marry in 1795 and others to act as witnesses to slave marriages in the parish.[51] Just as importantly, as an official the

[49] On this debate, see Bentancur and Aparicio, *Amos y esclavos* and Mario Rufer, *Historias negadas: esclavitud, violencia y relaciones de poder en Córdoba a fines del siglo XVIII* (Córdoba: Ferreyra Editor, 2005), 50ff and ch. 4.

[50] On shopkeepers as slave owners, see Socolow, *The Merchants of Buenos Aires 1778–1810*, 77.

[51] Díaz de Guerra, *Documentación relativa a esclavos*, 17–18.

Royal Treasury, he oversaw a wide array of economic activities in the area, including the sale of slaves.

Expectations about the use of slave labor in the colonists' communities became most evident beginning in the early 1790s as Padre Amenedo employed slaves to help with the construction of the parish church in San Carlos. Among other tasks, his slaves Juan, Joaquín, and Pedro opened wells to catch rainwater.[52] In 1799, Amenedo complained that he needed additional workers and wanted to purchase up to seven slaves for the purpose, so that he would not have to pay wage laborers.[53] In response, Pérez del Puerto petitioned the viceroy for permission to import from Brazil "various blacks" to Maldonado to work on the church.[54] During the project, their work went uncompensated; however, Amenedo did not fail to pay them because they were slaves, but because he was short on funds for the construction. At the end of the project in 1801, Padre Amenedo paid the slaves their back wages. Other slave owners contributed their slaves' labor to the church. One slave, Antonio Rodríguez, was provided by his master to work for six days for charity and then was paid four *reales* per day as a wage laborer. Still other slaves were lent to the construction of the church as payment for a debt owed by their master.[55] Through the ways that they used and managed slave labor, Padre Amenedo and Pérez del Puerto provided critical guidance about the expectations of slave ownership to the neophyte peasant slave owners.

In addition, the monarchy provided clear instructions about slavery and the proper functioning of a slave society. Just as the first colonists began purchasing slaves, the Crown issued an edict on slave treatment, *His Majesty's Royal Decree on the Education, Treatment, and Work of Slaves* (1789).[56] Twelve copies were distributed in Buenos Aires for the Viceroyalty.[57] There was little new in the *Decree*. For the most part, it

[52] Luis Astigarraga, "Un cura de 1800 (Apéndice 1)," *Boletín Histórico del Ejército* 259–262 (1978): 180. On the clergy and slave labor, see Carlos A. Mayo, *Los Betlemitas en Buenos Aires: convento, economía y sociedad (1748–1822)* (Sevilla: Publicaciones de la Excma. Diputación Provincial de Seville, 1991).

[53] Astigarraga, "Un cura de 1800 (Apéndice 1)," 190.

[54] Díaz de Guerra, *Documentación relativa a esclavos*, 24.

[55] Astigarraga, "Un cura de 1800 (Apéndice 1)," 185 and 181.

[56] The text of the *Real Cédula de su Magestad sobre la educación, trato y ocupaciones de los esclavos* is available at: www.afehc-historia-centroamericana.org/index.php?action=fi_aff&id=1109.

[57] See Manuel Lucena Salmoral, "La instrucción sobre educación, trato y ocupaciones de los esclavos de 1789: una prueba del poder de los amos de esclavos frente a la debilidad de la Corona española," *Estudios de historia social y económica de América* 13 (1996): 155–78, available at: http://dspace.uah.es/dspace/handle/10017/5928, and María Eugenia

was a compilation of previously issued royal edicts with few additions, promulgated in response to the French *Code Noir*. Noting the failure of many owners to follow earlier laws, the preamble admonished owners to comply with the "principles and rules dictated by religion, humanity, and the good of the State, compatible with slavery and public tranquility." Owners were expected to baptize their slaves when necessary and teach them Christian doctrine. They were not to work, either for themselves or for their masters, on holy days except during harvest time, and every day at the end of work, masters and or overseers were to pray the Rosary with them. The *Decree* asserted that the "primary" occupation of slaves was to be agriculture, and that they were to only engage in work appropriate to their ages and sex. Owners were to house, dress, and feed them appropriately. In order to prevent fornication, owners were supposed to encourage their slaves to marry. The last seven sections of the *Decree* attempted to regulate the use of punishment, especially violent punishment, by owners.

Authorities did not make any attempt to enforce the code, and faced with opposition from slave owners across Spanish America, it was suspended in 1795, but never revoked.[58] As a result, scholars agree that in general, the *Decree* had little impact on the treatment of slaves in the region. However, unlike other residents of the Río de la Plata, the colonists lived under the direct supervision of royal authorities and it is likely that they would have known about the *Decree* and that it would have had some influence on their behavior towards their slaves. As with all large groups of people, there probably was considerable diversity in individual colonists' sense of obligation to comply with legal norms and religious ideals. Some would have willingly fulfilled the expectations of the *Decree*, while others circumvented or ignored the demands made upon them by the Crown.

Parish records reveal that at least some of the colonists made efforts to ensure the spiritual well-being of their slaves from birth to death. Of course, it is impossible to know how many slave children were not

Chaves Maldonado, "Paternalismo, iluminismo y libertad. La vigencia de la Instrucción esclavista de 1789 y su impacto en la sociedad colonial," *Historia y sociedad* 21 (July–December 2011): 61–93.

[58] On the opposition from slave owners in Cuba, see Matt D. Childs, *The 1812 Aponte Rebellion in Cuba and the Struggle against Atlantic Slavery* (Chapel Hill, NC: University of North Carolina Press, 2006), 36–7; in Louisiana, Jennifer M. Spear, *Race, Sex and Social Order in Early New Orleans* (Baltimore, MD: Johns Hopkins University Press, 2009), 103ff; in the Río de la Plata, Bentancur and Aparicio, *Amos y esclavos*, 18.

baptized, but at least some colonists fulfilled their spiritual duties as masters. Matías Alonso, a colonist from Astorga, had Manuela Isabel, the child of his slaves, Joaquin and Juana, baptized. Similarly, José Borrego had two teenage male slaves *de nación Mina* baptized in 1804.[59] By "encouraging" their slaves to marry, colonists could fulfill both their duty to their king as well as their spiritual obligations. According to Castilian laws dating from the Middle Ages, all Christians, whether free or enslaved, had the right to marry, and they could marry either free or enslaved partners, although marriage to a free person did not change the enslaved person's status. Slaves could even marry when their owners opposed the marriage.[60] The 1789 decree reiterated the Catholic emphasis on marriage as a means to control sexual activity among slaves. It urged slave owners to encourage their slaves to marry, even if it meant purchasing the spouse or potential spouse from another owner. As a result, either out of personal piety or in response to the *Decree*, colonist Domingo de Castro, who had arrived from Asturias at the age of twenty-two and settled in Minas in 1783, had the parish priest administer the sacrament of marriage when his slave Antonio married another of his slaves, Juana. Among the witnesses were a slave of Doña Barbara Frontan and Isabel María, a free black woman.[61] These married slave couples in turn produced legitimate, although enslaved, children. In February 1799, colonist Santiago Fernández had the legitimate child of his two married slaves baptized.[62]

Other colonists did not seem to take as much care to ensure that their slaves' sexual activity took place within the confines of marriage. Tomás Corbo's slave Juana had at least two illegitimate children, one in 1797 and another in 1802. The first child was baptized by colonist José Borrego. Borrego also appears in the baptismal record of another illegitimate child of a colonist's slave in 1805.[63]

[59] San Carlos Borromeo, *Libro de bautismos.*

[60] The provisions on slave marriage appear in Partida 4 title V in *Las Siete Partidas*, ed. Robert I. Burns, S.J. and trans. Samuel Parsons Scott, vol. 4 *Family, Commerce, and the Sea: Worlds of Women and Merchants* (Philadelphia, PA: University of Pennsylvania Press, 2001), 901. On the multiple meanings of slave marriage in Mexico during an earlier period, see Herman L. Bennett, *Africans in Colonial Mexico: Absolutism, Christianity, and Afro-Creole Consciousness, 1574–1640* (Bloomington, IN: Indiana University Press, 2003), ch. 4 and 128–32.

[61] Catedral de la Inmaculada, Concepción de Minas, *Libro de matrimonios, 1784–1805*, fol. 15.

[62] San Fernando Maldonado, *Libro de bautismos* and Apolant, *Operativo Patagonia*, 301.

[63] San Carlos Borromeo, *Libro de bautismos.*

Of course, many factors militated against the colonists' ability and/or desire to regulate their slaves' sexual activity. Most importantly, these peasants came from a peninsular culture that tolerated non-marital sex and came to slave owning with very little, if any, religious education.[64] Thus, even if they understood the spiritual implications of non-marital sex (which is unlikely), they were likely unable to articulate those concerns to their slaves. Moreover, the relative ease with which slaves moved around the region meant that they were often outside of their owners' supervision. And finally, the fact that many other slave owners did not "encourage" marriage among their slaves must have served as permission for some of the colonists to ignore both the royal and the religious admonishments.

Death provided the colonists with one last opportunity to see to the spiritual welfare of their slaves. Early on, Domingo de Castro made a somewhat half-hearted attempt. When one of his slaves (unnamed) died, Domingo gave him a Catholic burial and had his death registered in the parish register, indicating that he had been baptized, although the priest noted that he was *bozal*, recently arrived from Africa. The priest also noted that the slave had not received last rites before his death, as his master had not let the priest know in time to administer the sacrament.[65] However, with time, more colonists, like Miguel Malo in San Carlos, ensured that their slaves received both last rites and Catholic burial.[66]

It is difficult to know the degree to which the presence of female slaves altered sexual relations within the colonists' communities. As we have seen, some of the colonists' slaves bore illegitimate children, but none of the parish registers name the non-slave fathers of illegitimate children or even designate the race of the child. Thus, all of the master/slave relationships within the colonizer cohort existed outside of the extant documentary record. In fact, there is documentary evidence of only two male colonists engaging in interracial relationships, one who married a free black woman and another who married two different Indian women.[67]

64 Allyson M. Poska, *Women and Authority in Early Modern Spain: The Peasants of Galicia* (Oxford: Oxford University Press, 2005) and *Regulating the People: The Catholic Reformation in Seventeenth-century Spain* (Leiden: Brill, 1998).

65 Catedral de la Inmaculada Concepción de Minas, *Libro de defunciones, 1783–1805*, fol. 18.

66 San Carlos Borromeo, *Libro de defunciones*, 1771–1821, 14 marzo 1797.

67 José Rodríguez Bobes and José Ventura Echavarría, Apolant, *Operativo Patagonia*, 338 and 295.

Slaves affected the lives of these colonizing families in less tangible ways, as they added to the rich cultural diversity of the frontier. Most of the slaves were from West Africa, by way of Brazil.[68] Tomás Corbo's slave Juana was from Angola.[69] José Borrego owned a Mina slave, from the Gold Coast or Bight of Benin.[70] African slaves brought their own foodways and ideas about household management to their owners' homes. In many cases, the slaves would have known more than their newly arrived owners about agriculture and livestock herding in the region.

The most striking effect of slave ownership on the peasant colonists was on the gendered division of labor, as slaves frequently took on work that peasant women traditionally engaged in on the peninsula. In northern Spain, most peasants worked small plots, sometimes supplemented by some small-scale animal husbandry. On those peasant farms, men and women worked side by side in the fields with very little evidence of any particular gendered division of labor.[71] Indeed, both Spanish Enlightenment thinkers and foreign travelers commented on northern peasant women's agricultural labor. Referring to women from Cantabria and other parts of northern Spain, royal minister Pedro Rodríguez Campomanes, noted that the women cared for livestock, guided the carts, harvested and gleaned, and even worked the land when men were absent.[72] Englishman Alexander Jardine was struck by the fact that "women in Galicia also drive the carts, steer the plow and spread manure, frequently by hand, instead of using a pitchfork." Moreover, in the parts of northern Spain that were marked by high rates of male migration, women did all or much of the agricultural labor and animal husbandry without male assistance.[73] Once they settled in the rural Río de la Plata, the colonists quickly came to understand that slaves, not women, were expected to do much of that work. So, for instance, by 1799, Francisco Aparicio and his wife Teodora de Plada raised some livestock on their small *estancia*.[74]

68 For more on the place of origin of slaves in the region, see Borucki, "The Slave Trade to the Río de la Plata," 93.

69 San Carlos Borromeo, *Libro de bautismos*, 1797.

70 San Carlos Borromeo, *Libro de bautismos*, 1804.

71 Serrana Mercedes Rial García, "Trabajo femenino y economía de subsistencia: el ejemplo de la Galicia moderna," *Manuscrits* 27 (2009): 86. On Asturias, Patricia Suárez Álvarez and Alberto Morán Corte, "Más allá del telar: el mundo del trabajo femenino en la Asturias del siglo XVIII," *El Futuro del pasado: revista electrónica de historia* (2011): 483–98.

72 Pedro Rodríguez Campomanes, *Discurso sobre la educación popular de los artesanos y su fomento* (1775) sección XVII, available at: www.cervantesvirtual.com.

73 Rial García, "Trabajo femenino," 86, 87.

74 Apolant, *Operativo Patagonia*, 273.

Once they purchased slaves, it is likely that they tended the animals, freeing Teodora for other tasks. Moreover, on the peninsula, it frequently fell to women to supplement family incomes by engaging in incoming generating activities such as spinning, selling surplus produce at market, or earning extra wages as harvesters, hoers, and weeders.[75] As noted earlier, many slaves in the Río de la Plata supplemented their owners' incomes by working as day laborers engaging in exactly those same tasks.

Within the home, the presence of female slaves would have significantly altered the work tasks of the peasant women.[76] Peasant women were accustomed to moving seamlessly between what Serrana Rial García has called the "floating frontier" between indoor and outdoor labor.[77] With slaves to wash the laundry or cook or help tend the garden, they could engage in less rigorous work. Colonial society viewed indoor work as more appropriate for women, and the presence of slaves may have allowed the colonists to restrict their work accordingly. In general, slave ownership gave peasant women some choices – they could decide what work they preferred and assign the less desirable tasks to slaves. Just as importantly, slave ownership gave these poor, peninsular women some status in their new communities and clearly differentiated them from other lower class women.

The project to colonize Patagonia brought around 1,900 northern Spanish peasants to the Americas late in the colonial period. They arrived in the Río de la Plata with little previous experience with racial difference and none with slavery. However, they quickly came to understand the complicated relationship between race, class, and slavery in the region. Over time, a number of colonists took the journey from peasants to slave owners, an experience that transformed their place in the colonial labor hierarchy. Female colonists were most affected by this change, as slaves altered the gendered division of labor in their families and gave them more choices about the type of work that they engaged in. Despite their poverty and dependence on the Crown, with their entry into slave-owning society, these peasant colonists would no longer be "slaves to all."

[75] Rial García, "Trabajo femenino," 85, 88. On women harvesting for wages, see Meijide Pardo, "La emigración gallega intrapeninsular," 532ff.

[76] On women and work on the frontier, see Socolow, "Women of the Buenos Aires Frontier," 76–8; Mayo, *Estancia y sociedad*, ch. 10.

[77] Rial García, "Trabajo feminino," 86.

PART III

LAW AND LETTERS

8

On Early Modern Science in Spain

María M. Portuondo

History, even the history of science, unfolds in a capricious fashion and only in retrospect can any rationality be discerned in its unfolding.[1]

In Miguel de Cervantes' *Coloquio de los Perros*, the dog Berganza recalls overhearing the conversations of a poet, an alchemist, a mathematician, and an *arbitrista* lying next to each other in a hospital. We pick up the story with the alchemist, who, having heard the poet's complaints, chimes in with his own:

"Personally," replied the alchemist, "I don't know much about poetry; and so I can't properly appreciate the misfortune you are complaining of. But, however great, it could not equal mine, which consists in the fact that, since I lack instruments, or a patron to support me and make available the things required by the science of alchemy, I am not, as I should be, rolling in riches and wealthier than all the Midases, Crassuses and Croesuses who ever lived." "Have you, Mr. Alchemist," said the mathematician at this juncture, "carried out the experiment of extracting silver from other metals?"

"Up to now," replied the alchemist, "I haven't managed to extract it; but I do know that it can be extracted, and I only need a couple of months to get the philosopher's stone, with which it is possible to make gold and silver out of stones." "You have certainly been exaggerating your misfortunes," said the mathematician at this point, "for after all, one of you has a book to dedicate to someone and the other is very near to extracting the philosopher's stone; but what about my misfortune, all alone and with no one to turn to? I've been looking for the fixed point for twenty-two years, and within an ace of finding it; and when I think I've

[1] William A. Wallace, "Domingo de Soto and the Iberian Roots of Galileo's Science," in *Hispanic Philosophy in the Age of Discovery*, ed. Kevin White (Washington, DC: Catholic University of America Press, 1997), 113–29 (p. 115).

found it and that it can't escape me come what may, all of a sudden I find I'm so far away from it that I'm absolutely amazed ... At times I think I'm on the point of arriving at the truth, and the next minute I'm so far from it that I'm climbing up the hill which I've just come down, with the stone of my labours on my back, just like Sisyphus.[2]

For Cervantes, the alchemist and mathematician were comical characters, hopelessly entangled in what he hinted were futile problems. For the alchemist, only the lack of instruments and a generous patron stood between him and the coveted philosopher's stone, while the mathematician believed tireless work would ultimately solve his theoretical problem. But Cervantes' readers knew that for thousands of years alchemists had unsuccessfully tried to transmute materials into gold and that since the discovery of the New World the best cosmographers in Europe – among them the royal cosmographers at the Casa de la Contratación – had been on a fruitless search for a way of determining longitude at sea, or finding the *punto fijo*: problems that in fact would not be solved until well into the eighteenth century. Writing in 1613 Cervantes portrayed these as chimerical projects ripe for mockery. And yet his nameless characters are also emblematic of something else: a deep-seated preoccupation with unraveling the natural world's most intractable secrets. One could argue that by choosing the alchemist and the mathematician Cervantes was also acknowledging the uncertain nature of the early modern scientific enterprise.

This chapter sketches the intellectual context of Cervantes' alchemist and mathematician. After surveying some historiographical questions, it considers the recent contributions of several analytical approaches to the history of science and explains how these have enriched our perception of the early modern scientific enterprise in Spain. Studies now include a broad spectrum of scientific practitioners (such as our alchemist), recognize the importance of practices considered dead-ends by previous historiographies (such as the "search" for the *punto fijo*), and emphasize the diversity of approaches and alternatives to scientific questions of the early modern scientific enterprise.[3] This chapter then surveys the

[2] Miguel de Cervantes Saavedra, "The Dog's Colloquy," Penguin Books, http://gateway.proquest.com.proxy3.library.jhu.edu/openurl?ctx_ver=Z39.88-2003&xri:pqil:res_ver=0.2&res_id=xri:lion-us&rft_id=xri:lion:ft:pr:Z201578918:2.

[3] As Richard L. Kagan often reminded his students, speaking of a single "Spain" during the early modern era can be inaccurate. Historians should not take for granted the questions: what was Spain; who was a Spaniard? Nonetheless, I find it a useful analytical category in the history of ideas. As Jorge J. E. García suggested, this field of investigation in Spain entails "the philosophy produced by a group of philosophers who span

panorama of early modern Spanish science, inquiring principally about the questions and motivations behind the formulation of alternative natural philosophies and their praxis. Finally, it draws on this panorama to argue that Spain's scientific enterprise wrestled with fundamentally the same philosophical concerns that drove the development of science elsewhere in Europe, but, by having tested the bounds of empiricism early on and having identified its shortcomings, sought – perhaps longer than most – a cohesive and comprehensive natural philosophy through which to interpret the natural world.

HISTORIOGRAPHICAL TRENDS IN THE HISTORY OF SPANISH SCIENCE

One of the principal developments in the field of history of science has been a shift in our current understanding of what constituted "science" during the early modern era. As a whole, the field now acknowledges the rich and variegated aspect of natural philosophy and its associated scientific practices without singling out for study only those that contributed to the rise of modern science. Long gone is the progressivism where the benchmark of scientific merit was taken to be whether a country contributed to the developments that culminated in the new experimental science of the seventeenth century – the New Science – the cornerstone of a historical narrative known as the Scientific Revolution. What the field finds interesting about the early modern scientific enterprise is the very diversity of responses to a broadly acknowledged "crisis of doubt" about explanations of the natural world that gradually overtook early modern European thought. This crisis was characterized – in broad strokes – by an increasing distrust of Aristotelian or Scholastic natural philosophy

diverse political, territorial, linguistic, and ethnic and racial boundaries, but who are closely tied historically," in García, "Hispanic Philosophy: Its Beginning and Golden Age," in *Hispanic Philosophy in the Age of Discovery*, ed. Kevin White (Washington, DC: Catholic University of America Press, 1997), 3–27 (p. 9). It is also problematic – if not anachronistic – to speak of "science" during the early modern period. Although the term "natural philosophy" was widely used at the time, it generally referred to theoretical postulates about nature. Yet, many activities carried out by early modern historical actors were directed towards understanding the workings of the natural world, making them therefore "scientific," but not "philosophical" in our modern understanding of the words. These could include writing detailed descriptions of plants and animals, carrying out trials with different chemical substances to understand how they interacted, or rolling balls down inclined planes. I prefer to call these activities "scientific practices," while using "scientific enterprise" to refer to both natural philosophy and scientific practices.

and a growing skepticism about the ability of this and other ancient philosophical systems, or even sensual perception, to explain the natural world and humanity's place in it.[4] Throughout this chapter I will refer to this with the shorthand "crisis of knowledge."

Surveys of the history of science in Spain have long noted the contrast between the vitality of the scientific enterprise in the sixteenth and seventeenth centuries.[5] The cosmographical and medical arts flourished during the earlier century, while during the one that followed it seemed as if the latest developments in astronomy and mechanics simply passed the peninsula by, so much so that, by the beginning of the eighteenth century, critics charged that Spain had been left out of the European Republic of Letters that discussed experiments, the tenets of the New Science, and the natural laws that resulted from them. By the 1740s Fray Benito Jerónimo Feijóo (1676–1764) had been remarking for years about the slow adoption of the Copernican thesis and Newtonian physics in Spain and the marginal status of Spain's scientific enterprise.[6] During the latter half of the twentieth century and with the invention of the historiographical concept of the Scientific Revolution, the same concerns were voiced: why did Spain not play a role in the Scientific Revolution?

The persistence of the flourish-then-decline narrative was also due largely to how nicely it fitted within the broader historiography of the seventeenth-century decline of Spain, a strand of which maintains that economic decline and the rise of religious orthodoxy hampered scientific activity. Historians have little difficulty finding many examples to illustrate both periods. The Casa de la Contratación (House of Trade in Seville) is a case in point. Whereas during the sixteenth century it was the locus of innovative solutions to cosmographical and cartographical problems, by the mid-seventeenth century these efforts were largely abandoned and were portrayed as having gradually waned in proportion to the severe decline in silver and other remittances from America. Furthermore, this narrative hints at the disproportionately negative influence of recalcitrant

[4] As general introduction to area, see Brian P. Copenhaver and Charles B. Schmitt, *Renaissance Philosophy* (Oxford: Oxford University Press, 1992), 22–4, 239–60. Richard H. Popkin frames it as a crisis that originated with the Reformation and quickly overtook philosophy, in Popkin, *The History of Scepticism: From Savonarola to Bayle* (Oxford: Oxford University Press, 2003), 15–16.

[5] José María López Piñero, *Ciencia y técnica en la sociedad española de los siglos XVI y XVII* (Barcelona: Labor, 1979), 372–3.

[6] Benito Jerónimo Feijóo, "Carta XXIII – Sobre los Sistemas Filosóficos (1745)," in "Cartas eruditas y curiosas," *Edición digital de las Obras de Feijoo*, ed. Biblioteca Filosofía en español and Fundación Gustavo Bueno – Oviedo (Filosofia.org).

religious authorities who, stoked by Counter-Reformationist zeal, kept Spanish natural philosophers from participating fully in the intellectual renewal associated with the New Science.[7] For the premier historian of Spanish science of the late twentieth century, José María López Piñero, the rescuers were the *novatores*, mostly physician advocates of the experimental science who began the difficult and gradual introduction of the New Science to Spain during the late seventeenth century and who found their most vocal advocate – if not a practitioner – in Feijóo.

As an explanation, the narrative outlined above proved very convincing, yet it is flawed. While it correctly identified linkages between a reduction in scientific activity and broader causes of the state's decline, its flaw lies in framing the seventeenth century as one of scientific decline *relative* to the New Science and the Scientific Revolution. As a result of this analytical perspective, earlier historiography tended to dismiss the alternate ways in which the Spanish scientific enterprise addressed the crisis of knowledge that had overtaken European natural philosophical thought. This was one of the conclusions that emerged from a pivotal international conference of historians of Spanish science titled "Beyond the Black Legend" held in Valencia in the fall of 2005.[8] One analytical approach singled out at the conference as especially productive entails adjusting the historian of science's frame of analysis such that the objective becomes explaining the thoughts, practices, and works of Spanish practitioners in their own right, rather than measuring Spain's scientific achievement against that of England or Italy, or even against that of Spain one hundred years earlier. Undergirding this approach is a fundamental change in our understanding of the origin of modern science that rejects the positivist view of the "New Science" as the inevitable outcome of the early modern crisis of knowledge. In fact, experimental natural philosophy was a highly contested enterprise harboring no guarantees of its eventual success.

By choosing not to limit the historian of science's work solely to members of an elite interested in natural philosophy, the field has also benefited by looking at how non-elites and marginalized groups understood

[7] This explanation has been examined by José Pardo Tomás and significantly qualified in his *Ciencia y censura: la Inquisición española y los libros científicos en los siglos XVI y XVII* (Madrid: CSIC, 1991).

[8] William Eamon and Víctor Navarro Brotóns, *Más Allá De La Leyenda Negra. España y La Revolución Científica (Beyond the Black Legend: Spain and the Scientific Revolution)* (Valencia: Instituto de historia de la ciencia y documentación López Piñero, Universitat de València, CSIC, 2007), 27–39.

the natural world. Pictorial representations, poetry, novels, music, consumption patterns, and other manifestations of popular culture all yield signals of how society understood and explained nature. The practices of empirics and healers, as well as alchemists, astrologers, and magi have been integrated into our understanding of early modern thought. In the case of Spain, the persistence of ideas about the natural world cultivated by marginalized groups – crypto-Jews and Protestants, *moriscos* and gypsies – suggests that Spanish thinkers had access to a rich spectrum of ways of interpreting the natural world, and that they enthusiastically, if not openly, engaged with them.[9]

THE CONTEXT: DISCOVERY, EMPIRE, AND SCIENCE

Finally, the "contextual turn" in historical studies has broadened the range of questions that are asked of historical actors beyond those associated with intellectual history. The life and work of doctor Francisco Hernández (1514–87) is a case in point. Fully understanding Hernández and his work, both while in Spain and later during his expedition to Mexico, has meant seeking answers to questions about patronage, court culture, and authority (both political and professional), as well as about social standing, friendship networks, moral, ethical, and racial biases, and even about a man's psychology. The goal has been to contextualize the doctor or natural philosopher as a person of his time and to embrace all the complexities his or her case presents, thereby considering the whole of the individual's work – not just those aspects that make that person a "precursor" or that smack of modernity.[10] Situating Hernández's written

[9] For some examples of this type of literature see Luis García Ballester, *Medicina, Ciencia y minorias marginadas: los moriscos* (Granada: Universidad, 1977); María José López Terrada, *Tradición y cambio en la pintura valenciana de flores (1600–1850)* (Valencia: Ayuntamiento de Valencia, 2001); Marcy Norton, *Sacred Gifts, Profane Pleasures: a History of Tobacco and Chocolate in the Atlantic World* (Ithaca, NY: Cornell University Press, 2008); Juan Martos Quesada, "Islam y ciencia en Al-Andalus," *Ilu. Revista de ciencias de las religiones* 16 (2006): 75–92; Mariano Gómez Aranda, "La ciencia de los judíos de Sefarad, ejemplo de intercambio de culturas," *Hesperia* 7 (2007): 73–84; John D. Slater, *Todos son hojas: literatura e historia natural en el Barroco español* (Madrid: CSIC, 2010); and Mar Rey Bueno, "Mágicos prodigiosos y verdades acrisoladas," *The Colorado Review of Hispanic Studies* 7 (2010): 49–66.

[10] For examples of this approach, see Francisco Hernández, *The Mexican Treasury: The Writings of Dr. Francisco Hernández*, ed. Simon Varey (Stanford, CA: Stanford University Press, 2000); Simon Varey, Rafael Chabrán, and Dora B. Weiner, et al., eds., *Searching for the Secrets of Nature: The Life and Works of Dr. Francisco Hernández*

work within humanistic culture reveals him wrestling with ill-fitting natural philosophical explanations based on Galenic or Aristotelian principles and Dioscoridian taxonomic systems that simply could not encompass – theoretically or methodologically – the richness of the vegetation found in Mexico.[11] While his mind was exquisitely informed by the classics and drove him to try to emulate Pliny the Elder in his writings, his hands were performing experiments that revealed the inability of existing natural philosophical systems to account for the reality of the nature of the New World he experienced firsthand. Hernández emerges as a prime example of a Spanish scientific practitioner seeking to find solutions in the intellectually tense environment of early modern natural philosophy.

An even broader contextual view makes the experience of the discovery of the New World central to any analysis of the Spanish scientific enterprise. The intimate contact with a new world and the desire to understand its nature was an unprecedented conceptual and methodological challenge in the history of the West, where the extraordinary work of Spanish and Portuguese naturalists was unparalleled. This enterprise dictated new questions that required new approaches, which often meant setting aside thematic models that had adequately served Europeans since antiquity. The need for a new paradigm is evident in the contrast between the scientific productions of Andrés Laguna (*c.* 1499–1559), who worked during the first half of the sixteenth century, and Nicolás Monardes (*c.* 1512–88), who flourished during the next generation. Both were practicing doctors trained as medical humanists with an empiricist bent that informed their compilation of medical herbaria. Although Laguna's survey of medicinal plants, *Pedacio Dioscorides Anazarbeo* (1555), mentions several American plants, he brought them to bear only when they added a new or an otherwise unavailable therapeutic effect to the collection of plants mentioned in Dioscoridies' treatise of the first century CE.[12]

(Stanford, CA: Stanford University Press, 2000); and José María López Piñero and José Pardo Tomás, *La influencia de Francisco Hernández (1515–1587) en la constitución de la botánica y la materia médica modernas* (Valencia: Instituto de Estudios Documentales e Históricos sobre la Ciencia, Universitat de València-CSIC, 1996).

11 Jesús Bustamante García, "Francisco Hernández, Plinio del Nuevo Mundo: Tradición clásica, teoría nominal y sistema terminológico indígena en una obra renacentista," in *Entre dos mundos. Fronteras culturales y agentes mediadores*, eds. B. Ares Queija and S. Gruzinski (Seville: Escuela de Estudios Hispano-Americanos, 1997), 243–68 (pp. 256–61).

12 José Luis Fresquet Febrer, "Terapéutica y materia médica americana en la obra de Andrés Laguna (1555)," *Asclepio* 44 (1992): 53–82 (p. 76); Teresa Huguet-Termes, "New World Materia Medica in Spanish Renaissance Medicine: From Scholarly Reception to Practical Impact," *Medical History* 45(3) (2001): 359–76 (p. 363).

Since Laguna's objective was to correct and enlarge this corpus using his personal experience – and having had limited access to American specimens – these new entries necessarily took a back seat to European plants he knew much better. Nicolás Monardes, however, was well acquainted with American medicinal plants; he grew many in his *hortus medicus* in Seville and ran a business with them. In contrast to Laguna, his *Historia medicinal de las cosas que se traen de nuestras Indias Occidentales* (1565–74, originally published in three volumes) sets aside the Dioscoridian scheme and instead presents a text organized around the medicinal benefits derived from American plants.[13] Monardes' proximity to American plants made it epistemologically feasible for him to write as an authority on New World *material medica*, although he never traveled to the Indies.

The enterprise of the Indies also coincided with attempts to consolidate an immense empire within the composite political structure of the Habsburg monarchy. The thrust of the scientific enterprise in Spain during the sixteenth century was directed *by* and *towards* understanding the New World and integrating this knowledge into the empire's *arcana imperii*.[14] The bureaucratic and economic structures put in place in response to the discovery created productive spaces for specialized scientific practitioners and defined the scientific disciplines that would receive the most patronage: astronomical navigation, cosmography, natural history, and medicine. Long before Francis Bacon (1561–1626) wrote about the *potential* utility of science, Spanish institutions demanded utilitarian results from these scientific practitioners, which resulted in adjustments to the methodologies and the epistemic criteria used to ascertain facts in these fields.[15] Barrera-Osorio has found that, in order to service this utilitarian mandate, scientific practitioners developed methodologies that relied on empirical practices that yielded

[13] Daniela Bleichmar, "Books, Bodies, and Fields: Sixteenth Century Transatlantic Encounters with New World Materia Medica," in *Colonial Botany: Science, Commerce, and Politics in the Early Modern World*, eds. Londa L. Schiebinger and Claudia Swan (Philadelphia, PA: University of Pennsylvania Press, 2005), 83–98 (pp. 87–9); Nicolás Monardes, *La historia medicinal de las cosas que se traen de nuestras Indias Occidentales (1565–1574)*, ed. José María López Piñero (Madrid: Ministerio de Sanidad y Consumo, 1989), 29–34.

[14] Richard L. Kagan, "Arcana Imperii: mapas, sabiduría y poder en la corte de Felipe IV," in *El atlas de Rey Planeta: La descripción de España y de las costas y puertos de sus reinos de Pedro Texeira* (Madrid: Editoral Nerea, 2002), 49–70.

[15] By 1894, if not earlier, this observation about the Spanish scientific enterprise had become commonplace: see Marcelino Menéndez y Pelayo, "Esplendor y decadencia de la cultura científica española," in *Edición nacional de las obras completas de Menéndez Pelayo*, ed. Enrique Sánchez Reyes (Santander: CSIC, 1941), 435.

actionable answers.[16] In the case of cosmography, the epistemological adjustments were significant and took place under the injunction that the compiled information and collection methodologies remain state secrets.[17]

This utilitarian emphasis proved determinative not just for government officials, but also for missionaries and laymen alike. Some of the most important natural historical and ethnographical work was carried out by members of religious orders for whom natural philosophy and the descriptive sciences were valuable explanatory and instructive tools; the missionary had to be well prepared in natural philosophy in order to make sense of the new surroundings. It was to this end that José de Acosta wrote his *Historia natural y moral de las Indias* (1590). Acosta compiled useful information about the New World for the benefit of Jesuit missionaries; gleaned from experience and backed with causal explanations drawn from the Ptolemaic–Aristotelian framework, his work attempted to make philosophical sense of the seemingly never-ending puzzles presented by the Americas. Yet in Acosta's work, when what was learned through experience threatened to clash with Aristotelian natural philosophy, what was learned through experience prevailed. Earlier, Franciscan missionaries had understood that Christian theology required a solid understanding of Aristotelian natural philosophy, particularly when Christian verities came face to face with the radically different cosmogony of the Aztecs. So when the sons of Tenochtitlan's indigenous elite attended the Colegio de Santa Cruz de Tlatelolco (est. 1536), their course of study included Aristotelian natural philosophy, although the success of the philosophical experiment might have left a lot to be desired.[18]

Taking the composite nature of the Habsburg monarchy into account and understanding its implications for scientific activity have also constituted an important development for the history of science in Spain.[19] The "macrocosm" of imperial Spain extended well beyond Castile to Flanders, the Kingdom of Naples, Portugal, "Spanish Rome," and to the furthest reaches of the Atlantic and Pacific Spanish world. Recent

[16] Antonio Barrera-Osorio, *Experiencing Nature: The Spanish American Empire and the Early Scientific Revolution* (Austin, TX: University of Texas Press, 2006).

[17] For cosmography in Castile, see María M. Portuondo, *Secret Science: Spanish Cosmography and the New World* (Chicago, IL: University of Chicago Press, 2009), 103–8.

[18] José María Kobayashi, *La educación como conquista (empresa franciscana en México)* (México: El Colegio de México, 1974), 315–20.

[19] Here John H. Elliott's perspective has been essential. John H. Elliott, "A Europe of Composite Monarchies," *Past & Present* 137 (1992): 48–71.

scholarship has made great strides in identifying networks of practitioners and philosophers spanning this first truly global empire, as well as the nature of information exchanged.[20] The historical lens has also turned to the "microcosm" of local practice, particularly in Seville, Catalonia, and Valencia. Some of the resulting historical narratives argue – in an important corrective to the decline narrative – that the scientific enterprise in these regions retained their vitality throughout the seventeenth century.[21]

So, what happens to the history of early modern Spanish science when the analytical perspectives and broad context outlined above are taken into account? These perspectives pose a powerful corrective to historical narratives that either ignore or purposely set aside the Spanish case when studying early modern science. Furthermore, they show how the enterprise of the Indies and its overarching monarchical structure proved a strong determinant of the scientific enterprise. From the brief examples considered thus far, it is evident that Spanish natural philosophers and scientific practitioners often acted in response to the perception that existing natural philosophical systems and associated practices no longer functioned as suitable tools for producing new knowledge, or that they failed to supply satisfactory explanations about the natural world. But how widespread was this reaction to prevailing natural philosophical systems? In what follows, I survey a range of Spanish proposals of the sixteenth and seventeenth centuries. My objective is to uncover or identify the underlying disquiet(s) that motivated this work and determine whether these doubts place the Spanish scientific enterprise in the midst of the pan-European debate about the need to develop new ways of explaining the workings of the natural world.

THE SPANISH DISQUIET

For some scientific practitioners, particularly those who stayed away from natural philosophical pronouncements or left no trace of them, the

[20] Thomas James Dandelet, *Spanish Rome, 1500–1700* (New Haven, CT: Yale University Press, 2001) and Karel Davids, "Dutch and Spanish Global Networks of Knowledge in the Early Modern Period: Structures, Changes and Limitations," in *Centres and Cycles of Accumulation in and Around the Netherlands During the Early Modern Period*, ed. Lissa Roberts (Munich: LIT Verlag, 2011), 29–52.

[21] José Pardo Tomás and Álvar Martínez Vidal, "Medicine and the Spanish Novator Movement: Ancients vs. Moderns, and Beyond," in *Más Allá De La Leyenda Negra. España y La Revolución Científica (Beyond the Black Legend: Spain and the Scientific Revolution)*, eds. William Eamon and Víctor Navarro Brotóns (Valencia: Instituto de historia de la ciencia y documentación López Piñero, Universitat de València, CSIC, 2007), 323–44 (pp. 332–41).

response to the crisis of knowledge becomes apparent only after careful examination of their proposals and determination of how they went about addressing the shortcomings they found in prevailing systems. Meanwhile for others, the inability of Aristotelian natural philosophy to give satisfactory explanations about the natural world became the very explicit motivation for their work. Their solutions consisted in proposals ranging from modest adjustments to Aristotelian philosophy to novel and idiosyncratic systems that sought to build a new natural philosophy on an empiricist epistemic criterion. Many of them took the form of conversations with ancient systems of thoughts – Neoplatonism, Atomism, Skepticism, Stoicism – which in Spain, as elsewhere in Europe, were seen as possible alternatives to Scholastic Aristotelianism. Early modern natural philosophers rarely adopted these alternatives blindly but rather wove them together with other ideas, often yielding highly idiosyncratic perspectives. I propose here that rather than simply slapping the label of "eclectic" or "syncretist" on them, it is far more productive to inquire about the nature of the problems these philosophers and scientific practitioners were trying to address, and to think of eclecticism or syncretism as the response to a perception of inadequacy in a particular philosophical program. This opens up a number of interesting avenues of inquiry: what did the author find inadequate and in need of reform? What informed the nature of the reforms or new postulates?

One aspect that many of the proposals put forth shared was a deep commitment to empiricism in all its gradations. Speaking generally, authority for an empiricist did not lie in the natural philosophical system that interpreted the observation or experience, but rather in the credibility of the observation and its witness. Whereas the historical Aristotle could be admired as a keen observer of nature, he was not considered infallible. The great skeptic Francisco Sanches wrote, "I believe Aristotle the most acute among the observers of Nature and to be the most extraordinary mind produced from our infirm human nature. But I also affirm he did not know many things and doubted many others ..."[22] Furthermore, Aristotelian natural philosophy had accrued a scholastic apparatus that had come to define it to its early modern students. Experience acquired by the senses and the resulting perception of reality had to be interpreted through the lens of Scholastic Aristotelianism to fit within a view of

[22] "Herculè Aristotelem inter acutissimos Naturae scrutatores plurmùm valere judico; unúmque esse praecipuum ex mirabilis humanae infirmitatis ingeniis. Nullibi tamen erase, non assererem: plurima ignorasse affirmo ..." Francisco Sanches, *Quod nihil scitur* (Lugduni: Ant. Gryphius, 1581), *Ad lectorem*.

reality determined by a set of a priori principles. After the Aristotelian machine had churned on an observation, what was ultimately held up as knowledge was a version of it that harmonized with universal principles. For someone who placed great value on experience and thought it a suitable foundation for new knowledge, the frustration was hard to dissimulate. As Oliva Sabuco explained, "I do not want them to believe me, but instead the experience and the truth of the thing."[23]

The empiricist also wanted to attain some measure of predictability and, perhaps, control over nature. This presented a problem to the more thoughtful empiricists; understanding particulars might account for a given instance, but it might not prove convincing when this experience was applied by extension to other cases. As a minimum, the empiricist sought to *dar regla* (create rules) about how something worked, better yet if the predictive power of mathematics and geometry could be brought to bear. The royal cosmographer, Andrés García de Céspedes (*c.* 1545–1611), explained that he had shown all the mathematical proofs necessary in his art of navigation because "he who makes the rules has the obligation of making them accurate and just."[24] On other subjects, such as the deviation of the compass needle (due to magnetic declination), he chose not to speculate: "Although some have sought to make rules about the deviation of the needle from North, because – as we will explain later – this is an uncertain thing, here I am simply giving some reports of what I know about it."[25]

The empirical approach to collecting, organizing, and making sense of natural knowledge was often accompanied with a pragmatic and utilitarian attitude. For a pragmatist, the criteria of truth lay in the merit of the practical outcome of a conceived notion, rather than on the logical integrity of its theoretical explanation. For example, if understanding the trajectory of a cannon ball as composed of mixed motions – a violation of Aristotelian dicta – helped to predict reliably the landing site of the shot – a good practical outcome – the pragmatist would find this the better notion.

[23] "[N]o quiero que me crean a mi, sino a la experiencia, y verdad de la cosa ..." Oliva Sabuco de Nantes y Barrera and Miguel Sabuco, *Nueva filosofia de la naturaleza del hombre* (Madrid: Pedro Madrigal, 1588), 205r.

[24] "[E]l que haze las reglas, tiene obligación de darselas ciertas y Justas." Andrés García de Céspedes, *Regimiento de navegación e hydrografía* (Madrid: Casa de Juan de la Cuesta, 1606), 73v.

[25] "Aunque algunos han querido dar reglas de lo que nordestea, o noroestea la aguja, porque esto es cosa incierta, como adelante diremos, aqui no se trata sino de cierta advertencia, segun lo que della tengo conocido." García de Céspedes, *Regimiento de navegación e hydrografía*, 73v.

Spanish artillery manuals of the era offer clear examples of this attitude.[26] What we find is that most scientific practitioners who were pragmatists operated in ways, if not entirely divorced from Aristotelian natural philosophy, certainly far removed from its scholastic version, or quite simply in the theory-free zone of radical empiricism. This pragmatic attitude could be attributed to utilitarianism and a preference for empirical practices, but it can also be ascribed to dissatisfaction with the systems of knowledge available to address – let alone answer – fundamental questions about nature. If so, this places Spanish scientific practitioners alongside their European counterparts wrestling with the crisis of knowledge that had overtaken early modern Europe. In fact, Jeremy Robbins has argued that this pragmatism prepared the way for the open reception of the experimental sciences in Spain during the latter half of the seventeenth century.[27]

Natural philosophers and theologians friendly to the empiricist approach nonetheless pointed to several problems. Some complained that it was untenable to have everyone come up with their own explanation. Equally problematic were empiricists who simply collected facts without any other purpose than finding some utility in them: danger lurked in knowledge for its own sake or solely for material gain, which could make the soul teeter towards materialism. Among the critics of a purely empiricist approach was Juan Luis Vives (1493–1540), who, although uncertain about the human capacity to unravel the secrets of nature, called for a moderate empiricism in *De disciplinis* (1531):

> The first precept in the contemplation and discussion of nature, is that since we cannot gain any certain knowledge from it, we must not indulge ourselves too much in examining and inquiring into those things which we can never attain, but that all our studies should be applied to the necessities of life, to some bodily or mental gain, to the cultivation and increase of reverence ... Nor does such excessive application leave time for any thought of God, and even if a man attempts it, his own investigation, in which he is totally wrapped up, presents itself instead before his eyes. Thus the contemplation of nature is unnecessary and even harmful unless it serves the useful arts of life, or raises us from a knowledge of His works to a knowledge, admiration and love of the Author of these works.[28]

[26] Just to name a few, see Luis Collado, *Plática manual de artillería* (Milan, 1592); Diego de Prado y Tobar, *Enciclopedia de fundicion y Platica manual de artillería* (1603); and Diego Ufano, *Tratado de Artilleria* (Brussels, 1612).

[27] Robbins' exploration of the cultural manifestations of this pragmatic attitude is evident in the use of *ser/parecer* and *engaño/desengaño*: Jeremy Robbins, "The Arts of Perception," *Bulletin of Spanish Studies* 82(8) (2005): 1–288 (pp. 5, 18, 225–53).

[28] Juan Luis Vives, *Vives: On Education; A Translation of the De tradendis disciplinis of Juan Luis Vives; Together with an Introduction*, ed. Foster Watson (Cambridge: The University Press, 1913), 166–7.

Conversely, Vives (and Montaigne) also reminded Scholastics that "the world of empirical reality could not be ignored without disastrous results to knowledge and science."[29] Thus Vives urged teachers to guide students through the thicket of ancient natural philosophy using experience as the basis of sound judgment, "So the studious will become accustomed to give assent to reason, rather than to human authority."[30] A natural philosophical preparation was necessary, insisted Vives, to learn how to balance knowledge of nature and knowledge of the Divine.

Although Vives spent his most productive years outside Spain, many others, especially those associated with Spanish universities, sought to correct ambiguities or shortcomings in the established natural philosophical system. Rather than simply adding a layer of empirical data unsupported by axioms derived from deductive logical reasoning, their goal was to build upon the epistemic and methodological apparatus of Scholastic Aristotelianism and thus retain the search for causality as the desired objective. In its Scholastic embodiment, the search for causality in natural phenomena – inherent to Aristotelian natural philosophy – ultimately ended with evidence of the actions of the Creator. The one had become inexorably tied to the other and thusly turned natural philosophy into a legitimate and virtuous pursuit, countering the threat posed by a systemless quest into the secrets of nature, or, worse yet, by one guided by an unsound natural philosophy. These perils could distance man from God and undermine the religious truths upon which Christianity was based. The stakes were very high.

In Spain, as in the rest of Europe, the most authoritative explanation about the purpose of science came from Saint Augustine: natural philosophy was needed to understand the Bible, often called the "science as handmaiden" justification for science.[31] St. Augustine, however, was pessimistic about postlapsarian humanity's ability to fully comprehend the "book of nature," a view illustrated in Vives' quote above.[32] Notwithstanding this, Spanish natural philosophers found a much more optimistic view in the Catalan Raymond Sebond's (1385–1436) *Theologia naturalis seu Liber*

[29] Hiram Collins Haydn, *The Counter-Renaissance* (New York: Grove Press, 1960), 238–9.

[30] Vives, *On Education*, 213.

[31] David C. Lindberg and Ronald L. Numbers, eds., *When Science and Christianity Meet* (Chicago, IL: University of Chicago Press, 2003), 18–19.

[32] Harrison sees this as the prevailing view in Protestant nations, while he argues Catholics followed St. Thomas' more optimistic view about humanity's ability to read the book of nature. See John R. Harrison, *The Fall of Man and the Foundation of Science* (Cambridge: Cambridge University Press, 2007).

creaturarum (1484).[33] This work put forth the idea of a book of nature open for all people to read according to their abilities, and, through this activity learn about their own nature, the nature of the Divinity, and how to live in concert with natural and Divine law. It was a course of study that stood proudly apart from the Scholastic curriculum and offered a pleasant – but also orthodox – path to the reconciliation of faith and reason. Moreover, Sebond's natural theology suggests a way to endow theory-free empiricism with a just purpose, for it was through the contemplation (or study) of nature and the arduous process of deciphering the Creator's writing therein that humanity could understand its place in Creation. Thus began a journey, buttressed by grace and faith, that would lead to greater knowledge of God. Through the 1549 Castilian edition published by Francisco Fernández de Córdona, *Violeta de Anima*, Sebond's book reached a popular audience, despite the censorship of its prologue in 1558.[34]

In an era scarred by the Reformation, wary of doctrinal dissention, and grappling with an evangelical effort unprecedented since the days of Constantine the Great, Catholic thinkers saw the empirical turn and revival of some ancient natural philosophies as potential threats to humanity's salvation. By the mid-sixteenth century these concerns motivated some formidable intellects to revisit Aristotelian Scholasticism and to try to address some of its shortcomings: Alejo de Venegas (1498–1562), Domingo de Soto (1494–1560), Diego de Zúñiga (1526–98), Benito Pereria (1535–1610), and Francisco Suárez (1548–1617) – the *Doctor eximius* – whose work is finally receiving the attention it deserves by historians of philosophy.[35] The Aristotelian rehabilitation effort continued throughout the seventeenth and into the eighteenth century, if with somewhat less momentum.

For the Jesuit Francisco Suárez, the goal was to reestablish Aristotelian natural philosophy on a metaphysical foundation that was in concert with reality as experienced through the senses, which pointed towards a theology where a better understanding of God was achieved through his role as

33 Thomas Woolford, "Natural Theology and Natural Philosophy in the Late Renaissance," PhD dissertation (University of Cambridge, 2011), 150–3.

34 Jesús Martínez de Bujanda, "L'influence de Sebond en Espagne au XVIe siècle," *Renaissance and Reformation* X(1974): 78–84 (p. 80); Jaume de Puig, *Les sources de la pensée philosophique de Raimond Sebond (Ramón Sibiuda)* (Paris: Honoré Champion, 1994), 65.

35 "Suárez seems to have slipped between the cracks in our historiographical taxonomy," Benjamin Hill and Henrik Lagerlund, eds., *The Philosophy of Francisco Suárez* (Oxford: Oxford University Press, 2012), 2.

Creator.[36] This perspective offered the proper point of departure for interpreting the natural world. By putting it forward, Suárez was also responding to what he saw as a dangerous misinterpretation of the Aristotelian corpus and radical departures from that corpus that seemed to be overtaking natural philosophy. As Hill points out, "Indeed much of Suárez's philosophical work is best viewed as a reaction against the philosophical naturalism prominent in the Averroist interpretations of Aristotle that were bouncing around the universities in Italy during the sixteenth century."[37] Understanding metaphysics to be the starting point of any philosophical program, Suárez composed his most influential work, the *Metaphysical Disputations* (*DM*, 1597), as a program of study that broke with the commentary tradition, and in the style of a modern philosophical treatise.[38] At first glance the profuse citations of authorities in the *DM* and its intractable scholastic language might suggest another Thomistic recitation, but under its scholastic trappings are some of the most radical reformulations of Aristotelian natural philosophy to take place during the early modern era. Suárez proposed restricting metaphysics to the study of real beings (both finite and infinite and who have a real essence), privileging the efficient and material causes over the formal (while making God the Creator always the final cause) and denying a real distinction between essence and existence.[39] Furthermore, he restricted the operation of the substantial form to bodies and thusly reduced the formal cause to an operation: the joining of the substantial form with the body.[40] Suárez used common experiences as supporting evidence for most of these changes in an effort to bring Scholastic Aristotelianism in line with the empiricism that characterized his era. The *DM* was widely read by Catholics and Protestants alike and had immense influence on the natural philosophers who addressed the metaphysical aspects of the New Science, such as Descartes and Leibniz.[41]

[36] The purpose of his metaphysics is to lay the foundation setting the relationship between being (*ente*) and creation. Emmanuel J. Bauer, "Francisco Suárez (1548–1617): Scholasticism after Humanism," in *Philosophers of the Renaissance*, ed. Paul Richard Blum (Washington, DC: Catholic University of America Press, 2010), 236–55 (pp. 244–5).

[37] Hill and Lagerlund, *Philosophy of Francisco Suárez*, 3.

[38] John P. Doyle, *Collected Studies on Francisco Suárez, S.J. (1548–1617)*, ed. Victor M. Salas (Leuven: Leuven University Press, 2010), 7.

[39] Doyle, *Collected Studies on Francisco Suárez*, 8–11.

[40] Helen Hattab, "Suárez's Last Stand for the Substantial Form," in *The Philosophy of Francisco Suárez*, eds. Benjamin Hill and Henrik Lagerlund (Oxford: Oxford University Press, 2012), 101–20 (p. 118).

[41] José Luis Abellán, *Historia crítica del pensamiento español*, vol. III (Madrid: Espasa-Calpe, 1979), 529; Doyle, *Collected Studies on Francisco Suárez*, 18–20.

The Galenic medicine taught at the universities was another target for reformers. One trend among the critics of Galen was to challenge the Aristotelian theory of four elements, such as Pere d'Oleza (*c.* 1495–1531) did with his version of Averronian corpuscularianism in his *Summa totius philosophiae et medicinae* (1536),[42] while Gómez Pereira relied on a rejection of Aristotelian prime matter to support the controversial ideas on mechanistic animal behavior in his *Antoniana Margarita* (1555).[43] Later in the century, Oliva Sabuco's *Nueva Filosofia* (1587) proposed a complete reformulation of theories of the body and disease that ran counter not only to Galenic medicine as taught at the Schools but also to some fundamental aspects of Aristotelian natural philosophy.[44] The author found that only two elements (earth and water) of the Aristotelian four sufficed, and this because of their analogous function to the sun and moon for maintaining life.[45] Although Galen bore the brunt of these reformers' critiques, many of them also brought into question the ability of Aristotelian natural philosophy to explain the fundamental building blocks of the human body.

As elsewhere in Europe, some Spanish natural philosophers also turned to natural philosophical systems rediscovered during the Renaissance as alternatives to Scholastic Aristotelianism. Renaissance Platonism is often used as a case in point, with the principal mode of transmittal of Neoplatonic ideas as Marsilio Ficino's *Platonic Theology* (1482).[46] In Spain, however, this was not the case. Several reasons seemed to mitigate the Italian's influence and that of his Neoplatonism, among them Vives' poor opinion of Ficino, the fact that Neoplatonic ideas were already integrated into the religious mysticism prevalent in many Spanish religious orders, and familiarity with an alternative in the operative Platonism of Llulian philosophy. Not surprisingly, the three figures cited as exemplars of Renaissance Platonism in Spain flourished outside Spain: the exiled

42 Josep Lluís Barona, *Sobre medicina y filosofía natural en el Renacimiento* (Valencia: Seminari d'Estudis sobre la Ciencia, 1993), 77–103.

43 Gómez Pereira, *Antoniana Margarita*, eds. J. L. Barreiro Barreiro, Concepción Souto García, and Juan Luis Camacho Lliteras (Santiago de Compostela: Universidad de Santiago de Compostela, 2000), 111.

44 The question of Oliva's authorship continues to be debated with some scholars attributing the work to her father Miguel Sabuco, an apothecary. Gianna Pomata has recently discovered that Sabuco took her/his thematic cues from Francisco Valles' (1524–96) *Controversiae Medicae et philosophicae*. Oliva Sabuco de Nantes y Barrera and Miguel Sabuco (?), *The True Medicine. Edited and Translated by Gianna Pomata* (Toronto: Iter, 2010), 35–7.

45 Sabuco de Nantes y Barrera and Sabuco (?), *Nueva filosofia*, 328–30.

46 Copenhaver and Schmitt, *Renaissance Philosophy*, 161–2.

Sephardim León Hebreo (Yehuda Abrabanel [1460? –1521?]) and his influential *Dialoghi d'amore* (1535), the heretic Miguel Servet (1509/11–53), and Sebastián Fox Morcillo (1528–60).

Fox Morcillo, Sevillian by birth and educated in Alcalá and Louvain, is considered Spain's most committed Platonist, but his flourishing career was cut short when his ship foundered as he returned to Spain as tutor to Philip II's son Don Carlos.[47] His few works give us some indications of how "rediscovered" philosophical systems were recruited to address what the philosopher saw as shortcomings of the prevailing natural philosophy. His project, as put on display in his *De naturae philosophia seu Platon et Aristoteles consensione* (1554), was to reconcile Plato and Aristotle by evaluating and extirpating what he considered were emendations and misinterpretations accrued through the ages. In the dedicatory *proemio* to Philip II, Fox Morcillo complained about the inadequate, complicated, and boring way the natural philosophical works of Plato and Aristotle were taught, and yet how necessary this knowledge was for the republic. He placed the blame of having to undertake his project squarely on the shoulders of instructors who were servile and dim-witted (*serviles y tozudos de ingenio*) and who made the subject unnecessarily complicated.[48] The purpose of his compendium was to put forth the philosophers' original ideas clearly and explain how these had been corrected by Church doctrine. Furthermore, he claimed that in this exercise he had found a universal philosophy (*universam philosophiam*) that could reconcile many of the warring schools of thought.[49]

His poorly dissimulated objective was quite a trifecta: to build a unified philosophical system on the strength of the original ideas of the philosophers, one that was in concert with empirical observation and also did not challenge Church doctrine. For example, in an attempt to reconcile their metaphysics, he mapped Plato's agents, the Demiurge, Idea, and Receptacle to Aristotle's Privation, Form, and Matter, explaining

[47] Marcelino Menéndez y Pelayo, "Platonismo en España," in *Ensayos de crítica filosófica* (Madrid: Est. Tipográfico "Sucesores de Rivadeneyra", 1892), 30, 65–74; José Luis Abellán, *Historia crítica del pensamiento español*, 2 vols. (Madrid: Espasa-Calpe, 1979), II, 136–43; Pedro Urbano González de la Calle, *Sebastián Fox Morcillo: Estudio histórico-crítico de sus doctrinas* (Impr. del Asilo de huérfanos del sagrado corazón de Jesús, 1903), 24.

[48] As cited in González de la Calle, *Sebastián Fox Morcillo*, 95.

[49] Sebastián Fox Morcillo, *Sebastiani Foxii Morzilli Hispalensis De natvræ philosophia, seu, De Platonis, & Aristotelis consensione, libri V nunc denuò recogniti, & à me͞dis, quibus anteà scatebant, sedulò repurgati* (Parisiis: Ex officina Iacobi Puteani, 1560), *praefatio*.

differences in their capabilities and functions, but ultimately settling for compromises.[50] When the opinions of the two philosophers could not be reconciled, Fox Morcillo turned to the Bible and to Church doctrine. In another example, we see him trying to reconcile Plato's claim that the heavens are made of fire with Aristotle's notion of ethereal spheres. As a way of solving the apparent contradiction, he investigated the meaning of the Hebrew רָקִיעַ, "raqiya" or "firmament" as used in Genesis 1:6. He found the word could also mean "diffusion," "extension," or "solidity," so he settled on "*naturae tenuis*," an observable quality shared by both fire and the purported quintessence as the best descriptor of celestial matter.[51] (It also helped that this view was corroborated by St. Basil.) His was an attempt to propose a solution to what he perceived was the poorly understood and intellectually fraught mélange of ideas about the working of the natural world typical of his era. He gave empiricism an important epistemic role in his philosophy, but his untimely death robs us of knowing how he would have engaged with the utilitarianism so characteristic of Philip II's court.

The work of Portuguese *converso* Francisco Sanches and his *Quod nihil scitur* (1581) are often taken as the point of origin of early modern Spanish skepticism. It is a strident manifesto – a rant, really – that grew out of the dissonance between experience and what natural philosophical systems taught. Whereas Robbins attributes the pragmatic attitude towards knowledge in early modern Spain to an interest in the rediscovered Pyrrhonist Skepticism and Neostoicism,[52] I suspect that – with the caveat that more work is needed in this area – rather than guided by Sextus Empiricus, *Quod nihil scitur*'s many readers embraced Sanches' rant because it echoed their frustration with Scholastic Aristotelianism and the confusion caused by the many alternative natural philosophical systems circulating at the time.

50 Fox Morcillo, *De natvræ philosophia*, 9.

51 Fox Morcillo, *De natvræ philosophia*, 47v; González de la Calle, *Sebastián Fox Morcillo*, 115–16.

52 Robbins, "The Arts of Perception," 7–8. For whereas the influence of Neostoic moral philosophy in Spain is undeniable, Stoicism's influence on natural philosophy is less clear. The principal authority was Seneca, and to a lesser extent Cicero and Diogenes Laertius. Yet, only certain sections of the *Questiones naturales* survived where Seneca describes Stoic natural philosophy and these give only a cursory view into Stoic cosmology. Despite this limitation, Seneca's message concerning the relationship between humanity and nature comes across clearly: humanity must live according to the rules of nature and face it with fortitude. For Stoics, as a consequence, the rules set by nature had to be investigated and understood.

Not even the burnished ideas of the ancient philosophers sufficed for others who understood the grave implications of the crisis of knowledge, principal among them the theologian and polymath Benito Arias Montano (1527–98).[53] Arias Montano sought guidance for a new natural philosophy in the philological study of Sacred Scriptures, the careful observation of nature, and, ultimately, in Divine revelation. He was well aware of what a cohesive natural philosophical system entailed and therefore developed in his project epistemic approaches and methodological tools that allowed him to formulate a new metaphysics, cosmology, physics, and natural history. His approach to the study of nature framed "nature" as unfolding from a series of historical events, which when properly interpreted described the workings of the natural world. He found these historical episodes in the Bible, often obscured in the *arcano sermone*, or arcane language, of the Mosaic texts. Therefore, he marshaled Hebrew, Greek, and Latin philology and an impressive command of the biblical text to find a new set of first metaphysical principles (mostly in the Book of Genesis) and clues to a complete natural philosophical system in the rest of the Bible. A careful analysis of the Montanian natural philosophy resulting from this exercise reveals that he never strayed too far from the comfort of accepted metaphysical structures – as in the case where his four causes of nature neatly map onto the four Aristotelian causes[54] – but also that he was motivated by the desire to present a whole natural philosophical system where the only arbiters were experience and the Word of God.

Also competing with these scholarly natural philosophies were alternative knowledge systems that addressed one of the vexing questions behind the crisis of knowledge we have been discussing: the nature of

[53] Appreciation of Arias Montano as a philosopher is a rather recent development in Spanish historiography. Abellán recognized Arias Montano's skepticism about rational philosophical systems, *Historia crítica del pensamiento español*, II, 234. Also see the studies by Juan José Jorge López, *El pensamiento filosófico de Benito Arias Montano: una reflexión sobre su Opus magnum* (Mérida: Editora Regional de Extremadura, 2002); Luis Gómez Canseco's introduction to Benito Arias Montano, *Libro de la generación y regeneración del hombre, o, Historia del género humano. Primera parte de la Obra Magna, esto es, Alma. Estudio preliminar de Luis Gómez Canseco*, trans. Fernando Navarro Antolín et al. (Huelva: Universidad de Huelva, 1999) and Benito Arias Montano, *Historia de la naturaleza. Primera parte del Cuerpo de la Obra Magna*, ed. Fernando Navarro Antolín, Bibliotheca Montaniana (Huelva: Universidad de Huelva, 2002). I discuss the influence of Arias Montano's philosophy of nature at the library of the Escorial, in María M. Portuondo, "The Study of Nature, Philosophy, and the Royal Library of San Lorenzo of the Escorial," *Renaissance Quarterly* 63(4) (2010): 1106–50.

[54] Portuondo, "The Study of Nature, Philosophy, and the Royal Library," 1126–7.

bodies and their energies. The alchemist (whether preparing medical remedies or seeking the philosopher's stone), the astrologer (practicing either licit or judiciary astrology, or both), and the magus or professors of secrets (from the Hermetist to the charlatan) relied on systems of knowledge that, once mastered, helped to make sense of the world and perhaps control it. Recent historical work has made it quite clear that Spain had a significant community of alchemical practitioners, often working under royal patronage and building on an autochthonous tradition that held Raymond Llull (or rather the pseudo-Llull), Arnau of Vilanova, and John of Rupescissa as preceptors. These practitioners operated loosely within a number of philosophical systems of thought and sought in their emananantist forces and occult properties the ideas that gave their alchemical trials coherence.[55] In pharmacy, the role of American plants and the gardens of simples at Aranjuez and the Escorial brought together alchemical distillation practices, the iatrochemical practices of empirics, and those of the most learned physicians.[56]

No longer relegated to the dustbin of history, astrology and natural magic, which had a long tradition in Spain and were invigorated by the translations of the *Picatrix* and other works during the reign of Alfonso X, had by the sixteenth century emerged as viable sources for alternative explanations about nature.[57] In particular, the tradition of Arabic and Hebrew alchemy and magic had a strong presence and had become an important alternative to Scholastic Aristotelianism for some practitioners. In magic, the dialectic between advocates and denouncers came to a head during the seventeenth century. While manuscript copies of the *Clavicula salomonis* circulated widely in Spain,[58] the popular books of Martín del Río (1551–1608) and Hernando Castrillo (1585–1667) tried to bring magic and witchcraft out of the shadows – and to the attention of the inquisitors.[59] As Del Rio remarked in his the prologue

[55] Miguel López Pérez, "Ciencia y pensamiento hermético en la Edad Moderna Temprana," in *Más Allá De La Leyenda Negra. España y La Revolución Científica (Beyond the Black Legend: Spain and the Scientific Revolution)*, eds. William Eamon and Víctor Navarro Brotóns (Valencia: Instituto de historia de la ciencia y documentación López Piñero, Universitat de València, CSIC, 2007), 57–72 (pp. 65–9).

[56] Mar Rey Bueno, *Los señores del fuego. Destiladores y espagíricos en la corte de los Austrias* (Madrid: Corona Borealis, 2002), 59–90.

[57] Laura Fernández Fernández, "El "arte mágica" en el scriptorium Alfonsí: del Picatrix al Libro de astromágia," in *De cuerpos y almas en el judaísmo hispanomedieval: entre la ciencia médica y la magia sanadora*, eds. Yolanda Moreno Koch and Ricardo Izquierdo Benito (Cuenca: Ediciones de la Universidad de Castilla-La Mancha, 2011), 73–109 (p. 77).

[58] Julio Caro Baroja, *Vidas mágicas e inquisición*, 2 vols. (Madrid: Taurus, 1967), I, 138–47.

[59] Caro Baroja, *Vidas mágicas e inquisición*, I, 168–70.

to the *Disquisitionum magicarum* (1595), "demons might find a home in heretics."[60] His motivation was driven by "how widespread magic has become in recent times. Evil spirits are on the loose seeking to take possession of foolish and deluded souls ... I have read that, as a result of the Moorish occupation of Spain, the magical arts were virtually the only subjects being taught in Toledo, Seville and Salamanca ..."[61]

By the beginning of the seventeenth century, all the systems of knowledge discussed above were in frank competition with additional systems joining the fray as the century progressed. The system-building efforts of Spain during the sixteenth century were matched only by Italy's, where the Naturalism of Bernardino Telesio (1508–88) (and its empiricist variants), the Platonism of Francesco Patrizi (1529–97), and several versions of reformed Scholastic Aristotelianism were widely discussed. Later in the century the systems hatched in Italy and Spain would be joined by the Christianized atomism of Gassendi, Boyle's Corpuscularianism, Cartesian mechanism, and the New Science of experimental natural philosophy. The range of proposals was astounding, from the conservative – mostly Aristotelian – natural philosophy of the Jesuits' *Ratio studiorum* and its course of study as laid out in the *Cursus conimbricensis* (1591–2) to the magical naturalism of Paracelsianism as interpreted by Robert Fludd (1574–1637). These were either very exciting or very troubling times to be a natural philosopher. But only when we divest ourselves from the notion that the triumph of the New Science was inevitable can we appreciate the variety and richness of these system-builders' projects and appreciate the uncertainty of the whole enterprise.

It is undeniable that by the 1640s the broader economic and political context affecting the scientific enterprise in Spain had changed. The troubles that plagued the composite monarchy of the Habsburgs caught up with the practitioners who had made the focus of their work the natural history and cosmography of the New World. By then it was clear that American flora harbored no panacea and that its seemingly inexhaustible silver mines were showing signs of depletion. Even a monarch's lack of interest could hamper the scientific enterprise, as was the case in Philip III's disinterest in botanical gardens and alchemical laboratories, which

[60] "*dæmones in hæreticis ... habeant domicilium.*" Martin Antoine Del Rio, *Disquisitionum magicarum libri sex,. Quibus continetur accurata curiosarum artium, & vanarum superstitionum confutatio, utilis theologis, iurisconsultis, medicis, philologis* (Moguntiae: Apud Johannem Albinum, 1612), Prologue.

[61] Martín Del Río, *Martin Del Rio: Investigations into Magic* (Manchester: Manchester University Press, 2000), 27–8.

ultimately drew patronage away from these activities.[62] The history of this disinterest (*desinterés*) in natural history is yet to be written, but John Slater is developing an interesting thesis that sees a growing unease in the ambiguity caused in various constituencies by different models of natural philosophical authority and in what "usefulness" meant in the context in which it undermined the formation of a "broad culture of natural historical investigations."[63] The result was the development of an approach to natural specimens that saw them as emblematic types and therefore more useful to rhetoric than as scientific specimens. American novelties found places of honor in the virtuosi's cabinet of curiosities and in the literature of the Spanish Baroque, but not in natural history books. The publication of natural history books on American flora practically stopped after the 1620s and did not revive until the eighteenth century. By then American natural productions had somehow crossed the threshold between wonders and commodities: tobacco, chocolate, and some medicinal plants, such as *mechoacán*, sarsaparilla, and *Cassia fistula* underwent the process of demystification described by Marcy Norton.[64]

The bureaucratic structures established to support the enterprise of the Indies remained in place, but by the mid-seventeenth century the Casa de la Contratación had stopped training ship pilots, and cosmographical productivity slowed down significantly. In 1628 the post of cosmographer was transferred to the professors of the *Reales Estudios* of the Jesuit Colegio Imperial de Madrid,[65] and the Council of Indies only half-heartedly required yearly works from its chroniclers of the Indies.[66] The most productive of these was Antonio de León Pinelo (1595–1660). As a long-time affiliate to the Casa de la Contratación, León Pinelo had access to the wealth of descriptions of the Americas collected by the Council of Indies over the previous hundred years or more. He used some of this

62 Mar Rey Bueno, "Juntas de herbolarios y tertulias espagíricas: el círculo cortesano de Diego de Cortavila (1597–1657)," *Dynamis* 24 (2004): 243–67 (p. 260).

63 John D. Slater, "The Green Gold Fallacies: Myth and Reality in the Transatlantic Trade in Medicinal Plants," in *Geografías médicas. Fronteras culturales de la medicina hispanoamericana, siglos XVI y XVII*, eds. Mauricio Sánchez-Menchero and José Pardo Tomás (Mexico: CEIICH-UNAM, 2014).

64 Norton, *Sacred Gifts*, 141–2, 235–6.

65 José Simón Díaz, *Historia del Colegio Imperial de Madrid* (Madrid: CSIC, Instituto de Estudios Madrileños, 1952); Víctor Navarro Brotóns, "Momentos y lugares de la ciencia española Siglos XVI–XX. El Colegio Imperial de Madrid," *Historia 16* 287 (2000): 49–64.

66 Ernesto Schäfer, *El Consejo Real y Supremo de las Indias: su historia, organización y labor administrativa hasta la terminación de la Casa de Austria*, 2 vols. (Madrid: Junta de Castilla y León, Marcial Pons, 2003), I, 255; II, 362–6.

material – selecting only the most renowned authors – and his impressive command of ancient and modern sources to argue in his *El paraíso en el Nuevo Mundo* (1650) that Paradise had been located in America. Interwoven within the eighteen principal reasons he gives to support his thesis is a natural history of the New World.[67] It is a highly selective natural history, where the fantastic and marvelous are magnified to reinforce the argument. It is emblematic of the scientific climate of the time that this was deemed by León Pinelo to be the best use for the descriptive material that lay securely guarded at the Council of Indies. (The Council seemed to have been of like mind as they granted him the post of chronicler of the Indies after the outline of the manuscript was published in 1568.) León Pinelo's regard for and use of empirical information was diametrically opposite to the utilitarian pragmatism that had guided the cosmographer–chronicler majors in the previous century, yet in concert with the Baroque fascination with curiosities and its obsession with displays of erudition.[68]

At the Colegio Imperial de Madrid, however, Jesuits from all over Europe came to teach natural philosophy, astronomy, mathematics, and mechanics. Although the educational program faltered after about a decade,[69] Navarro Brotóns considers its contributions the sole bright spot for the development of these disciplines during this period.[70] The Jesuits' focus on applied mathematics and their eclectic approach to cosmological questions opened a space for the discussion of novel ideas.[71] Yet their attitude towards the new discoveries was pragmatic, for although they rejected radical empiricism, they decided on the merits of new natural philosophical propositions by their logical coherence and adherence to religious orthodoxy. Figures such as Charles della Faille, Jean-Charles Van Langren, and José de Zaragoza comprised a significant node in the network of Jesuit scholars centered at the Collegio Romano and the towering figure of Athanasius Kircher. Perhaps more than any other area of early modern studies of science, the scientific activities of the Jesuits during the seventeenth century has undergone a reassessment; the field's

[67] The editor remarked about his "tremenda erudición e implacable citomanía." Antonio de León Pinelo, *El paraíso en el Nuevo mundo: comentario apologético*, ed. Raúl Porras Barrenechea, 2 vols. (Lima: Imprenta Torres Aguirre, 1943), I, xxvii.

[68] His treatment of antediluvian giants illustrates this. León Pinelo, *El paraíso*, I, 208–15.

[69] Adolfo Carrasco Martínez, "Los estudios reales del Colegio Imperial de Madrid y otros proyectos educativos de Olivares," *Cuadernos de investigación histórica* 28 (2009): 99–121 (p. 112).

[70] Navarro Brotóns, "Momentos y lugares de la ciencia española," 50–1.

[71] Navarro Brotóns, "La ciencia en la España del siglo XVII: el cultivo de las disciplinas físico-matematics," *Arbor: Ciencia, Pensamiento y Cultura* 153(604–5) (1996): 197–252.

bibliography is now extensive and sheds light on how mathematics, experiments, and knowledge system-building took place in the Catholic world, even though we still await a comprehensive study of science at the Colegio Imperial.[72]

Jesuit science's sense of purpose was aligned with the natural theological concerns of the previous hundred years, but it also had to contend with the sensibilities of the times. For Juan Eusebio Nieremberg (1595–1658), this concern is evident in his natural histories, which often nimbly situate the curious and wondrous in nature within a number of traditions: ancient philosophies, Sacred Scriptures, Patristic commentaries, as well as the opinions of contemporary observers in a process that both perpetuated what was marvelous and yet sought ultimately to demystify.[73] Pimentel has described his approach as Baroque in spirit and style.[74]

In the culture of display characteristic of the Baroque, collectors' fascination with curiosities increased, not as the objects of systematic natural historical study, but instead as vanity.[75] Antonio Maravall observed how the pursuit of novelty during this time was encouraged within cultural productions in the form of "artistic and poetic caprice"[76] (while at the same time it was rejected by elites when these *novedades* threatened social norms and hierarchies). Nieremberg built on the Baroque fascination with novelties with his exhaustive exercises of erudition but did so to illustrate how their thoughtful study could lay out a path through

72 Peter Dear, *Discipline and Experience: the Mathematical Way in the Scientific Revolution* (Chicago, IL: University of Chicago Press, 1995); M. T. Borgato, ed., *Giambattista Riccioli e il merito scientifico dei gesuiti nell'età barocca* (Florence: L. S. Olschki, 2002); Mordechai Feingold, ed., *Jesuit Science and the Republic of Letters* (Cambridge, MA: MIT Press, 2003); Mordechai Feingold, *The New Science and Jesuit Science: Seventeenth Century Perspectives* (Dordrecht: Kluwer Academic, 2003); Paula Findlen, ed., *Athanasius Kircher: The Last Man Who Knew Everything* (New York: Routledge, 2004); John Edward Fletcher, *A Study of the Life and Works of Athanasius Kircher, 'Germanus Incredibilis'* (Leiden: Brill, 2011).

73 See the editor's contributions in Domingo Ledesma and Luis Millones Figueroa, *El saber de los jesuitas, historias naturales y el Nuevo Mundo* (Madrid: Hispanoamericana/Vervuert, 2005).

74 Juan Pimentel, "Baroque Nature: Juan E. Nieremberg, American Wonders and Preterimperial Natural History," in *Science in the Spanish and Portuguese Empires, 1500–1800*, eds. Daniela Bleichmar, et al. (Palo Alto, CA: Stanford University Press, 2009), 93–114 (p. 99).

75 For a prime example, see the essays on the Baroque collector Vincencio Juan de Lastanosa (1607–81), in Miguel López Pérez and Mar Rey Bueno, eds., *The Gentleman, the Virtuoso, the Inquirer: Vincencio Juan de Lastanosa and the Art of Collecting in Early Modern Spain* (Cambridge: Cambridge Scholars, 2008).

76 José Antonio Maravall, *Culture of the Baroque: Analysis of a Historical Structure* (Minneapolis, MN: University of Minnesota Press, 1986), 126–9.

virtuous curiosity and ultimately to salvation. His was an effort to redirect the interest in curiosities towards natural theology and away from vanity, greed, or the purely empirical study of nature.

While Nieremberg was not seeking to build a novel natural philosophical system, others of his peers were. Sebastián Izquierdo (1601–81), who taught for twenty years at the Colegio Imperial, addressed what he perceived was at the root of the crisis of knowledge: the serious disjunction between empirical knowledge and metaphysics. The *Pharus scientiarum* (1659) – his only philosophical work – was an ambitious effort to create a new and universal "general art of knowledge" (*scientia de scientia* or *arte general del saber*) that would lead from empirical observations to knowledge of universal truths.[77] In his survey of other general arts of knowledge, he mentioned those of Francis Bacon. And although he sided with Bacon's call for experimentation and the investigation of the nature of things, Izquierdo faulted him for having dismissed too swiftly those disciplines that relied on purely intellectual exercises to produce new knowledge, such as geometry, arithmetic, and metaphysics.[78] Izquierdo was also uneasy with Bacon's search for universal rules using particular experiments, which for Izquierdo only proved individual cases and could not be extrapolated into larger truths.[79] He positioned his proposals as contributing to the growing and rapidly unfolding activity of science of the latter half of the seventeenth century, but what is most relevant to our current discussion is that he found his motivation in what he took to be a serious epistemic problem with the New Science: its new doctrines had been posited without a proper basis (*fundamentum*) or even a metaphysics, something reprehensible and that had be corrected.[80]

CONCLUSION

As we have seen, the utilitarian purpose for science proposed by Francis Bacon in *The New Organon* (1620) was not a novel idea in Spain.

[77] For what follows, I rely on José Luís Fuertes Herreros, *La lógica como fundamentación del arte general del saber en Sebastián Izquierdo: estudio del "Pharus scientiarum" (1659)* (Salamanca: Universidad de Salamanca, Instituto de estudios albacetenses, 1981), 225–6.

[78] Sebastián Izquierdo, *R.P. Sebast. Izquierdo, ... Soc. Iesu ... Phari scientiarum: pars altera: vbi, quid, quid ad cognitionem humanam humanitùs acquisibilem pertinet, vbertim iuxtà, atque succintè pertractatur: scientia de scientia, ob summam vniuersalitatem vtilissima, scientificisque incundissima, scientific ...* (Lugduni: Sumptibus Claudii Bourgeat & Mich. Lietard, 1659), 290.

[79] Fuertes Herreros, *La lógica*, 250–1.

[80] Izquierdo, *Phari scientiarum*, 271.

Likewise, neither were natural philosophical proposals that emphasized empiricism as part of a knowledge system. The radical aspect of Bacon's proposal – the aspect that brought Izquierdo such alarm – was that laws of nature could be induced from an accumulation of empirical facts (learned by means of Bacon's albeit loose definition of experiment) *without* the guide of a philosophical system based on first principles.[81] By arguing that the natural philosophy that would result from his program still lay in the future as the ex post facto result of the systematic collection of verifiable observations, Bacon avoided the thorny problem of laying down a metaphysical proposal that not only accounted for natural phenomena but was also in concert with religious mysteries. For the Anglican, the notion that nature was a reflection of God's glory sufficed, thereby opening a space where the seemingly philosophically free quest for knowledge could take place.

Yet, for many of the Spanish natural philosophers discussed in this chapter, knowledge claims based on experience *had to be* embedded in a philosophical system that stood on firm metaphysical and logical ground, did not contradict Catholic doctrine, and was universal. For them, adopting Bacon's empiricist approach was seen as negligent or, at the very least, epistemologically sloppy. Furthermore, the purpose of the scientific enterprise assigned to it by natural theology (to know God through nature) could not – *should not* – be divested from the scientific enterprise in favor of pure utilitarianism. Despite these caveats, the proposals by Spanish natural philosophers and scientific practitioners that we have surveyed shared with the broader community of European natural philosophers in the resolve to address a perceived crisis of knowledge about the natural world. In this sense, the early modern Spanish enterprise moved in lockstep with Europe. But while responding to the same disquiet, Spanish thinkers remained committed to an earlier understanding of natural philosophy and to the circumscribed role of empirical knowledge therein. Their commitment was not so much to the content of a particular natural philosophy, but rather to how well it could encompass all there was to know about the natural world. Little did they suspect that the seeds of modern science lay in explaining particulars and setting aside many of the lofty goals of natural philosophy.

[81] Bacon explained, "As the sciences in their present state are useless for the discovery of new works, so logic in its present state is useless for the discovery of sciences" (Book 1, Aphorism XI). Francis Bacon, *The New Organon*, eds. Lisa Jardine and Michael Silverthorne (Cambridge: Cambridge University Press, 2000), 35.

9

Spanish Inquisitors, Print, and the Problem of Publication

Kimberly Lynn

In 1575, Diego de Simancas – who had then been in Rome for several years as a representative of the Spanish Inquisition – published the final revision of his treatise on inquisitorial law. He opened it with a dedication to Pope Gregory XIII, which began:

> Never had I considered, most blessed Father, that this book would reach your sacred hands; or, if it fell into them by chance, that it would not immediately displease your most wise judgment. For how could I have suspected or hoped for it? When some time ago I had composed the first, then the only one, then, twenty-three years ago, with youthful heat, I issued it without consultation, by no other better method than by alphabetical order, filled with a scornful multitude of authors, and crammed with many extra ornaments and superfluous things, which could certainly be easily condemned in any place by any rigid censor.[1]

Simancas appealed to the pope to see the worth of his book, forgive its failings, and understand the constraints imposed on the author by time and by his desire to serve an audience of legal experts who valued works packed with citations. These words were the rhetorical flourishes of an elite ecclesiastic seeking patronage and influence; yet they also deliberately revealed a Spanish inquisitor's conviction that an inquisitor's manual was quite likely to become the object of ecclesiastical book censorship. It is well known that early modern inquisitors and inquisitions censored the written and spoken word, and that they created spectacles aimed to discipline various populations – from the infamous *auto de fe* to the publication of edicts of faith to the spectacle-in-reverse of the secrecy

[1] Diego de Simancas, *De catholicis institutionibus* (Rome: in aedibus Populi Romani, 1575), fol. a2.

of their judicial proceedings. Less well known are other ways inquisitors interacted with the processes of writing and publication, including the censoring of inquisitors' words. This essay aims to call attention to how Spanish inquisitors evaluated and utilized the varied modes of communication available to them: writings in print and in manuscript; institutional correspondence and decrees; sermons and speech at court; processions and other forms of symbolic action, more or less choreographed.

Recent scholarship has made the case that public opinion was extant and a meaningful factor in sixteenth-century politics, and that varied forms of communication affected political processes. With a persistent, dynamic, interplay between the oral and the written, pamphlets, news reports, rumors, and social interactions were increasingly perceived as politically meaningful, altering the shape of disputes and decision-making, changing the valence of physical presence at an event or before a figure of authority, and raising the political value of theoretically privileged information.[2] The religious transformations and conflicts of the Reformation era have, moreover, long been recognized as bound up with shifting technologies of – and attitudes towards – information and communication.[3] And the persuasive and disciplinary mechanisms of the early modern Catholic world are receiving ever more scholarly attention, as, for example, in the increasingly nuanced vision of the workings of ecclesiastical censorship.

The core elements of book censorship – the drafting of indices and procedures for expurgation, the commissioning of censors and inspectors – developed significantly between the 1550s and 1590s, intertwined with a Catholic perception of a definitive break with Protestants and the reforming ethos of the Council of Trent. The increasing involvement of the Spanish Inquisition in censorship coincided with broader processes of institutionalization and expansion, as it sought to regularize

[2] See, in particular, Filippo de Vivo, *Information and Communication in Venice: Rethinking Early Modern Politics* (Oxford and New York: Oxford University Press, 2007); Michele Olivari, *Entre el trono y la opinión: la vida política castellana en los siglos XVI y XVII*, prologue by Ricardo García Cárcel, trans. Jesús Villanueva (Valladolid: Junta de Castilla y León, 2004). See also Elizabeth Horodowich, "Introduction: Speech and Oral Culture in Early Modern Europe and Beyond," *Journal of Early Modern History* 16 (2012): 301–13, and the five subsequent essays in that special issue, as well as Daniel Bellingradt, "The Early Modern City as Resonating Box: Media, Public Opinion, and the Urban Space of the Holy Roman Empire, Cologne, and Hamburg ca. 1700," *Journal of Early Modern History* 16 (2012): 201–40.

[3] For a useful recent overview, Andrew Pettegree, "Catholic Pamphleteering," in *The Ashgate Research Companion to the Counter-Reformation*, eds. Alexandra Bamji, Geert H. Janssen, and Mary Laven (Farnham and Burlington, VT: Ashgate, 2013), 109–26.

its procedures and encompass new territories and criminal targets. As those appointed as inquisitors came to be primarily secular clerics with law degrees, those jurist-judges in turn commissioned theologians (often members of the regular clergy) to consult on trials and weigh the contents of books. Censorship initiatives, moreover, were simultaneously elaborated in Spain and in Rome (as well as at Trent, and elsewhere), and the relationship between different decrees, indices, and censoring authorities was not clear. Thus scholars are revealing a climate of censorship in which there was extensive contestation between Catholic authorities who had differing ideological commitments and political affiliations; in which prominent ecclesiastics were also the objects of censorship; in which there were struggles between the Roman Congregations of the Inquisition and the Index (then newly founded arms of papal government); and in which, throughout the territories under the jurisdiction of the Spanish and Roman Inquisitions, there were substantial gaps between edicts and their application and ongoing disputation of both the mechanisms and content of censorship by a variety of actors.[4]

At the same time, scholars have produced a new vision of the Spanish Inquisition's place in early modern communications – as an institution that sought to build its authority by manipulating a multiplicity of persuasive tools, and as an institution whose actions provoked dissent and debate across the social spectrum and throughout the Hispanic world.[5] In such ways, the contingency of the Spanish Inquisition's actions is being recaptured, and with that an earlier historiographical tradition is

4 For avenues into this field, in the Spanish case: Martin Nesvig, *Ideology and Inquisition: The World of the Censors in Early Mexico* (New Haven, CT and London: Yale University Press, 2009); Virgilio Pinto Crespo, *Inquisición y control ideológico en la España del siglo XVI* (Madrid: Taurus, 1983); Patricia W. Manning, *Voicing Dissent in Seventeenth-Century Spain: Inquisition, Social Criticism, and Theology in the Case of El Criticón* (Leiden: Brill, 2009). On the Roman case: Gigliola Fragnito, ed., *Church, Censorship and Culture in Early Modern Italy*, trans. Adrian Belton (Cambridge: Cambridge University Press, 2001); Marco Cavarzere, *La Prassi della Censura nell'Italia del Seicento* (Rome: Edizioni di Storia e Letteratura, 2011); Amnon Raz-Krakotzkin, *The Censor, the Editor, and the Text: The Catholic Church and the Shaping of the Jewish Canon in the Sixteenth Century*, trans. Jackie Feldman (Philadelphia, PA: University of Pennsylvania Press, 2007); Thomas F. Mayer, *The Roman Inquisition: A Papal Bureaucracy and Its Laws in the Age of Galileo* (Philadelphia, PA: University of Pennsylvania Press, 2013).

5 See the fundamental approaches in Francisco Bethencourt, *The Inquisition: A Global History, 1478–1834*, trans. Jean Birrell (Cambridge and New York: Cambridge University Press, 2009); Stefania Pastore, *Il Vangelo e la Spada. L'Inquisizione di Castiglia e i Suoi Critici (1460–1598)* (Rome: Edizioni di Storia e Letteratura, 2003); Stuart Schwartz, *All Can Be Saved: Religious Tolerance and Salvation in the Iberian Atlantic World* (New Haven, CT: Yale University Press, 2008).

being thoroughly revised. The early nineteenth-century historian of the Inquisition, Juan Antonio Llorente (himself a former Inquisition secretary), wrote, for example, about the royal councilors of the Inquisition, "that these ministers of persecution have been the chief causes of the decline of literature, and almost the annihilators of nearly all that could enlighten the people, by their ignorance, their blind submission to the monks who were qualifiers, and by persecuting the magistrates and the learned who were anxious to disseminate information."[6] Without in any way discounting the potential power of the Spanish Inquisition or its very real effects on persecuted individuals and communities, this essay aims to further complicate the understanding of inquisitorial thought and actions. My aim here, quite simply, is to observe how Spanish inquisitors developed a spectrum of relationships to the written forms available to them; in particular, I propose to explore how inquisitorial writings themselves hint at the contingency of – and contestation surrounding – inquisitorial actions and to suggest how communication around (and about) inquisitorial trials was perceived as a strategic political tool. To do so, this essay proposes a close reading of how Simancas' involvement in the highly charged atmosphere of the end of the Inquisition trial of the archbishop of Toledo, Bartolomé Carranza, illuminates both some inquisitorial attitudes towards the politics of communication and some of the motivations which drove writing about Inquisition cases.[7]

In 1575, Simancas (1513–83) was in his early sixties and although he was the bishop of Badajoz, he had been in Rome for eight years, sent there to pursue Carranza's Inquisition trial on charges of Protestant heresy, when it was revoked from Spain to the papal court; educated as a jurist, Simancas had experience in universities, as a royal appellate judge, and on the *Suprema* (the royal council of the Spanish Inquisition). He was involved in the prosecution of Carranza from its start in 1559 until

[6] Juan Antonio Llorente, *A Critical History of the Inquisition of Spain* (1826; repr. Williamstown, MA: The John Lilburne Company, 1967), xvi. On such emphases in nineteenth- and earlier twentieth-century scholarship on the Hispanic world, see Richard L. Kagan, "Prescott's Paradigm: American Historical Scholarship and the Decline of Spain," *American Historical Review* 101 (April 1996): 423–46. There is a voluminous amount of sophisticated revisionist scholarship on the Spanish Inquisition; for an overview of some of its emphases, Kimberly Lynn Hossain, "Unraveling the Spanish Inquisition: Inquisitorial Studies in the Twenty-First Century," *History Compass* 5(4) (2007): 1280–93.

[7] This essay builds on and draws from my study: Kimberly Lynn, *Between Court and Confessional: The Politics of Spanish Inquisitors* (New York: Cambridge University Press, 2013), chs. 1, 2, and 6.

Gregory XIII concluded the case in 1576. Also trained as a jurist, the pope had been a legate to Spain in the early years of the trial; during his papacy, among other initiatives, he spurred development of the Roman Inquisition, created the Congregation of the Index, and sponsored a new printed edition of the corpus of canon law and a substantial collection of legal commentaries to gloss it.

The dedication of Simancas' 1575 manual was thus bound up in the contentious climates of the end of Carranza's trial and the early implementation of Tridentine reforms. It captures the inquisitor's career-long investment in the written word. He published three revisions of his compendium of Inquisition law, and another several editions of an abridged version of it, along with Latin volumes on other legal issues, ecclesiastical administration, and political advice. He had written correspondence and official memoranda and would go on to compose an autobiographical *Vida*. He printed some of his writings, left others in manuscript, published under a pseudonym, and chose between Latin and Spanish. He had revised his treatises, incorporating things he had read, adding and deleting portions over time; he had fretted about the quality of different printers, his competition in the print marketplace, and the organization of his books. And his books had advertised his authority, in at least two instances including a laudatory poem in the prefatory material.[8] Simancas is not representative of all Spanish inquisitors – in fact, the shape of his career was rather odd, consumed by the Carranza trial for nearly two decades – but he can be used to illustrate the variety of ways in which such judges engaged with the written word.[9] In the useful reminder of one scholar, "The world of the law, remember, was a world based on the book."[10] Formed by university legal study and by an office that required them to judge potentially heretical speech and writing and then to explain their judgments (sometimes in speech, sometimes in writing), the men who held the post of inquisitor were a variegated and complex group, not infrequently at odds with one another, and who operated in institutions with a variety of constraints on their powers. Despite ongoing anxieties about the written word – and especially about print – memoranda,

[8] A verse epistle by Juan de Verzosa in *De Republica Libri IX* (Venice: Bolognino Zaltieri, 1569); an epigram (and an approbation) by Simón Cuña Ribera (identified as Portuguese and Alcalá's printing corrector) in *De catholicis institutionibus* (Alcalá de Henares: Andrea de Angulo, 1569).

[9] See the study of Simancas' career in Lynn, *Between Court and Confessional*, ch. 2.

[10] Rodolfo Savelli, "The Censoring of Law Books," in Fragnito, *Church, Censorship and Culture*, 251.

pamphlets, and treatises also became some of the most useful tools available to inquisitors to seek to advertise the fight against heresy or to try to shape their contemporaries' perceptions of a particular trial, to shore up the authority of their institution, to influence a particular case, or to attempt to build their own careers.

As a compendium to aid the practice of Inquisitions, Simancas' manual contained a chapter specifically about books. There, he defended the condemnation and burning of heretical books. As was his customary method, he began that chapter with an historical justification for the practice; thus he aligned civilization and censorship. Starting from Plato, he alleged precedents in Athens, imperial Rome, among the Hebrews and the Christians, in order to assert that it was characteristic of all recognizable republics to condemn harmful books; he noted that even peoples he classified as impious or in error offered proof of the universal good of rooting out harmful reading material. He proceeded to treat some points of procedure and contested elements of the practice of book censorship. Simancas also made a clear distinction between written and oral communication. He reasoned, "for if we should shun the distorted conversations of the impious, how much more their books, for that which they have written can do more harm, because they are premeditated, and have stronger forces." He asserted that things read took hold more strongly than those heard.[11]

Simancas' writings might thus offer a glimpse of inquisitorial attitudes and actions at a crucial moment in the development of Inquisitions and mechanisms of censorship. His warnings against heretics and their books correlate well with the customary picture of Spanish inquisitors, whose image has long been tied to censorship, among other repressive forms of persecution, intellectual fettering, and social control. Yet his treatise also hinted at the unease of Spanish inquisitors, among other Catholic authorities, about what kinds of writings they should approve or be involved in producing, and for what potential audiences. Such judges understood deciding what to suppress and how to do so as a complex problem and as only part of an equation; fighting heresy also required the production

[11] Simancas, *De catholicis institutionibus*, 287. Martin Nesvig has recently made sophisticated use of the second half of this paragraph (where Simancas claimed that books were especially powerful because they lasted over time and could travel widely) to open his study of censorship in colonial Mexico. *Ideology and Inquisition*, 1. For Simancas' doubts about what it meant to read or own a particular book, see Pinto, *Inquisición y control ideológico*, 252.

of theological, legal, political and historical literature to strengthen the Church.[12]

THE WAR AROUND CARRANZA

A publicity war swirled around Carranza's trial. The prosecution of the archbishop of Toledo – an esteemed protagonist of Catholic reform suddenly charged with promoting Protestant heresies – immediately attracted intense interest. And the Carranza case can be approached, in retrospect, as a microcosm of some of the key tensions of early modern Catholicism in the era of Trent. Among those, the *cause célèbre* can be read as a testing ground for judicial responses to heresy, for the roles of inquisitions in early modern states, and for varied approaches to inquisitorial procedure and process. Moreover, it can reveal much about inquisitorial attitudes towards the problem of publication – which in sixteenth- and seventeenth-century Castilian Spanish connoted the act of declaring something in public, and was intimately related to language about fame and notoriety.[13]

The case spanned seventeen years and two peninsulas, begun under the Spanish Inquisition in Valladolid in 1559 and moving to Pope Pius V's Rome in 1567. Before his dramatic nighttime arrest, Carranza had already led a remarkable career. Famous within the Dominican order and at the Habsburg court, Charles V had sent him as a theologian to the early sessions at Trent, and he had accompanied Philip II to England and the Low Countries. Carranza had distinguished himself as an advocate of episcopal residency, a preacher, a confessor, and as a combatant against the Protestant heresies to which he would be accused of adhering. The trial was bound up with longtime enmities between prominent courtiers as well as substantive arguments over what was potentially dangerous to say or write. The case involved judging whether or

[12] See the analysis of bibliography as missionary activity (and the description of Simancas as 'vir doctissimus' in Possevino's bibliography) in Luigi Balsamo, "How to Doctor a Bibliography: Antonio Possevino's Practice," in Fragnito, *Church, Censorship and Culture*, 64. Adriano Prosperi – playing upon a seventeenth-century manual title – has usefully termed inquisitors' law books their "arsenal." Moreover, scholars are increasingly analyzing the most important early modern Catholic intellectuals – such as Robert Bellarmine or Cesare Baronio – as objects of censorship and as complex thinkers whose treatises are full of complicated stances on contemporary debates. See, inter alia, Mayer, *The Roman Inquisition*; Stefania Tutino, *Shadows of Doubt: Language and Truth in Post-Reformation Catholic Culture* (Oxford: Oxford University Press 2014).

[13] Sebastián de Covarrubias Orozco, *Tesoro de la lengua castellana o española*, ed. Felipe C. R. Maldonado (Madrid: Editorial Castalia, 1995), 839.

not Carranza's *Commentaries on the Christian Catechism*, published in Antwerp in 1558, contained heretical content. It incapacitated the archbishop of Toledo at a critical juncture, and so resulted in a lengthy vacancy of Spain's primacy during the aftermath of the Council of Trent. Pope Gregory XIII concluded the proceedings in 1576, when he sentenced Carranza to abjure sixteen gravely suspicious heretical propositions. Carranza did so, and died in Rome a scant two and a half weeks later, on May 2, 1576, in the Dominican church and convent of Santa Maria sopra Minerva.[14]

News of the Carranza trial appeared in official letters; it became part of diplomatic correspondence and inquisitorial precedent in the *Suprema*'s archives. It appeared in gossip and rumor that circulated at the papal court and on the streets of Rome.[15] Roman *avvisi*, the well-known Italian manuscript news sheets, included notes about the trial's progress.[16] Following his release, one sympathetic dispatch described how Carranza made a tour through Rome, with a large entourage of his supporters, on one day stopping at seven churches where he was received "with great love and courtesy," and shown relics – enacting in public "how such a great prelate has come out clean of heresies, as they hoped." Attempts to interpret the true nature of the trial and of the archbishop only proliferated after his death. Carranza's admirers and detractors fired off letters advancing their positions and folded partisan accounts of the trial into a variety of manuscripts. Brief manuscript accounts of Carranza's sentence,

14 For an introduction to the scholarship on Carranza: John Edwards and Ronald Truman, eds., *Reforming Catholicism in the England of Mary Tudor. The Achievement of Friar Bartolomé Carranza* (Aldershot and Burlington, VT: Ashgate, 2005), and the bibliography in José Ignacio Tellechea Idígoras, *El Arzobispo Carranza, "Tiempos Recios"* (Salamanca: Publicaciones Universidad Pontificia, Fundación Universitaria Española, 2003–7), I, 495–506. There is also analysis of the historical significance of the case in, inter alia, Geoffrey Parker, *Imprudent King: A New Life of Philip II* (New Haven, CT: Yale University Press, 2014); Pastore, *Il Vangelo e la Spada*; Olivari, *Entre el trono y la opinion*; Henry Kamen, *The Spanish Inquisition: A Historical Revision*, 4th ed. (New Haven, CT: Yale University Press, 2013).

15 On rumors circulating about the trial, Simancas, *Vida y Cosas Notables del Señor Obispo de Zamora Don Diego de Simancas*, in Manuel Serrano y Sanz, ed., *Autobiografías y Memorias* (Madrid: Bailly, Bailliére, S. B., 1905), II: 174. For some of Simancas' letters from Rome to Madrid, Tellechea Idígoras, "Cartas inéditas de un inquisidor por oficio. El Dr. Simancas y el proceso romano de Carranza," in *Homenaje a Julio Caro Baroja*, eds. Antonio Carreira, Jesús Antonio Cid, Manuel Gutiérrez Esteve, and Rogelio Rubio (Madrid: Centro de Investigaciones Sociológicas, 1978), 965–99.

16 For example, reports of the arrival of Carranza and the Spanish delegation to Rome in 1567, and a few details about the trial procedure to follow, BAV, MS Urb. Lat. 1040, fols. 32r, 46v–47r, 66v, 411r-v, 413r, 438v.

last days, and death quickly circulated.[17] A pair of letters appeared, purportedly a posthumous vindication from the prominent jurist who had defended Carranza, Dr. Martín de Azpilcueta (popularly known as Dr. Navarro), and a rebuttal – reasserting Carranza's multiplicity of heresies – from an anonymous "friend."[18] Azpilcueta and Simancas – participants in the trial – folded their experiences of it into Latin manuals for legal practice and for prayer. An account of Azpilcueta's life published in 1575 enumerated his defense of Carranza as one of his many praiseworthy and virtuous achievements.[19] Historians also grappled with how to record the trial: mention of it appeared in royal chronicles, histories of the Dominican order, and biographies of Pope Pius V.[20] Despite the obviously exceptional nature of the Carranza trial, these writings provide a suggestive case study of how an Inquisition trial and its meanings was argued about outside the courtroom. Not only the trial itself, but also accounts of the trial became sites of conflict about some of the most contentious issues of Catholic reform: questions about episcopal residence, fraternal correction, jurisdiction, inquisitorial procedure, and the repelling of heresy.[21]

Closer study of two accounts of Carranza's case, circulated after his death, can illuminate the qualities of such texts more fully. Written by Simancas and Pedro Salazar de Mendoza, both were in Spanish and,

[17] For example, *Sentencia y últimas palabras de Carranza*, BNE, MS 6035, no. 16, fols. 200–21. For a series of such materials – Latin copies of Gregory XIII's sentence and Carranza's abjuration; a report of the "news of the case" in Castilian, composed just after its conclusion and full of praise of Carranza (quotations here from this text, fol. 385r); a letter opposing the archbishop – BAV, MS Reg. Lat. 385 (2), fols. 360r–389v.

[18] Simancas, *Vida*, 206–10. See also the analysis of biographies of Carranza in Ronald Truman, "Pedro Salazar de Mendoza and the First Biography of Carranza," in Edwards and Truman, eds., *Reforming Catholicism*, 181.

[19] Simon Magnus Ramlotaeus, *Vita Excellentissimi Iuris Monarchae Martini ab Azpilcueta Doctoris Navarri* (Rome: Victorius Elianus, 1575), fol. 6r.

[20] It appears in Joannes Antonius Gabutius, *De Vita et Rebus Gestis Pii V. Pont. Max. Libri Sex* (Rome: Aloisius Zannetti, 1605), 91–3 (a defense of Carranza's virtues); Girolamo Catena, *Vita del Gloriosissimo Papa Pio Quinto* (Rome: Vincenzo Accolti, 1586), 97–8 (an ambivalent treatment, not even mentioning the sentence); D. Antonio de Fuenmayor, *Vida y Hechos de Pío Quinto Pontífice Romano. Con algunos notables sucessos de la Christiandad del tiempo de su Pontificado. Tercera Edición.* (Madrid: widow of Juan Sánchez, 1639), fol. 103r-v (1st ed., Madrid, 1595; 2nd ed., Zaragoza, 1633; it cast Carranza's comportment in the trial and his reconciliation as virtuous). See also analysis of the depictions in Ambrosio de Morales' notes as an official royal chronicler of Philip II and in the Dominican Friar Juan López's history of his order, printed in Valladolid in 1615 in Truman, "Pedro Salazar de Mendoza."

[21] Stefania Pastore has argued, for example, that the Carranza affair caused Simancas to harden his stance on fraternal correction, *Il Vangelo e la Spada*, 241–51.

notably, issued in manuscript rather than print. Both authors were jurists, and were obviously well aware that their narratives were part of a field of conflicting reports; they wrote explicitly to contest other versions of the trial, and they often used the same descriptive categories, frequently emphasizing the disjuncture between secular and regular clergy, between jurists and theologians, and between Italian and Spanish jurisprudence.

THE LIFE OF SIMANCAS

Simancas centered his autobiography on his involvement in Carranza's trial. Begun in November 1577, after his return to Spain, it sought to demonstrate his virtue and his authority as a jurist, as well as the hardships he had suffered during a career in royal and ecclesiastical service.[22] And he was careful to note having received Gregory XIII's permission to communicate about the case. Despite the recuperation of Carranza's image following the trial's conclusion, Simancas used his *Vida* to systematically attempt to demonstrate the archbishop's guilt, and so to reassert the justice of the trial. He focused on the quality of Carranza's speech, making it show his inclination to heresy and his lack of repentance. He claimed that observers remarked that the criminal (*reo*) – as he continued to refer to Carranza, offering his own verdict – had read his abjuration very dryly, as if it did not pertain to him at all. Instead he praised Melchor Cano (1509–60), another prominent Dominican. Carranza's supporters had frequently depicted Cano as a principal instigator of the archbishop's trial, attributing the theological opinions he rendered in the case to personal enmity and thus seeking to posthumously disqualify him. Simancas, predictably, defended him from these charges and cast him as a friar and qualifier of heretical propositions notable for his judgment and learning. He maintained that Carranza and Cano had once been friends, and that this amplified even further the latter's virtue in following his conscience. Thus Simancas also converted Cano into a model for fraternal correction, judging his brother but forgiving what he could. Again, Simancas attended closely to particular forms of communication, crediting Cano with discovering in Carranza's *Catechism* "the language of the heretics."[23]

In Simancas' account, Carranza and his supporters made a mockery of the inquisitorial process. Following the sentencing, they greeted the archbishop as if he had been absolved. The archbishop then attempted

[22] Lynn, *Between Court and Confessional*, 111–18.
[23] Simancas, *Vida*, 167.

to compel a Spanish secretary – recently employed in his trial – to record a final rebuttal, a testament to a Catholic death: "I was never a heretic, before I preached against the heretics, if not as the best, not as the worst, and I wrote against them, if not as the best, not as the worst."[24] In this telling, Carranza stoked the passions of his partisans and suborned the inquisitorial process even as he lay dying. In retort, Simancas glossed the papal sentence, emphasizing its gravity. He reasserted the heretical material revealed in Carranza's trial, as an inquisitor with privileged access to its contents, and he noted that Gregory XIII had granted him special permission to communicate about the case, breaching the customary secrecy. When he contended that Carranza had publicly used the same words as heretics, he sought to combat those who treated Carranza's death as that of a martyr, and his body as something holy. He heaped scorn on accounts that circulated just after Carranza's death, attacking – as ignorant or foolish – those who believed friars who said they had dreamt that Carranza was in heaven alongside St. Dominic and St. Peter Martyr (that is, in the company of those often considered the inquisitor-saints of his order).[25]

Rome, too, became suspect, as Simancas reported being told even before he was sent there that "in Italy they took the Spanish jurists for barbarians."[26] And so did its inquisitorial regime; while the Spanish Inquisition's judges were increasingly jurists and secular clerics, those in the Roman Inquisitions were more often theologians, members of the regular clergy, and frequently Dominicans. His portrayal of both Pius V and Gregory XIII was ambivalent, at times dissatisfied, sometimes admiring or seeking to show his proximity to the pope. Yet when Simancas remembered his departure from Rome in 1576, he claimed a whole cohort accompanied him to the port, attesting to the fame and good reputation he had acquired there. His reported experience thus mirrored the retinue who had accompanied Carranza on his release. Simancas created a cast of characters. He sorted both the jurists and the theologians, many of them Dominicans, into camps of the learned and virtuous, on one side, and, on the other, the perverters of justice. Refuting his reputation as an enemy of theologians, he frequently invoked them – and

[24] *Ibid.*, 183.

[25] *Ibid.*, 184. Or the timing and manner of Carranza's death was likened to the Dominican Pius V's (beatified in 1672, canonized in 1712), *ibid.*, 206. On the political resonances of dreams, Richard L. Kagan, *Lucrecia's Dreams: Politics and Prophecy in Sixteenth-Century Spain* (Berkeley, CA: University of California Press, 1990).

[26] Simancas, *Vida*, 153.

especially prominent Dominicans – to establish his authority as a jurist. He reported how the friar Domingo de Soto, the great theologian of the University of Salamanca, had praised him when he was a student, and how Marco Medici, "the Inquisitor General of Venice, Dominican friar of great authority and doctrine," praised his publications on inquisitorial law.[27] He made himself a model inquisitor, asserting: "I alone won the case in the time of Gregory XIII, with the greatest honor for Spain, and especially for the Catholic King and the Holy Office, in which all the work I have given has been well employed, although in this life they have not shown me gratitude."[28]

He described those he pursued as a judge. He classified Domingo de Rojas as "the most notable heretic" condemned in the 1559 *auto de fe* in Valladolid in which elite courtiers had been revealed as dangerous secret Lutherans, an event that had opened Simancas' career on the *Suprema*. His notoriety – in Simancas' explanation – stemmed from his noble birth as the son of a marquis, from his status as a Dominican friar and the last disciple of the archbishop of Toledo, from his wide dissemination of heretical teaching to relatives and friends, and from his pertinacity nearly until the bonfire of the *auto de fe*.[29] Simancas cast Carranza as the inverse of the good Dominican, the good archbishop, and the good theologian. He noted how in the opening audience, "We made him the ordinary admonitions, and he was negative to all of them, saying that he had no more guilt than Saint Dominic."[30] Simancas made Carranza's invocation of the saint a mark of his perversity, lacking reverence for his order and mocking the inquisitorial process from his opening audience.

Then, in the subsequent paragraph of his *Vida*, Simancas described how friar Diego de Chaves, a Dominican (and former royal confessor), produced an exhaustive assessment of the bad propositions in Carranza's *Commentaries*, working with a Franciscan, a Benedictine, and a Hieronymite. He juxtaposed Carranza to Chaves – another Dominican, another theologian – who worked diligently alongside the jurist-inquisitors, tirelessly and carefully judging the heresies of Carranza, his false brother. Simancas created a lesson about inquisitorial procedure: theologians were required to judge theological propositions with precision and care, as the foundations upon which any good case must rest. The theologians might assess with "prolixity" (preventing future

[27] *Ibid.*, 151, 180–1.
[28] *Ibid.*, 189.
[29] *Ibid.*, 156–7.
[30] *Ibid.*, 159.

questioning of their authority), in order to allow the inquisitors to act with good judicial brevity. Simancas even reported that he had petitioned especially for Chaves' presence in Rome. But, he complained, in Rome everyone had attributed the detailed nature of these theological assessments to ill will, partisanship, and enmity, and thus sought to disqualify them. Simancas claimed to have foreseen this conflict and urged moderation in the theologians' work, but protested that such detailed assessments were requisite because "they have to assess everything, that the Judges take from it what they wish." Carranza, on the other hand, negated process constantly, "justifying himself in everything and by everything," thus creating a muddle. And Simancas recounted how the king, with his "great judgment," had preferred his arguments to those of Azpilcueta, Carranza's defender, rejecting Azpilcueta's initial attempt to have the case transferred to the papal court in Rome.[31] Simancas wrote his jurist opponent as his foil: Azpilcueta cloaked bad reasoning in eloquence; he was consumed by passion.

THE LIFE OF CARRANZA

In 1586, Pedro Salazar de Mendoza (*c.* 1550–1629), a canon in Toledo's cathedral and legal consultant to the Inquisition, began composing an account of Carranza's life. He credited Gaspar de Quiroga (1512–94) with spurring him to write it.[32] A contemporary of Simancas, Quiroga was also a jurist who had spent time in Rome, Carranza's immediate successor as archbishop of Toledo, and the Inquisitor-General when Carranza's case closed; in 1578, Gregory XIII made him a cardinal and in 1583, Quiroga issued a newly expanded Index. Salazar's *Vida* of Carranza sought to rehabilitate Carranza, the archdiocese, and the Spanish Inquisition, following charges that it had behaved scandalously in Carranza's trial,

[31] *Ibid.*, 159. See also *ibid.*, 176.

[32] I have consulted the eighteenth-century copy (bound with an account of the Froilán Díaz case) in the Spanish History manuscripts at the Lilly Library, Indiana University: *Vida, causa, y sucesos, prósperos, y adversos del Illustrísimo y Reverendísimo Señor Don Fray Bartolomé de Carranza, y Miranda, Arzobispo de Toledo, de la esclarecida Orden de Predicadores, Escrita por Don Pedro de Salazar y Mendoza, Canónigo Penitenciario de dicha Santa Iglesia á instancia del Excelentísimo Señor Cardenal de Quiroga su dignísimo Arzobispo de Toledo año de 1586* (pages unnumbered). See the analysis of this biography in Truman, "Pedro Salazar de Mendoza"; and of Salazar in Richard G. Mann, *El Greco and His Patrons: Three Major Projects* (Cambridge: Cambridge University Press, 1986), 112–20; Richard L. Kagan, "Pedro de Salazar de Mendoza as Collector, Scholar, and Patron of El Greco," *Studies in the History of Art* 13 (1984): 85–93.

succumbing to the envy of the Inquisitor-General Fernando de Valdés or of other members of the Dominican order. Salazar justified his interpretation by reference to another history – recently printed in Latin – that termed the trial a complete invention and affront, even though he protested that he would not go so far as that; he cast the prosecution as an abuse of justice, even as he simultaneously reinforced the rectitude of Gregory XIII's sentence and the inquisitorial process in general.[33] And he rendered Salazar as an exemplary figure, with qualities reminiscent of holy martyrs: "those who dealt with him admired the tranquility and serenity of his spirit in the midst of such terrible misfortunes."[34]

Salazar deployed similar details to Simancas, as each sought to advance the truth of their interpretation of events. The narrative's opening counted the length of Carranza's confinement in Valladolid and in Rome down to the days, noting the time in transit between the two, in order to total the years, months, and days of imprisonment the archbishop had suffered; Simancas had also counted the days of his Roman sojourn. Salazar provided authoritative figures to question Carranza's guilt. He reported that a serious Franciscan friar had claimed to long for Judgment Day, in order to know the truth of the matter. A decorated Dominican had mused that if an angel were to say that either he or Carranza was a heretic, he would sooner have believed himself the heretic. Salazar framed the archbishop's life as an object lesson in the vicissitudes of fortune and the impossibility of discernment in this world. Nowhere were the contradictions of Carranza's story more apparent, he implied, than in the epitaph on his sepulcher. There, the archbishop ordered to publicly abjure sixteen heretical propositions was publicly memorialized as illustrious both for preaching and for his Christian life.

From the start, Salazar emphasized Carranza's renown as a theologian. As a further proof of his authority and orthodoxy, he stressed his involvement in inquisitorial business as an expert theologian; he emerged as the model bishop, friar, and heretic-fighter. Salazar found signs of Carranza's holiness, recounting the archbishop's conversations about pious works and his suffering at his lack of access to the sacraments during his long inquisitorial trial. His faithful companion friars also suffered diminished access to the sacraments. Salazar emphasized how Carranza "prayed the canonical hours by the Breviary of St. Dominic, like the rest of the friars, the same as he did being free," but he added even more verses

33 He referenced the history of Gabutius (called there Graviso); see note 20 above.

34 Salazar, *Vida, causa, y sucesos*, [fol. 2v].

of the psalms while imprisoned, prayed tenderly and with tears.[35] He depicted Carranza's imprisonment as sowing scandal and aiding heretics; Carranza became the adept arguer, who marshaled historical precedents, divine and human, to show the injustice of his trial. Simancas appeared in Salazar's account only in passing, in describing the arrangement of the participants in the Valladolid hearings, the crossing to Rome, and those involved in the case there.[36] Such silences are always tempting; Salazar may have meant to silently indict Simancas, or perhaps he contradicted his version of the trial merely by rendering him fairly irrelevant.

Like Simancas, Salazar recounted Carranza's contention – in the first audiences of his trial – that he "had no more guilt than his Father Saint Dominic." In this context, however, Carranza appeared the hero: he "always justified himself with great conscience" against the opinions of Chaves and the other theological consultants.[37] Salazar alluded to the machinations of envious members of Carranza's own order as a potential cause for the trial, but depicted Cano ambivalently. He recounted how the trial had jeopardized Cano's reputation and position of authority in his order. He depicted Cano as receiving permission to confer with two other learned and respected friars (including Domingo de Soto), in preparing his judgment of the *Catechism*. He reported that Cano had reasserted his respect for Carranza, and for the important dignities he held.[38] This time, Ugo Boncompagni – the future Gregory XIII – became a model jurist. Part of the 1564 papal delegation to Spain, unsuccessful in settling the Carranza affair then, Salazar described him as a "prelate of great opinion, and authority, jurist by profession," and so the ideal judge to issue the final sentence.[39] Salazar's account was often an inversion of Simancas': the inquisitor recounted how his health failed during the long years in Rome; Salazar depicted Carranza's health – and implicitly, perhaps, the health of the trial – as improved in Rome. If polemicists often used metaphors of health or sickness to assess the body politic and its meting out of justice, so too might they illuminate the nature of a trial through the physical affect of its protagonists.

They invoked different biblical templates. In summarizing the meanings of Carranza's career, Salazar called up Jonah. He likened the archbishop's ascent to the plant that grew to shade Jonah, which flourished

35 Salazar, *Vida, causa, y sucesos*, ch. 23.
36 *Ibid.*, chs. 22, 25, 27.
37 *Ibid.*, ch. 23.
38 *Ibid.*, ch. 20.
39 *Ibid.*, ch. 24.

in one night and perished in another. He reinforced this with a gesture to the *flor de la maravilla*, whose beauty lasted only a day. This imagery gestured to divine providence, and to the fleeting nature of the temporal life. Paired with an assertion of Carranza's quietude and the serenity of his soul throughout the trial – reportedly remarked upon by those he encountered – these allusions rendered Carranza a virtuous penitent. Salazar chose language laden with ambivalent meanings. The message might invoke Jonah's history as a preacher against immorality, or it might be Christological, as Jonah was one of the figures of the Hebrew Bible depicted as prefiguring Christ. Alternately, the description of Carranza might offer a hint of something more sinister: some seventeenth-century uses of the *flor de la maravilla* gestured not only to its ephemeral nature, but also to dissimulation, as something of beauty which contained the seeds of sorrow. Salazar's critique of the Inquisition's actions emphasized the difficulties of discernment and the limitations of human judgment.[40]

Simancas, likewise, had selected authorities who implied the virtue of his actions and the hope of divine reward, knitting together a church father, a pope, and a patriarch to make his case. He referred to Jerome's comments that those nearing death – or who thought their mortal end was near – more easily renounced temporal things, suggesting that he no longer sought higher worldly charges. He claimed that, as he prepared to leave Rome, his competitors had hurled the same calumny at him that had once been directed at Pope Innocent III (1198–1216). He thus forged a link to a papacy associated with a council of notable doctrinal import (perhaps akin to the era of Trent) and with the initial authorization of both the Dominican and Franciscan orders; he reminded a learned audience of a prelate known as a skillful jurist and theologian – associated with the universities at Paris and Bologna – and for his staunch pursuit of heretics. He intimated numerous parallels

[40] Salazar, *Vida, causa, y sucesos* [fol. 2v]. Jonah 4: 6–7. The identity of this plant – frequently rendered as gourd or as a varietal of ivy in English – has been the subject of debate since the early Church. See Natalio Fernández Marcos and Emilia Fernández Tejero, *Biblia y humanismo. Textos, talantes y controversias del siglo XVI español* (Madrid: Fundación Universitaria Española, 1997), 18. Tirso de Molina employed this image in his Corpus Christi *auto sacramental*, entitled *No le arriendo la ganancia* (*c.* 1612–13), to similar effect, to demonstrate the fleeting nature of the mortal life. Ann Nickerson Hughes, *Religious Imagery in the Theater of Tirso de Molina* (Macon, GA: Mercer, 1984), 14. A cursory database search indicates that other early modern playwrights and poets (with Tirso, also Luis de Góngora, Pedro Calderón de la Barca) employed the imagery of the *yedra de Jonás* and the *flor de la maravilla*. *El inobediente o la ciudad sin Dios* of Félix Lope de Vega y Carpio (1562–1635) included a conversation between Jonah and God, drawing from the same biblical passages; www.cervantesvirtual.com.

with his own day. He chose a biblical citation with even more polemical force. He asserted that he too could say to his opponents what Joseph had to his brothers, that they had thought to do him ill, but God had converted it into good. He thus selected a familiar figure from the Hebrew Bible to associate himself with a story of divine vindication and reward, effected even in the temporal life, a tale also illustrative of showing mercy to vanquished opponents, and one steeped in an aura of fraternal correction and forgiveness.[41]

PUBLICITY BATTLES AND INQUISITION TRIALS

The various narratives produced about the Carranza trial – both during and after – reveal attempts to influence perceptions of the actions of the Spanish Inquisition, to build precedent and institutional memory in order to shape future decisions, as well as potentially to sway trials in process.[42] Although the Spanish Inquisition was a legal institution whose trials took place within a specific framework of law and procedure, inquisitors perceived their work as having important extrajudicial and symbolic components. Their office embroiled them in debates with their colleagues and superiors and in an ongoing negotiation of authority with other royal and ecclesiastical officials; moreover, the conduct of inquisitorial business required the ongoing cultivation of an image of inquisitorial authority. Thus the writings produced surrounding trials can be seen as substantive pieces of political communication, one of a complex of persuasive tools inquisitors and their associates employed.[43]

Yet, as Simancas' career suggests, sixteenth-century inquisitors also saw great risks to circulating their opinions in writing, and they seem to have been particularly wary about the use of print. In 1567, not long after his arrival in Rome, Simancas was part of a joint Spanish–papal commission to review the censure of a theological work that had been published in Latin in Louvain.[44] Later, he sought to silence the Discalced

[41] Simancas, *Vida*, 195–6. Gen 50:20.

[42] Members of the *Suprema* referenced Simancas' *Vida* during another high-profile trial in 1643; it was recopied in the seventeenth century and maybe printed. Salazar's life continued to be recopied in manuscript for more than two centuries; it was printed in the eighteenth century and widely known.

[43] See Andrew Pettegree, *Reformation and the Culture of Persuasion* (Cambridge: Cambridge University Press, 2005).

[44] The censored work was even seemingly a treatise by an inquisitor, Michael Baius, Inquisitor-General in the Low Countries. On his career, Violet Soen, *Geen pardon zonder Paus!: studie over de complementariteit van het koninkrijk en pauselijk generaal pardon*

Franciscan friar Alonso de Lobo, who was preaching in Rome in the 1570s against the Toledo cathedral's purity of blood statutes. Simancas propelled the Roman Inquisition to try Lobo, who then publicly abjured his errors in Rome in late 1572. This was not the end of the affair, however. Written narratives followed. And Lobo's career as a popular preacher who traveled throughout Italy – favored by figures as famous as Carlo Borromeo – continued as well, through the 1580s. One short, anonymous, manuscript account of Lobo's abjuration circulated decades later, after his death in 1593, listing his three errors, but also describing him as responsible for many good works, as one of Spain's best preachers, well loved by Gregory XIII, and "with no less holiness in his death than he had in his life."[45] Simancas, on the other hand, persisted in his opposition to Lobo, writing a defense of the statutes and having it printed in Latin in Antwerp, under a pseudonym, in 1573 and 1575.[46] This was, seemingly, one solution to an ongoing problem confronted by inquisitors: how to publicly rebut something that they claimed it was harmful to publicly debate. There are also later manuscript copies of that treatise, in Spanish.[47]

Significantly, Simancas' extensive corpus of printed writings was all in Latin. And yet, as Simancas' dedication to Gregory XIII indicated, these were also seen to be potentially controversial. He circulated some compositions only in manuscript – not only his vernacular *Vida*, but also with a small Latin legal treatise on financial interests and abuses he entitled *De ambitu*, which he noted that he sent to correspondents.[48] Simancas folded

(1570–1574) en over inquisiteur-generaal Michael Baius (1560–1576) (Brussels: Paleis der Academiën, 2007).

45 BNE MS 6035, fols. 101–3; Lynn, *Between Court and Confessional*, 105–6, 128–9; Pastore, *Il Vangelo e la Spada*, 414–15. On Lobo's full career, see Fedele Merelli, "P. Alfonso Lupo Cappuccino e San Carlo Borromeo," *L'Italia Francescana* 64(2–3) (1989): 138–348.

46 Didacus Velásquez, *Defensio statuti Toletani a Sede Apostolica saepe confirmati, pro his, qui bono et incontaminato genere nati sunt* (Antwerp: Christophe Plantin, 1573; 1575). A manuscript copy in Spanish: "Libro en favor de la Iglesia de Toledo Compuesto en latin por el lic. Diego Velázquez (1575) y traducido Madrid, agosto de 1592," BNE, MS 2592. See also Robert Aleksander Maryks, *The Jesuit Order as a Synagogue of Jews: Jesuits of Jewish Ancestry and Purity-of-Blood Laws in the Early Society of Jesus* (Leiden: Brill, 2009), ch. 1. A copy of the 1573 Latin treatise has written in as a sort of appendix – still in Latin – an account of Lobo's revocation following the proceedings of the Roman Inquisition, BAV, shelfmark Stamp.Barb.FF.I.126.

47 Similarly, inquisitorial treatises consistently struggled with the problem of how and when to dispute heretics.

48 Simancas, *Vida*, 153–4. Guy Lazure has found evidence that Simancas gave it to at least one correspondent.

opinions about the evolving Carranza case into the revised editions of his published manuals, even as Azpilcueta had had a Latin treatise of his, published in 1572, denounced for precisely that reason, for incorporating into it "a few sentences in defense of Carranza."[49]

Nor was Simancas immune to such denunciations. The concerns about censorship that Simancas aired in the dedication of his 1575 Inquisition manual to Gregory XIII took definitive form in a series of events in Rome which Simancas related in his *Vida*. He painted himself as the virtuous persecuted, his book attacked as a result of his opposition to both Carranza and Lobo:

> The partisans of the Archbishop and of Lobo, not being able to suffer that I should leave Rome with so much honor, with rabid hatred and diabolical malice thought how they could darken about my name; and not being able to put into question my life and customs, some of them agreed to publish (*publicar*) that in my book the *Instituciones Católicas*, so examined in Spain and in Rome, and so approved and praised in all Italy, there was a great error, and they gave a petition about this to the Pope.

He reported that Gregory XIII, in turn, handed the matter over to Cardinal Madruzzo for review (according to Simancas the latter was a "very good theologian"). Simancas proceeded to describe how he went to visit the cardinal in his house, and how – through that visit, further conversations in his own residence, and writing an explanatory brief – he successfully defended his book before the charge at hand, which hinged on his use of a quotation of Pope Innocent III. Simancas, moreover, explicitly tied the episode back not only to the trials he had propelled but to other events of the day, reporting how *conversos* in Toledo and – according to the cardinal – the delegates at Trent had been heatedly debating the interpretation of Innocent.[50]

* * *

And there is suggestive evidence of the persuasive efforts that surrounded Spanish Inquisition cases even in the decades preceding the Carranza affair. The renowned Franciscan friar Francisco Ortiz preached in Toledo against the Inquisitor-General's prosecution of Francisca Hernández in the early 1530s, before being prosecuted himself.[51] Manuscript news of

[49] This was his *Finibus humanorum actuum*, Pinto, *Inquisición y control ideológico*, 33. See also references to Carranza in Azpilcueta, *Enchiridion sive Manuale Confessariorum, et Paenitentium* (Venice: heirs of Francesco Ziletti, 1589), 887, 910.

[50] Simancas, *Vida*, 188–9.

[51] Lu Ann Homza, "How to Harass an Inquisitor-General: The Polyphonic Law of Friar Francisco Ortíz," in *A Renaissance of Conflicts: Visions and Revisions of Law and*

Spanish Inquisition trials and their results circulated in the sixteenth century, for example in reports of the demonic possession of a Cordoban nun who was tried in the 1540s.[52] In the seventeenth century, the quantity of print material generated by Spanish Inquisition tribunals increased markedly.[53] Then, print was used more extensively to publicize edicts, in broadsheets about heresies to avoid, denounce, and confess.[54] There survive numerous printings of copies of the constitutions of the confraternity for the Inquisition's officials, dedicated to Saint Peter Martyr.[55] There had been sixteenth-century printings of the collected procedural instructions of the Spanish Inquisition, but there was seemingly a concerted effort in the late 1620s to reissue them in print as more useful and accessibly organized reference tools; while most of the legal commentary was printed in Latin, the instructions were in Spanish.[56] In that era, as the dissemination of pamphlet literature increased across European societies, regional Spanish Inquisition tribunals also used a mixture of manuscript memoranda and printed briefs to assert their jurisdictional authority and sway public opinion. Murcia's inquisitorial court used these tools to complain about the way the local bishop referred to himself as an inquisitor and meddled in their public displays of authority, and to defend inquisitorial personnel against charges, issuing pamphlets arguing that two familiars charged with murder should be under inquisitorial jurisdiction (which laid claim to the civil and criminal affairs of its officials), that they should be absolved and released from their imprisonment by secular justice.[57]

Society in Italy and Spain, eds. John A. Marino and Thomas Kuehn (Toronto: Centre for Reformation and Renaissance Studies, 2004), 299–336.

52 On Magdalena de la Cruz, see Lu Ann Homza, ed. and trans., *The Spanish Inquisition 1478–1614. An Anthology of Sources* (Indianapolis, IN and Cambridge, MA: Hackett, 2006), 168–75. For the insight that Simancas referenced that case in his Inquisition manuals, see Anne Jacobson Schutte, *Aspiring Saints: Pretense of Holiness, Inquisition, and Gender in the Republic of Venice, 1618–1750* (Baltimore, MD: Johns Hopkins University Press, 2001), 62.

53 See Fernando Bouza, *Communication, Knowledge, and Memory in Early Modern Spain*, trans. Sonia López and Michael Agnew (Philadelphia, PA: University of Pennsylvania Press, 2004).

54 Of many examples of this, there are close to thirty printed broadsheet edicts issued by the Inquisition tribunal in Mexico between the 1610s and 1660s in one file at the Lilly Library, AC 86. M6 I5 1646 Mendel.

55 For example, the pamphlet dated Murcia 1607 in AHN, Inq., libro 500, fols. 107–26v.

56 For example, the edition prepared by the secretary of the *Suprema: Instruciones del Santo Oficio de la Inquisición sumariamente antiguas y nuevas puestas por abecedario por Gaspar Isidro de Argüello* (Madrid: Imprenta Real, 1627; 1630).

57 The inquisitors requested their prosecutor, Juan Adam de la Parra, be reimbursed for the costs of printing a pamphlet, May 15, 1634, AHN, Inq., leg. 2810. For the jurisdictional

Reports of *autos de fe* – those infamous, highly choreographed displays of the Inquisition's sentenced criminals – had long been a staple of the Inquisition's internal correspondence. Inquisitors described them as a method of instructing the public, and used them to demonstrate their diligent work to their superiors, usually with a request for extra funds. Manuscripts and printed pamphlets had reported news of the 1559 Valladolid *auto de fe*, some making it an example of necessary defense of the faith and others of horrifying persecution.[58] In the seventeenth century, the reports developed as an identifiable print genre, spreading the calculated display of the *auto de fe* far beyond those present at the event. They narrated, sometimes in ample detail, elements of the ceremony, the names of convicted heretics, and synopses of what had been found to be their crimes.[59] In addition, printed versions of a few sermons delivered at *autos* or at the proclamation of inquisitorial edicts began to appear.[60] The Spanish Inquisition appeared as news in other ways as well – as, for example, a rallying cry in pamphlets penned by early modern Protestants. Its policy changes also could become fodder for pamphleteers, as when English and Dutch reading audiences were quickly apprised of the Inquisition's newly moderated regulations for how to treat Englishmen who might appear before the tribunals.[61]

Inquisition tribunals and the men who were Spanish inquisitors – at various times in their careers – could be patrons of cheap print; they might select younger clerics or officials to preach sermons at their public events or write pamphlet reports of *autos de fe*. Inquisitor-General

conflict with the bishop of Cartagena, BNE, MS 6157, fols. 47r-51v, 85r-96r; on the importance of pamphleteering in his career, Lynn, *Between Court and Confessional*, ch. 4.

[58] At least twelve editions in Italian, Dutch, and German appeared in print in 1559, Emil van der Vekene, *Bibliotheca Bibliographica Historiae Sanctae Inquisitionis*, 3 vols. (Verduz: Topos Verlag, 1982–92), I: 191–3, III: 36–7. Also: BNE, MSS 721, 883, 6176. On the circulation of the manuscript accounts, David Vázquez Santiago, "Propedéutica de los autos de fe de Valladolid de 1559: Espiritualidad y poder en Castilla" (memoria de licenciatura, Universidad Complutense de Madrid, 2000); Elias Amezaga, *Auto de Fe en Valladolid* (Buenos Aires: Gráficas Ellacuría, 1966).

[59] Bethencourt, *The Inquisition*; Alejandro Cañeque, "Theater of Power: Writing and Representing the Auto de Fe in Colonial Mexico," *The Americas* 52 (1996): 321–43.

[60] A cursory sense can even be given of those cataloged in Spanish libraries at www.mcu.es/ccpb – of, among others, sermons published in 1621, 1624, 1633, 1652, 1654, 1655, 1688, 1695, or of reports of *autos de fe* of 1610, 1624, 1625, 1632, 1639, 1649, 1659, 1680.

[61] Appearing in pamphlets of 1605 and 1609, respectively; Gary K. Waite, "Empathy for the Persecuted or Polemical Posturing? The 1609 Spanish Expulsion of the Moriscos as Seen in English and Netherlandish Pamphlets," *Journal of Early Modern History* 17 (2013): 95–123 (here p. 107).

Quiroga was an early patron of Salazar de Mendoza, who later wrote various legal commentaries and histories, and of the inquisitor Luis de Páramo, who composed a lengthy Inquisition history (and Valdés had patronized Simancas' early career). The example of Simancas also proposes the consideration of inquisitors as writers, despite the anachronism of that category. It is quite clear that inquisitorial work required a substantial amount of writing, missives, and memoranda with specific rhetorical conventions. And there is evidence that a small but still significant minority of inquisitors published treatises, both before and after their appointment to that office. Teresa Sánchez Rivilla's prosopography indicates that of the forty-six Inquisitors-General who took office between 1483 and 1818, sixteen – distributed across three centuries – produced books of some kind, if indices of prohibited books and Inquisition instructions are included. Many of these were texts on law or ecclesiastical power. Of the 363 appointees to the *Suprema* she counted between 1488 and 1819 (excluding the thirteen who would eventually become Inquisitors-General), she noted some treatise by twenty-four of them (Simancas was one). They published in Latin and in Spanish, generally on legal or theological topics. All the councilors described as writers were appointed to the *Suprema* in the sixteenth or seventeenth centuries, and mainly in the latter.[62] And a handful of inquisitors in the twenty-one district tribunals also published treatises in manuscript and print.[63]

In sum, looking at the spectrum of ways Spanish inquisitors (and other officials and consultants of tribunals) interacted with the written word can contribute to the effort to build a more nuanced sense of the intricacies of Catholic intellectual, political, and religious culture, especially

[62] Teresa Sánchez Rivilla, "Inquisidores Generales y Consejeros de la Suprema: documentación biográfica," in *Historia de la Inquisición en España y América*, eds. Joaquín Pérez Villanueva and Bartolomé Escandell Bonet (Madrid: Biblioteca de Autores Cristianos, Centro de Estudios Inquisitoriales, 2000), III, 228–437.

[63] Of the seventy-two judges appointed to Sicily's court between 1487 and 1712, Arnaldo Albertí and Juan de Rojas had written about inquisitorial jurisprudence while they were inquisitors in Mallorca and Valencia, respectively, before each was transferred to Palermo. Besides his history of the Inquisition, Páramo published on Sicily's jurisdictional issues, as did Vincenzo Percolla. In the mid-seventeenth century, Diego García de Trasmiera (who was promoted to the *Suprema*) published on similar themes, in addition to writing hagiographies. Such treatises, like those of Simancas, included commentary on recent inquisitorial actions. Manuel Rivero Rodríguez, "La Inquisición Española en Sicilia," in *Historia de la Inquisición*, III, 1212–20. Lesser Inquisition officials also published treatises, including on aspects of Inquisition procedure; sometimes this was part of a strategy to pursue higher offices. Gabriel de Quemada, for example, published a treatise on the Inquisition's fiscal affairs in the 1560s; see also the examination of affiliates of the Mexican tribunal in Nesvig, *Ideology and Inquisition*.

in the wake of the Council of Trent. Inquisition trials took place in a climate of debate and contestation that remains to be more fully recaptured. These cases were sometimes part of the news that circulated in manuscript or print, and, in certain circumstances, inquisitors blurred their institution's conventions about secrecy and publicized particular visions of their activities.[64] Moreover, as scholarship like Martin Nesvig's is showing about the censors and theological consultants of the Spanish Inquisition, or as scholars of the Roman Inquisition and the Roman Congregation of the Index are increasingly demonstrating about the cardinals and officials of those bodies, the regulators of orthodox speech and writing were a fractious and varied group. They themselves held a range of ideological commitments which they not infrequently expressed in their own writings, they were frequently at odds with one another, and even busy censoring one another.[65] Spanish inquisitors must also be included in this revised vision, reconsidered as part of networks of intellectuals and circles of ecclesiastics. Publishing could be an avenue of career advancement; yet it could also have its prices. Simancas' hopes for rewards were largely frustrated, and none of his contemporaries on the *Suprema* printed writings as he did. Gigliola Fragnito has suggested, for example, how ecclesiastics at the papal court in the later sixteenth century, after seeing the investigation and correction of esteemed colleagues and their writings, decided not to risk issuing their own manuscripts in print.[66] The involvement in inquisitorial activities could also dissuade intellectuals from publication. All this suggests the possible ambivalences of inquisitorial careers in the crucial later sixteenth century, how inquisitors' attitudes about what speech and writing should be suppressed was based in part in their own reading, writing, and investment in books as a form, how inquisitors might seek not just to suppress supposedly erroneous print but also to create a renewed Catholic literature, and how inquisitors themselves were subject to correction and political and ideological constraints. Moreover, as inquisitors and other jurists and clerics reported on and argued about inquisitorial actions, their communications and controversies yield a picture of a Spanish Church and a Spanish Inquisition that was anything but univocal.

[64] Cf. de Vivo, *Information and Communication in Venice*.
[65] See also Pastore, *Il Vangelo e la Spada*.
[66] Fragnito, "The Central and Peripheral Organization of Censorship," in *Church, Censorship and Culture*, 48–9.

10

"An Immense Structure of Errors": Dionisio Bonfant, Lucas Holstenius, and the Writing of Sacred History in Seventeenth-Century Sardinia*

A. Katie Harris

On June 10, 1614, Gavino Manca de Cendrelles (1562–1620), archbishop of the northern Sardinian see of Sassari, ordered excavations within an ancient basilica located in the nearby seaside town of Torres (now called Porto Torres).[1] By June 22, workers had recovered the long-lost relics of the paleo-Christian martyr Gavinus and his companions Protus and Januarius, together with many inscriptions, grave goods, and human remains that the archbishop and his advisors judged to be the sacred relics of ancient Christian martyrs.[2] The discoveries were greeted with immense enthusiasm, and miraculous healings effected through invocation of the three saints proliferated in towns and villages throughout the

* Research for this paper was supported by a Franklin Research Grant from the American Philosophical Society, a Research Grant from the Renaissance Society of America, and by the Academic Senate of the University of California, Davis.

1 For the notarial records for the excavations, see Archivio Storico Diocesano di Sassari (ASDS), Fondo Capitolare (Fon. Cap.), Serie Q4. I am very grateful to Mons. Giancarlo Zichi, director of the ASDS, for facilitating access to the manuscript materials related to these finds.

2 Gavino Manca de Cendrelles, *Relación breve de la invención de los cuerpos de los ilustríssimos Mártires San Gauino y San Proto, y San Ianuario, Patrones de la Iglesia Metropolitana Turritana, que se han hallado con otros muchos Santos por el mes de Iunio del año 1614. en el templo dedicado a los mismos Santos de la ciudad antigua de Torres del Reyno de Cerdeña* (Madrid, 1615), fol. 19v-20r. In 1695, Archbishop Juan Morillo Velarde (1652–99) and a council of theologians assigned names drawn from the Roman Martyrology to the otherwise anonymous bones. ASDS, Fon. Cap, Serie Q4, fol. 83r–84r. For modern archaeological assessments of the seventeenth-century finds, see Guglielmo Maetzke, *Monte Agellu. Le origine della Basilica di San Gavino di Porto Torres secondo le testimonianze archeologiche* (Sassari: Chiarella, 1989); Fernanda Poli, *La Basilica di San Gavino a Porto Torres. La storia e le vicende architettoniche* (Sassari: Chiarella, 1997).

surrounding region.[3] Archbishop Manca de Cendrelles informed King Philip III of Spain about the finds in a short pamphlet published in 1615 in Madrid, and in the years that followed, other publications continued to spread the news.[4]

To the south, in Cagliari, Sardinia's largest and most populous city, the finds in Sassari prompted not piety but competition. While local patriotism and civic rivalries were common throughout the early modern Mediterranean, the competition between Sassari and Cagliari, the two most prominent cities on the island, was particularly intense.[5] Sardinia was one of many territories possessed by the Spanish Crown, and Sassarese and Cagliarese elites jockeyed with one another for the advantage in relations with Madrid and the institutions of Spanish government on the island. Sassari contested Cagliari's claim to be Sardinia's capital and viceregal seat, and in the early seventeenth century, Sassarese nobles insisted that they should meet and vote separately from their Cagliarese counterparts in the island's representative assembly, the Cortes.[6] The cultural sphere, too, became an arena for conflict, as the Cagliarese worked hard to undermine Sassari's efforts to found Sardinia's first university.[7]

The rancorous rivalry between the two cities was intimately linked to a bitter struggle between Sardinia's most prominent religious leaders. By the late sixteenth century, the archbishops of Cagliari and Sassari were engaged in an acrimonious dispute over which of the two could legitimately lay claim to the title of Primate of Sardinia and Corsica.[8]

[3] For the miracles associated with the finds in Torres, see ASDS, Fon. Cap, Serie Q3.

[4] Manca de Cedrellas, *Relación breve*; Francisco Bastelga, *Relación sumaria y verdadera de todo lo que ha sucedido, y de la multitud de cuerpos de Santos, que se han hallado en la Iglesia de san Gavino de Torres, que está situada, y plantada fuera de la Ciudad de Sacer, distante de doze millas, junto a la mar, y puerto de Torres de la dicha Ciudad, azia la parte que el sol se pone, en el Reyno de Cerdeña* (Barcelona, 1615).

[5] The literature on local and national identities and loyalties in early modern Spain and Italy is extensive and growing. For a good discussion of identification with the local *patria*, see James Casey, "Patriotism in Early Modern Valencia," in *Spain, Europe, and the Atlantic World: Essays in Honour of J.H. Elliott*, eds. Richard L. Kagan and Geoffrey Parker (Cambridge: Cambridge University Press, 1995), 188–210. On the struggles between Sassari and Cagliari in the early modern period, see Francesco Manconi, *Tener la patria gloriosa. I conflitti municipali nella Sardegna moderna* (Cagliari: CUEC, 2008); Gian Giacomo Ortu, "Centralismo e autonomia nella Sardegna di Filippo III," *Rivista storica italiana* 102(2) (1990): 302–39.

[6] Manconi, *Tener la patria gloriosa*, 24–34, 46–57.

[7] Raimondo Turtas, *La nascita dell'università in Sardegna. La politica culturale dei sovrani spagnoli nella formazione degli Atenei di Sassari e di Cagliari (1543–1632)* (Sassari: Dipartamento di Storia, Università degli Studi di Sassari, 1988).

[8] On Sardinia's relationship with the Spanish crown, see Franceso Marconi, "The Kingdom of Sardinia: A Province in Balance Between Catalonia, Castile, and Italy," in *Spain in*

While the history of the primacy is murky, it seems to have had a local, rather than papal origin. A frustrating paucity of written sources for both the ancient and medieval eras, coupled with Sardinia's turbulent history of repeated invasions and incursions by Roman, Vandal, Byzantine, Moorish, Berber, Genoese, Pisan, and Aragonese forces made documenting it and many other aspects of the island's early history quite difficult for would-be historians.[9] By the early sixteenth century, the prelates of both Sassari and Cagliari were styling themselves primates, and by the end of the century, the question had made its way to the judges of the Rota, the papacy's high court of appeals.[10] While the title itself was largely honorific, the possessor of the primacy stood to benefit by a closer relationship to Rome, spiritual jurisdiction over the whole of the island, and, perhaps, even some monetary advantage in tithes and other incomes. The newly uncovered relics from Torres were immediately drawn into this high-stakes dispute, since the tangible evidence they provided of the antiquity of Christianity in Sassari and its region lent crucial support to Sassari's claims to primacy.[11]

The Cagliarese recognized the challenge posed by the Torres relics and wasted no time in responding in kind. On the morning of November 6, 1614, Archbishop Francisco de Esquivel (1554?–1625) inaugurated the excavation of the basilica of St. Saturninus, one of Cagliari's most ancient holy sites.[12] By early afternoon, the laborers had begun to uncover what

Italy. Politics, Society, and Religion 1500–1700, eds. Thomas James Dandelet and John A. Marino (Leiden: Brill, 2007), 45–72.

9 Modern scholars suggest that it may have originated in a lesser title conceded to the archbishop of Pisa in the twelfth century. On the lack of sources for ancient and medieval Sardinian history, see the comments of Raimondo Turtas, *Storia della Chiesa in Sardegna dalle origini al Duemila* (Rome: Città Nuova, 1999), 32; Damiano Filia, *La Sardegna cristiana, 1, Dalle origini al sec. XI* (Sassari: Carlo Delfino, 1995), 12–13.

10 Turtas, *Storia della Chiesa in Sardegna*, 374–82; Damiano Filia, *La Sardegna cristiana, 2, Dal periodo giudicale al 1720* (Sassari: Carlo Delfino, 1995), 263–70.

11 While the origins of Sassari are obscure, by the thirteenth century the city was the capital of the Giudicato of Logudoro, one of the island's four indigenous principalities. Though the archdiocese of Torres was not formally moved to Sassari until 1441, the town had been in decline for centuries. Turtas, *Storia della Chiesa in Sardegna*, 290. On similar disputes over primacy in the Iberian Peninsula, see José Godoy Alcántara, *Historia crítica de los falsos cronicones* (Madrid, 1868; reprint, Granada: Universidad de Granada, 1999), 10–14; Ramón Sánchez González, *Iglesia y sociedad en la Castilla Moderna: el Cabildo catedralicio de la Sede Primada (siglo XVII)* (Cuenca: Universidad de Castilla-La Mancha, Excmo. Ayuntamiento de Toledo, 2000), 54–6.

12 For a day-by-day account of the excavations in Cagliari and surrounding areas, see Archivio Storico Diocesano di Cagliari (ASDC), Culto dei santi, 16. Many thanks to Mons. Tonino Cabizzosu, director of the ASDC, for facilitating access to the manuscript materials related to these finds.

were immediately assumed to be the bodies of ancient martyrs for Christ. The sanctity of the finds was confirmed the next day with the discovery of a fragmentary inscription that, to the delight of the attending clerics, appeared to promise *sancti innumerabile*, innumerable saints. Over the next few years, Esquivel and his agents recovered dozens of martyrs' bodies in Cagliari and in the surrounding towns and villages. In an elaborate ceremony in 1617, Archbishop Esquivel installed the relics in a sumptuous shrine inside Cagiliari's cathedral.[13] He trumpeted his city's extraordinary sacred antiquity in a lengthy treatise, published in Naples in 1617, that he sent to both Philip III and Pope Paul V.[14] Further publications advertised the finds, which continued to grow as excavations proceeded under Esquivel's successor. A local Capuchin friar, Serafín Esquirro, described each of the finds in exacting detail in a history of the discoveries published in 1624, but he died before completing the promised second volume.[15]

By 1629, Dionisio Bonfant (d. 1637), a local theologian and jurist, had completed his own history of the martyrs of Cagliari and sent the manuscript to the printer. His *Triumpho de los santos del reyno de Cerdeña* (Cagliari, 1635) would eventually become the best known of the texts and treatises that celebrated Cagliari's relic finds, though not for the rigor of its scholarship or the quality of its prose. Instead, the book and its author became notorious among specialists in ecclesiastical history and Christian antiquity as a particularly egregious example of scholarly ignorance, biased interpretation, and outright fraud. Among the very first scholarly readers to offer a thorough and sharply negative critique of Bonfant's *Triumpho de los santos* was Lucas Holste (1596–1661), better known both to his contemporaries and to posterity as Holstenius. This erudite German convert frequented the influential Roman scholarly circle of Cardinal Francesco Barberini (1597–1679), whom he served as

[13] Sergio Bullegas, *L'effimero barroco: festa e spettacolo nella Sardegna del XVII secolo. Il* Santuario de Caller *di Serafino Esquirro e la* Relación verdadera *di Antonio Sortes* (Cagliari: CUEC, 1995).

[14] Francisco de Esquivel, *Relación de la invención de los cuerpos santos, que en los años 1614. 1616. y 1616. fueron hallados en varias Yglesias de la Ciudad de Caller y su Arçobispado* (Naples, 1617). For the cover letter that accompanied the papal presentation copy of Esquivel's treatise, see Biblioteca Apostolica Vatican (BAV), Barb. lat. 8574, fol. 31Ar-v. The pope's bland and non-committal reply is reprinted in Dionigi Scano, *Codice diplomatico delle relazione fra la Santa Sede e la Sardegna, 2, Da Gregorio XII a Clemente XIII* (Cagliari: Arti Grafiche, 1941), 418.

[15] Serafin Esquirro, *Santuario de Caller y verdadera historia de la invención de los cuerpos santos hallados en la dicha ciudad y su Arçobispado: parte primera* (Cagliari, 1624).

personal librarian.[16] Holstenius' acid commentary, carefully penned into the margins of his patron's copy of the book (now part of the Barberini collection of printed works held by the Vatican Library), established the jeering tone and much of the content of the criticisms leveled against Bonfant by later critics.[17]

Bonfant's text and Holstenius' response offer a window into differing ideas at work in the seventeenth century about methods and standards of evidence in the writing of sacred history. In the sixteenth and seventeenth centuries, writers of *historia sacra*, a genre that included ecclesiastical history, hagiography, martyrology, and other forms of historical writing that centered upon religious topics, took up the new standards and approaches pioneered by humanist scholars working on more secular themes. Armed with new tools of textual criticism, a growing attentiveness to historical context, and an increasingly skeptical perspective, sacred historians explored legends surrounding the saints and their remains, giving equal or greater weight to evidence drawn from manuscripts, printed materials, and inscriptions as to miracle stories and other traditional sources.[18] Cardinal Cesare Baronio's (1538–1607) *Annales ecclesiastici* (1588–1607), for example, integrated new documentary and material testimony to chronicle the earliest centuries of Christian history, while the Jesuit hagiographers known as the Bollandists reassessed and revised the ancient and medieval manuscript sources for the lives of hundreds of saints, rejecting those *vitae* they found to be spurious.[19] But not all sacred historians approached their task with the same

[16] On Holstenius and his connections to Barberini, see Peter Rietbergen, *Power and Religion in Baroque Rome: Barberini Cultural Policies* (Leiden: Brill, 2006), 256–95.

[17] BAV, Barberini. U.IV.72. Holstenius seems to have preferred to mark up his patron's copy of the book, since his own copy of *Triumpho de los santos*, now in Rome's Biblioteca Angelica, is free of annotations. See Biblioteca Angelica, X.15.7.

[18] On hagiography and the development of modern historical method, see Simon Ditchfield, *Liturgy, Sanctity and History in Tridentine Italy: Pietro Maria Campi and the Preservation of the Particular* (Cambridge: Cambridge University Press, 1995); David J. Collins, *Reforming Saints: Saints' Lives and Their Authors in Germany, 1470–1530* (Oxford: Oxford University Press, 2008); Alison Knowles Frazier, *Possible Lives: Authors and Saints in Renaissance Italy* (New York: Columbia University Press, 2005).

[19] The literature on Baronio and the Bollandists is extensive. For an entry into Baronio's method, see Giuseppe Antonio Guazzelli, "Cesare Baronio and the Roman Catholic Vision of the Early Church," in *Sacred History: Uses of the Christian Past in the Renaissance World*, eds. Katherine van Liere, Simon Ditchfield, and Howard Louthan (Oxford: Oxford University Press, 2012), 52–71; *Baronio storico e la Controriforma. Atti del convegno internazionale di studi, Sora, 6–10 ottobre 1979*, eds. Romeo de Maio et al. (Sora: Centro di studi Soriani "Vincenzo Patriarca," 1982); Stefano Zen, *Baronio storico. Controriforma e crisi del metodo umanistico* (Naples: Vivarium, 1994). For an introduction to the

perspective. A comparison between text and commentary reveals that the divergence between Holstenius and Bonfant hinged upon differing understandings about the proper use of historical sources in constructing the past. Grounded in local attitudes and attentive to local needs, Bonfant deployed a wide array of historical sources – manuscripts and printed texts, images and local traditions, but especially inscriptions and other archaeological remains – not in order to question Cagliari's sacred traditions but rather to construct for his fellow citizens an imaginary heroic Christian past thickly peopled with martyrs for Christ. Holstenius' view, by contrast, was centered upon Rome and demanded a critical approach to the same evidence, one that eschewed pious conjecture and bound historical interpretation to the available sources. Thus, the conflict over the proper use of sources was also one that pitted the periphery against the center, the province against the metropole, local traditions against those of Rome and the universal Church.

Very little is known about Dionsio Bonfant beyond the handful of hints contained within his *Triumpho*. He apparently held doctorates in theology and in both civil and canon law, and served as *provisor general* of Cagliari's university in the late 1630s.[20] He probably frequented scholarly or clerical circles close to Cagliari's archiepiscopal leadership, since he claimed to have been a participant in and eyewitness to many of the discoveries.[21] What is certain, however, is that he was a very ardent patriot of his native city. At the heart of his project lay the exaltation of Cagliari's newly discovered ancient Christian history, the vindication of its saints, and the confounding of its rivals in Sassari. Unlike Esquivel and Esquirro, who limited themselves to a day-by-day account of the excavations, Bonfant erected a complete history of the saints of Cagliari. Using Esquivel and Esquirro's published accounts, together with the original notarial registers of the finds, he sketched out the lives and deaths

Bollandists, see David Knowles, *Great Historical Enterprises: Problems in Monastic History* (London: Nelson, 1963), 1–32; Flor Van Ommeslaeghe, "The 'Acta Sanctorum' and Bollandist Methodology," in *The Byzantine Saint. University of Birmingham Fourteenth Spring Symposium of Byzantine Studies*, ed. Sergei Hackel (London: Fellowship of St. Alban and St. Sergius, 1981), 155–63; Robert Godding, Bernard Joassart, Xavier Lequeux, François De Vriendt, and Joseph Van der Straeten, *Bollandistes. Saints et légendes: Quatre siècles de recherche* (Brussels: Société de Bollandistes, 2007).

[20] He describes himself as "doctor en theología, y ambos derechos, provisor general de la Universidad de Caller, y Cerdeña" on the title page to his *Breve tratado del primado de Cerdeña, y Corcega en fauor de los arçobispos de Caller, y del real patronasgo de su Magestad, que le tiene fundado en la dignidad primacial de la santa Yglesia de la ciudad de Caller, cabeça de todo el reyno de Cerdeña* (Cagliari, 1637).

[21] Bonfant, *Triumpho*, unp. (prologue).

of hundreds of saints, the vast majority of them previously unknown martyrs for Christ whose existence and careers were known only through inscriptions uncovered in the course of the excavations conducted by Esquivel and his successors. The vagaries of history made these sources crucial: because of "the incursions and invasions of so many princes by which this kingdom and city has been devastated and destroyed, and its archives burned, that ruined everything and extinguished the memory of its saints (which in these recent times their discovery has resuscitated)," inscriptions and other archaeological remains stood as the most important evidence of Cagliari's ancient Christian past, and the indispensable foundation for Bonfant's reconstruction of his city's history.[22]

Contained within thirteen lengthy books spread over more than 600 pages, Bonfant's narrative stretched from the earliest origins of Sardinian Catholicism to about the twelfth century. His primary focus, however, lay squarely upon the first centuries of Christian history, and especially on the persecution of Christians under the Roman emperors before the conversion of Constantine in the early fourth century. For each martyr, Bonfant provided a description of an inscription and its discovery, using the often fragmentary text as a source for reconstructing the outline of an exemplary life and death. His reading of these stone witnesses to Cagliari's early Christian past rested upon a presumption that, given that all were found within the structure or surrounding necropolis of the basilica of St. Saturninus, the city's oldest church, or within other ancient churches, the inscriptions must perforce refer to Christian martyrs.[23] Pointing to an allegedly ancient inscription that documented the building of the structure, he argued that the basilica had been created at the command of Constantine, who commemorated Rome's martyrs with the construction of grand pilgrimage churches. Given that the church of St. Saturninus had been known as a basilica since at least the early sixth century, and given that the emperor supposedly had ordered that only martyrs could be buried within these structures, Bonfant reasoned that all of the remains found there must belong to Christians who died

[22] "Ninguna cosa particular desta santa [Victorina] nos ha dexado la antigüedad, como ni la tenemos de innumerables santos martyres, y si algo dexó escrito … los incursos, e inuasiones, con que este reyno y ciudad ha sido de tantos príncipes vastada, destruyda, y quemados sus archiuos, ya todo lo arruynaron, y han del todo extinguido las memorias de los santos, que estos últimos tiempos nos han resuscitado sus inuenciones." Bonfant, *Triumpho*, 178.

[23] On the history of the basilica of St. Saturninus, see Tatiana K. Kirova, *La basilica de S. Saturnino in Cagliari. La sua storia e i suoi restauri* (Cagliari: Editrice Minipress, 1979).

in defense of the faith. The *sancti innumerabile* inscription discovered by Archbishop Esquivel, he contended, supported this claim, since it referred to all of the bodies within the structure.[24]

These arguments informed Bonfant's interpretations of the many funerary inscriptions discovered at the excavation sites. The Latin word *memoria*, for example, was to be understood as a word used only to commemorate martyrs and "proves that the body that is there is of a saint and martyr"; similarly, the letters B. M., which frequently adorned the graves, he rendered as *bonae memoriae*, of good memory. This wording, he argued, indicated the presence of saintly remains; for this reason, he regularly rendered them as "blessed martyr."[25] For example, a typically laconic grave inscription for a certain Rustica read as follows:

> × HIC IACET B.M.
> RVSTICA ANCILL
> A DI. QVIE BIXIT ANNI
> S PL. M. XII REQVIEBIT
> IN PACE SV. D. XVI. KAL. MAI. IN
> DICT OCTAB

For more erudite readers, Bonfant fleshed out the inscription, which was set apart on the page with an ornamental border, with a fuller version in italicized Latin: *Hic iacet Beata Martyr Rustica Ancilla Domini quae vixit annis plus minus duodecim, requievit in pace sub die sexto decimo Kalendae Maii, indictione octava.* A Spanish translation followed for readers not conversant in the ancient language. Beneath this evidence, Bonfant retold the martyr's story. From the bare-bones memorial inscription, he crafted a noble Cagliarese maiden, some twelve years old, who was so dedicated to the service of God that her fellow Christians gave her the honorific of *ancilla Domini*, handmaid of the Lord. During the

[24] In a separate preface titled "De la canonización de los santos," Bonfant argued that all of the martyrs were canonized "de la manera que la Yglesia acostumbraua al tiempo que murieron" – that is, canonized by Cagliari's bishops by authority delegated from the pontiff. The *sancti innumerabile* inscription also proved the martyrs had been canonized, since it could only have been placed there by order of the city's bishops. His insistence upon this point is very likely a response to decrees published in 1625 by Rome's Congregation of the Holy Office of the Inquisition that prohibited veneration of individuals reputed to be saints unless previously authorized by the pope. See Miguel Gotor, *I beati del papa. Santità, Inquisizione e obbedienza in età moderna* (Florence: Olschki, 2002), 285–356. Bonfant, *Triumpho*, unp.

[25] "El primero es la palabra *memoria*, título tan illustre, y tan calificado, que hallándose gravada esta palabra en un mármol junto de una sepultura, prueva que el cuerpo que allí está es de santo, y mártir." Bonfant, *Triumpho*, 20.

reign of the emperor Trajan, Rustica was called before the pagan authorities, who urged her to give up her service to Christ. The saint responded with an impassioned (and wholly imaginary) speech that praised the Incarnation and Christ's sacrifice for humanity. The "impious judge," seeing that she would not be dissuaded, ordered her decapitated.[26]

Other words and abbreviations Bonfant likewise read through a lens of presumptive martyrdom. The ubiquitous *requiescit in pace* (rest in peace), for example, was also a sure indicator of one dead in defense of the faith.[27] So too was *innocenti*, a word commonly used in Roman funerary inscriptions for dead children. According to Bonfant, *innocenti* "is the same as a saint *sine noxa, sine culpa*, without sin ..."[28] Other abbreviations might be taken as signs of Christianity. The letters D. M., an abbreviation found on both pagan and Christian inscriptions and usually deceiphered as *dis manibus* (literally, "to the ghost-gods," but understood as "in memory of"), Bonfant sometimes rendered instead as *Deo Maximus*, "to the greatest God."[29] He frequently read the insignia and symbols on gravestones as the tools of martyrdom. Hearts pierced by arrows on the inscriptions that marked the tombs of Sts. Lucifera and Vitalia, Egidius and Ananias, for example, signified martyrdom at the hands of archers, while an anchor on the grave of the priest St. Antiochus signified death by drowning.[30]

Bonfant's readings of incomplete inscriptions likewise reflected a presumption that the dead to whom they referred were Christian martyrs. Sometimes he reproduced the missing and damaged text – a problem that he regularly blamed on the malevolence and avarice of marauding Vandal and Saracen invaders – and then provided his own rendition; sometimes

[26] Bonfant, *Triumpho*, 119–20. On this inscription, see *Corpus Inscriptionum Latinarum*, X, pt. 1, ed. Theodore Mommsen (Berlin: George Reimer, 1883), 1360*.

[27] "Tambien la palabra *requiescit in pace*, es significatiua de que el alma de aquel cuerpo a quien sirue de título es santa, y que goza de Dios, la Yglesia (en prueua desto) de los que passaron desta vida, *optatiue*, dize *requiescant in pace*, significando el entrañable desseo, que tiene de que gozan de Dios. Luego si de todos los que tratamos se dize affirmatiue *requiescunt in pace*, no queda occasion de duda, de que este título *requieuit in pace*, es vno de los que a solos los santos, y mártyres se deue dar ..." Bonfant, *Triumpho*, 23.

[28] "Aquella palabra innocenti lo mesmo es, que santo *sine noxa, sine culpa*, sin pecado ..." Bonfant, *Triumpho*, 284. See Hanne Sigismund Nielsen, "The Value of Epithets in Pagan and Christian Epitaphs from Rome," in *Childhood, Class and Kin in the Roman World*, ed. Suzanne Dixon (New York: Routledge, 2001), 173.

[29] Bonfant, *Triumpho*, 345–6.

[30] Bonfant, *Triumpho*, 140–2 (Lucifera and Vitalia), 441–2 (Egidius and Ananias), 212–15 (Antiochus Presbyter).

he gave only a reconstructed version.[31] In either case, Bonfant's inventive analysis regularly found enough information in the broken and fragmentary inscriptions to reconstruct names, family relationships, careers, and deaths of the supposed ancient Christian martyrs. In one characteristic case, he filled in the many missing letters in a "muy destruyda" gravestone to reveal a St. Quintinia and her three sons, Ideresius, Isoveus, and Ippopeus.[32] In another, that of a St. Anthony, Bonfant transformed what was most probably a pagan funerary inscription into the gravestone of an eleven-year-old martyr named Anthony Appelles, drowned with an anchor around his neck for his constancy in the faith and buried with his nurse, Venusta.[33] For a certain St. Julius, Bonfant filled in not only the missing text, but reconstructed an excised Chi-Rho from what appears to be a single stray chisel mark.[34] In some cases, Bonfant created details of a martyr's life and death from the appearance of the human remains or the presence of grave goods. The bones of St. Bincema, for example, provided the clues needed to assign an approximate age: "Here lies Bincema of eternal memory and glory, who lived thirteen years, more or less. I say thirteen because that's what the framework of the bones indicated."[35]

These creative readings of the archaeological evidence allowed Bonfant to fill his city's past with newly discovered martyrs. He distributed them liberally throughout the earliest Christian centuries, which he organized in traditional form around numbered cycles of imperial persecution. An imperial indiction (tax levy) or other date found on one inscription provided a date for other, nearby remains, while the physical disposition of the bones signaled whether a grave was a body's original burial site or whether it had been moved there at a later time. "If it is the first burial, not made by a translation from another site, and the bodies are found together," he argued, "it proves that they are coevals and contemporaries in martyrdom."[36] This kind of evidence permitted Bonfant to

[31] On this complaint as an oft-repeated trope among early modern historians of Spanish history, see Katrina B. Olds, *Forging the Past: Invented Histories in Counter-Reformation Spain* (New Haven, CT: Yale University Press, 2015).

[32] Bonfant, *Triumpho*, 86–8. Cf. Bonfant, *Triumpho*, 209–11.

[33] Bonfant, *Triumpho*, 76–7.

[34] Bonfant, *Triumpho*, 240–1.

[35] "Aquí jace Bincema de eterna memoria, y gloria, que biuió 13. años poco más, o menos, digo 13. porque tantos denotaua el armasón de los guessos," Bonfant, *Triumpho*, 230–1.

[36] "... si es primero entierro, no hecho por traslación de otro lugar, y los cuerpos se hallan juntos, prueua ser coetáneos, o contemporáneos en el martyrio." Bonfant, *Triumpho*, 65. Cf. Bonfant, *Triumpho*, 48.

fix the otherwise undated remains in time and confect a history of their veneration. In one representative case, he found that Sts. Longinus and Saturninus (not to be confused with the later saint of the same name, to whom the basilica was dedicated) had received the palm of martyrdom around the year AD 70, but that their relics had been translated and reburied in the church fifty years later, under the reign of Hadrian.[37]

While the inscriptions and other remains recovered from the excavation sites were Bonfant's sole source for the vast majority of the hundreds of Cagliarese martyrs he described, other forms of evidence played an important auxiliary role. For a few saints he found supporting testimony such as the existence of nearby churches under the advocation of like-named saints, old retables and other images, and local traditions. A select handful of the inscriptions and remains could be identified as the relics of well-known saints, including St. Ephesius, St. Luxorius, the child martyrs Sts. Cesellus and Camerinus, St. Saturninus, and the controversial St. Lucifer of Cagliari (d. *c.* 370), a renowned theologian and ardent opponent of the Arian heresy. For these saints, Bonfant supplemented inscriptions with ancient commentators, medieval hagiographies, and some recent scholarship.[38] Such works did not, however, bear the full weight of Bonfant's new history of his city's ancient past; rather, archaeological evidence extracted from Esquivel, Esquirro, and the notarial registers remained the primary or even sole source for the preponderance of the martyrs whose lives and deaths filled Cagliari's earliest Christian centuries.

Such arguments were less than convincing for the erudite Holstenius, who filled the margins of Cardinal Barberini's copy of *Triumpho de los santos del reyno de Cerdeña* with critical commentary written in Latin lightly spiced with Greek. While many of his attacks were *ad hominem* – his preferred epithets for Bonfant were *nugator* (trifler, joker) and *tricator* (trickster) – Holstenius' main focus was methodology, and his dry and biting sarcasm overlaid an attentive and scholarly analysis of Bonfant's foundational assumptions about the nature and use of material evidence in the writing of sacred history. He contested as a "hypothesis falsissima" Bonfant's argument that, because of their location in and around the basilica of St. Saturninus, all of the newly discovered bodies must necessarily be those of martyrs. This "slippery and empty foundation,"

[37] Bonfant, *Triumpho*, 71–3.

[38] Contemporary scholarship included works by Baronio, Pietro Galesino, Alfonso Chacón, Onfrio Panvinio, Laurentius Surius, and Ferdinando Ughelli, together with works by Sardinian writers, especially Giovanni Arca, Giovanni Francesco Fara, and Dimas Serpi.

he noted, underlay an "immense structure of errors."[39] Indeed, Bonfant's assumption that only saints were buried in the basilicas of the martyrs, never the common faithful who died after the era of Constantine, was "the principal source and origin of all of the errors of this book."[40] The inscription that supposedly proved that Constantine built the basilica of St. Saturninus was "recent and suppositious" (though the stone itself might be old).[41] Moreover, the ubiquitous B. M. carved upon the gravestones meant not *bonae memoriae*, he contended, but *bene memoriae* or *benemerente* (and, by extension, should not be rendered as *beatus martyr*, "blessed martyr"). Funerary expressions like *bonae memoria* were not the exclusive province of martyrs, but instead were used "freely by all Christians deceased in the Lord but without martyrdom or being inscribed in the number of the saints."[42] Similarly, *requiescit in pace* was a biblically derived expression reserved not just for the saints but for all faithful Christians.[43]

Holstenius' objections centered on Bonfant's insistence on approaching the material evidence through the lens of presumptive martyrdom and his ignorance of ancient epigraphy, architecture, chronology, and other auxiliary sciences of ecclesiastical history and sacred archaeology. While the tone of his responses to Bonfant's inventive interpretations of the fragmentary inscriptions ranged from acerbic mockery to outraged sputtering, most of his comments focused on Bonfant's implausible, imagined, or erroneous readings.[44] The ascetic St. Hierus and his seven companions, for example, were fabricated "because seven dead bodies were found in the same tomb." Nor could the bodies' proximity within a subterranean "chapel" be taken as evidence of veneration after martyrdom, since similar family tombs could be seen throughout the catacombs of Rome. Bonfant's ignorance of sacred antiquity, said Holstenius, was astonishing.[45] Similarly, the funerary inscription of St. Emerita contained

39 "Hoc capita [in which Bonfant laid out his ideas] fundamenta ponit lubrica et vana, quibus deinde immensam errorum machinem superstruit." BAV, Barberini. U.IV.72, 20, 456.

40 "... atque hic est principuus omnium errorum hujus libri fons et origo." BAV, Barberini. U.IV.72, 24.

41 "lapis forte antiquus sed inscriptio recens et suppositicia," BAV, Barberini, U.IV.72, 27.

42 "promiscue omnibus Christianis in domino defunctis etiam sine martyrio, neque sanctorum numero adscriptis," BAV, Barberini, U.IV.72, 21, 29.

43 BAV, Barberini, U.IV.72, 23

44 See, for example, Holstenius' commentary on Bonfant's deciphering of the gravestone of St. Stephana, which ranges from the sarcastic ("lepidum caput") to the irate ("putidissimiae ineptiae et plane pueriles"), BAV, Barberini, U.IV.72, 89–91.

45 "haec confingit, quia VII corpore mortua in eodem conditorio reperta," BAV, Barberini, U.IV.72, 82–83. f. BAV, Barberini, U.IV.72, 26.

nothing that would suggest a violent death in defense of the faith. "So why the glory of martyrdom? He does not know the formulas of ancient inscriptions."[46] Holstenius also frequently pointed out how anachronistic phrases and names called Bonfant's chronology into question. For example, he suggested that the use of the title of "archbishop" for St. Florus, whose martyrdom Bonfant ascribed to the reign of Hadrian, indicated instead a date in the sixth century or after, and noted that Sts. Frederick and Leonard, who supposedly perished with thirty-six companions during the reign of Commodus, bore Germanic names unknown in the second century. He dismissed the tale of their martyrdom, exclaiming "O foolish trifler!"[47]

In Holstenius' analysis, Bonfant's presumption of martyrdom was closely linked to an insufficiently critical and overly imaginative approach to his sources. Where information was lacking, Bonfant filled in the gaps with "dreams" and fantasies from his own imagination.[48] The life of St. Beneria, for example, "he weaves from his own brain like a spider weaves its web."[49] The result of Bonfant's inventive flights of fancy, argued Holstenius, was to undermine the credibility of both his narrative and his sources. The reconstructed inscription of St. Lucianus, for example, held no evidentiary weight. "The author completed it from his own [imagination], and therefore the mutilated text cannot be cited as undoubted evidence."[50] Much to Holstenius' frustration, Bonfant applied the same approach to even the most approved and accepted of his documentary sources. For example, according to Bonfant, the martyrs Aemilius, Priamus, Lucianus, and Felix were converted to Christianity by St. Boniface, one of the seventy-two evangelists. Holstenius noted that the four were mentioned in the martyrology of Usuard, one of the standard medieval martyrologies. "Usuard's martyrology simply attributes these martyrs to Sardinia. But this author invents their time, place, and status from his own [imagination]."[51] Bonfant deployed his authorities without

[46] "quae ergo gloria martyrii? ignorat formulas inscriptionum antiquarum," BAV, Barberini, U.IV.72, 273–4. Cf. BAV, Barberini, U.IV.72, 418–19.

[47] "o insulsum nugatorem!," BAV, Barberini, U.IV.72, 225–6. Cf. BAV, Barberini, U.IV.72, 104, 108, 112, 128, 140–2, 410.

[48] E.g. BAV, Barberini, U.IV.72, 43, 96, 228, 408.

[49] "Totam hanc narrationem ut aranea suam telam ex proprio cerebro texit," BAV, Barberini, U.IV.72, 107.

[50] "ergo auctor de suo supplevit, ac proinde in scriptio mutila pro indubitato testimonio citari nequit," BAV, Barberini, U.IV.72, 63.

[51] "Martyrologium Usuardi Sardinia simpliciter hosce martyres tribuit. sed hic auctor tempus, locum et dignitate di suo affingit," BAV, Barberini, U.IV.72, 47.

regard to their historical context or quality. His proofs for the primacy of the archbishop of Cagliari, for example, were a "mishmash of authors selected without order of preferable things, where he prefers the recent to the ancient, the worst to the best, and he mixes it all up like a cook makes a sauce."[52] Instead of diligently comparing accounts and the records and monuments of the early Church like careful scholars do, said Holstenius, Bonfant either misunderstood them or arbitrarily twisted them, and pinned his claims to manuscript sources that existed only in Utopia – that is, nowhere.[53]

Holstenius' sharp critique of Bonfant's failings found a ready audience, as later scholars repeated and amplified his criticisms. Jean Bolland (1596–1665), for example, noted many of the same scholarly shortcomings that had so irked Holstenius and, on the advice of Cardinal Barberini (who was himself very likely informed by Holstenius, his librarian), opted not to include the newly discovered Sardinian martyrs in his *Acta sanctorum*.[54] His successor, Daniel van Papenbroeck (1628–1714), mentioned Holstenius among the erudites of Rome who cast doubt on Bonfant and the relic finds in Sardinia, and pointed to many of the same methodological problems highlighted by the German librarian.[55] In the eighteenth century, Ludovico Antonio Muratori (1672–1750) and Luigi Gaetano Marini (1742–1815) joined the dogpile.[56] Marini, for example, heaped scorn upon Bonfant's *Triumpho*, deriding it as "the summit of imposture, imprudence, boldness, ignorance (and one could say worse)."[57] By the nineteenth century the celebrated classicist Theodor Mommsen

[52] "Ecce tibi farraginem authorum sine ordine delectu productorum; ubi recentes antiquis, pessimos optimis praeponit atque omnia confundit ut multi iura coquus," BAV, Barberini, U.IV.72, 50. The reference is to Plautus' *Mostellaria* I, 3, 277.

[53] BAV, Barberini, U.IV.72, 49, 65. Cf. BAV, Barberini, U.IV.72, 489.

[54] *Acta sanctorum*, February, 1 (Antwerp, 1658), xx–xxi. Bolland's informant, the French scholar Raphael Trichet du Fresne (1611–61), overlapped with Holstenius during his sojourn in Rome as the librarian to Queen Christina of Sweden, and enjoyed connections to Cardinal Barberini and especially to the cardinal's secretary, Cassiano del Pozzo (1588–1657). The books on the saints of Cagliari by Esquivel, Esquirro, and Bonfant all appear in *Catalogus librorum bibliothecae Raphaelis Tricheti du Fresne* (Paris, 1662).

[55] *Acta sanctorum*, May, 5 (Antwerp, 1685), 221ff.

[56] Ludovico Antonio Muratori, *Antiquitates Italicae Medii Aevi*, 5 (Milan, 1741), disseratatio LVIII, col. 17. Muratori cites Papenbroeck as his main source on Bonfant.

[57] "L'opera poi del Bonfante è il colmo della impostura, imprudenza, sfrontatezza ignoranza, e se si puó dir peggio." Marini appears to have directly consulted Holstenius' handwritten comments, which he describes as "bellissime." Reported by the celebrated archaeologist Giovanni Battista de Rossi (1822–94), in BAV, Vat. Lat. 10528, 8v.

(1817–1903) dismissed both Bonfant and Esquirro and condemned most of the inscriptions from Cagliari and Torres as false.[58]

By dint of repetition and ever shriller rhetoric, Bonfant's *Triumpho* came to stand as a star example of the most egregious excesses of scholarly ignorance and bad faith. But as Simon Ditchfield has noted, such execration was both excessive and anachronistic, and more recent scholarship has suggested that Bonfant, Esquivel, Esquirro, and others who wrote in praise of the martyrs of Cagliari were probably more faithful to their material sources than their critics allowed.[59] Bonfant was not wholly without critical judgment; for example, he mocked the Sassarese Jesuit Jacobus Pinto (1575–1650) and his *paisan*, Giovanni Francesco Fara (1543–91), for their incautious reliance upon the forged chronicles of Annius of Viterbo (*c.* 1432–1502), and his frequent laments about the loss of archival materials in the many invasions and wars that had beset Sardinia suggested an awareness of a need for more and better sources.[60] Interestingly, his historical narrative included very few miracles, and he

[58] For Mommsen's opinions on Bonfant and on the discoveries in Torres and Cagliari, see *Corpus Inscriptionum Latinarum*, X, pt. 2, ed. Theodor Mommsen (Berlin: George Reimer, 1883), 779–80.

[59] Simon Ditchfield, "Martyrs on the Move: Relics as Vindicators of Local Diversity in the Tridentine Church," in *Martyrs and Martyrologies*, ed. Diana Wood (Oxford: Blackwell, 1993), 292. For reappraisals of Bonfant and his Cagliarese colleagues, see Donatella Mureddu, Donatella Salvi, and Grete Stefani, *Sancti innumerabiles. Scavi nella Cagliari del Seicento: testimonianze e verifiche* (Oristano: Editrice S'Alvure, 1988); Marcella Bonello Lai, "Le raccolte epigrafiche del '600 in Sardegna," in *Arte e cultura del '600 e del '700 in Sardegna*, ed. Tatiana K. Kirova (Naples: Edizioni scientifiche italiane, 1984), 379–95; Donatella Mureddu and Grete Stefani, "La diffusione del mosaico funerario africano in Sardegna: scoperte e riscoperte," in *L'Africa romana. Atti del III convegno di studio Sassari, 13–15 dicembre 1985*, ed. Attilo Mastino (Sassari: Edizioni Gallizzi, 1986), 339–61.

[60] Bonfant, *Triumpho*, 2, 576. Pinto, a Jesuit professor of Holy Scripture who had been closely involved in the excavations in Torres, defended Sassari's claim to primacy and exalted the Sassarese martyrs in his *Christus Crucifixus* (1624). Bonfant devoted an extensive section of the final chapter of *Triumpho de los santos* to confuting his rival. See Bonfant, *Triumpho*, 566–600; Jacobus Pinto, *Christus Crucifixus: sive selectorum ex scriptura universa locorum in certas classes pro variis Christi titulis digestorum nova & accurata discussio, sacrorum interpretum, et concionatorum usui accommodata*, vol. 1 (Lugduni, 1624; Naples, 1859), 295–302 (lib. 3, tit. 4, l. 12). On Jacobus Pinto, S.J. (aka Jaime/Diego Pinto/Pintu/Pintus), see the short biography in Miquel Batllori, *L'università di Sassari e i Collegi dei gesuiti in Sardegna. Saggio di storia instituzionale ed economica* (Nuoro: Poliedro, 2012), 124–5. On Fara, see Enzo Cadoni and Raimondo Turtas, *Umanisti Sassaresi del '500: le "biblioteche" di Giovanni Francesco Fara e Alessio Fontana* (Sassari: Edizioni Gallizzi, 1988; Antonello Mattone, "Giovanni Francesco Fara giureconsulto e storico del XVI secolo," in *A Ennio Cortese*, vol. II, eds. Domenico Maffei, Italo Birocchi (Rome: Il Cigno, 2001), 320–48.

never cited miracles as proof for the sanctity of Cagliari's newly discovered martyrs. Bonfant's approach to his evidence appeared to be predicated less upon rigorous critique and strict fidelity to the sources and more upon a spirit of what contemporaries called "pious affection" or "pious generosity." This approach, less a coherent methodology than an outlook, required the historian to take what Katrina Olds has called an "attitude of courtesy or deference toward the holy."[61] In one formulation, that of humanist historian Ambrosio de Morales (1513–91), the writer of *historia sacra* must approach his task with three things in mind. "The first and most important is pious affection, in order not obstinately to resist what should be understood morally, when deduced from good evidence. The second is zeal and skepticism, moderated and governed with prudence, so that one is not persuaded without good reason. The third requires one to take note of many details, and to know how to consider them well with knowledge and experience in order to deduce from them as much as one can and should infer."[62] While perhaps Bonfant's posture of pious affection went farther than Morales might have thought proper, it was this view that shaped his approach to the sources and distinguished his understanding of the historian's task from that of Holstenius.

The writing of history was yet another arena in which the Cagliarese and Sassarese worked out their struggle for predominance, and Bonfant's posture of pious affection towards his sources was closely linked to the propagandistic aim that underlay his version of Cagliari's Christian history. His generous interpretation of the excavated remains crowded his city's past with untold martyrs, their numbers far superior to those of

[61] Katrina B. Olds, "The Ambiguities of the Holy: Authenticating Relics in Seventeenth-Century Spain," *Renaissance Quarterly* 65(1) (2012): 162.

[62] "Para todo lo que desto se tratara, conuiene mucho aduertir, como en la materia deste santo negocio no puede auer euidencia, ni argumentos que del todo concluyan, sino vna buena probabilidad moral deduzida de buenos principios y fundamentos, de donde se forman razones, que tienen toda la fuerça de que es capaz la materia. Ésto es lo que trate al principio de los discursos generales de las antigüedades, que puse en esta mi corónica, por el autoridad de Aristóteles y Marco Tulio: y tiene mejor lugar aquí, por ser estas cosas de suyo dificíles de aueriguar, y que se deue tener en mucho qualquier aparencia de buena razón y conueniencia, que se pueda hallar. Quanto màs que para procederse bien en este santo negocio, son menester tres cosas. Y la primera y muy principal es pia affecion, para no resistir con porfía a lo que moralmente se dexa entender, quando se deduze de buenos fundamentos. La segunda es zelo y recelo concertado y regido con cordura, para no dexarse persuadir sin buena razón. Lo tercero se requiere notar muchas particularidades, y con noticia y experiencia saberlas bien considerar, para deduzir dellas, lo mucho que se puede y deue inferir." Ambrosio de Morales, *Los cinco libros postreros de la Corónica general de España* (Cordoba, 1586), 272.

Torres. While the title promised a hagiography of the whole of Sardinia, in truth Bonfant dealt almost exclusively with the ancient Christian past of Cagliari and its region. Even the handful of martyrs that he grudgingly allowed to Torres and Sassari, he made disciples or even natives of Cagliari.[63] At every possible turn, he cast doubt upon Torres's antiquity, episcopal succession, and hagiographic traditions, and impugned the sources and methodology of its historians.[64] His efforts did not go unnoticed in Sassari. As we will see, the Sardinian tribunal of the Spanish Inquisition, which was located in Sassari, took an early interest in his work, and after publication, in 1636, the archbishop of Sassari, together with two of his suffragan bishops, wrote to the Council of Aragon, alleging that, because the book had whipped up animosities between the two cities, they dared not attend the meeting of the Cortes, Sardinia's representative assembly, which was to be held in Cagliari.[65]

Sitting in Cardinal Barberini's library in Rome, far removed from Sardinian squabbles, Holstenius brought to the finds from Cagliari and Torres not the pious affection of a provincial patriot, but a broad and Rome-centric erudition that demanded a more critical approach to the remains of the past. His commentary frequently insisted that Bonfant look beyond the confines of Sardinia to contextualize the inscriptions with examples from the Italian Peninsula, especially Rome. Bonfant had visited the Eternal City, but had clearly not absorbed from his time there the lessons that a good scholar should have. "If you had examined the urban cemeteries," Holstenius sniped, "you would have abstained from such delusions and fictions."[66] Holstenius applied standards of historicity that valued critical skepticism over partisanship, local patriotism, and pious

63 For example, Bonfant claimed St. Gavinus, one of the three martyrs whose bodies were discovered in Torres, as a native of Cagliari on the grounds that, as a Roman citizen, Gavinus must have hailed from ancient Caralis, the most prominent urban settlement on the island. He explicity denied that Torres, a much smaller town, could ever have been a Roman colony. Bonfant, *Triumpho*, 329–31, 566ff. In truth, Torres originated as Turris Libisonis, the sole full-fledged colony on the island.

64 See, for example, his lengthy digression on the 1620 publication of the Condaghe of St. Gavinus, a late medieval manuscript in Sardo that documented the history of the Torres saints and served as an important source for Pinto. Bonfant, *Triumpho*, 537–48. On the Condaghe of St. Gavinus, see *Il Condaghe di San Gavino*, ed. Giuseppe Meloni (Sassari: Centro di Studi Filologici Sardi; Cagliari: CUEC, 2005).

65 Francesco Manconi, "Storia di un libro di storia," in Francisco de Vico, *Historia general de la Isla y Reyno de Sardeña*, ed. Francesco Marconi (Sassari: Centro di Studi Flologici Sardi; Cagliari: CUEC, 2004), LI.

66 "si cimiteria urbica lustrasses [lustravisses?] abstinuisses tot deliriis et figmentis." BAV, Barberini, U.IV.72, 82. Cf. BAV, Barberini, U.IV.72, 411, 450.

affection. Sacred history called not for pious conjecture but for documentary evidence, "for in this kind of study I believe on that account, enlightened by faith."[67] He repeatedly demanded reliable historical authorities to support Bonfant's flights of hagiographic fantasy. He countered the martyrdom of St. Asuma, for example, with a skeptical "Where did you get this? Report the author, and we will believe you."[68] Moreover, all evidence should be approached with circumspection. Holstenius lauded the "modesty" (i.e. caution) with which Cesare Baronio, the foremost ecclesiastical historian of the age, assessed certain ancient writers, and condemned Bonfant's "intemperance" and "defensiveness."[69] He also praised the judiciousness with which the Sassarese historian Jacobus Pinto addressed the uncertainties of ancient history: "thus it is proper for a prudent and modest man to deal with doubtful and uncertain things, but not to affirm thoughtlessly [both] known and unknown things, [as is] your fashion."[70] The historian's imagination must not overstep his evidence; interpretation must operate within the bounds of the sources.

In sum, Holstenius and Bonfant approached the historian's task with very different notions of the use and abuse of sources, notions closely linked to their distinct perspectives at the center and periphery of the Catholic world, respectively. Holstenius' sarcastic commentary (and the venerable tradition of Bonfant-bashing it established) notwithstanding, the Sardinian's inventive reading of late-antique Christian inscriptions was not particularly unusual. Even among the select scholarly community of Rome, opinions differed as to how gravestone abbreviations should be interpreted and which insignia could be understood as signs of martyrdom.[71] Nor was Bonfant's local patriotism and the imagined

[67] "Sed Traiuanum Sardinia patrem appellatrum vix crediderim, nisi clariora afferantur documenta. nam in hoc studiorum genere credo quod ideo, fide oculata." BAV, Barberini, U.IV.72, 401.

[68] "unde haec hausisti? affer auctorem, et credemus tibi," BAV, Barberini, U.IV.72, 121.

[69] "vide quantum distet Baronii modestia ab intemperantia et praesidentia hujus censoris," BAV, Barberini, U.IV.72, 489. For similar critical attitudes expressed in marginalia penned by an anonymous critic in Rome – very possibly Holstenius himself – into another volume of local sacred history (an account of ancient martyrs' remains discovered Arjona, Spain) that was part of Cardinal Barberini's private library, see Olds, "The Ambiguities of the Holy," 159–60.

[70] "ita hominem prudentem et modestum in rebus dubiis et incertis agere par est: non autem nota ignotaque temere tuo more affirmare," BAV, Barberini, U.IV.72, 453.

[71] On the debate over the markers of martyrdom, see Giovanni Battista de Rossi and Antonio Ferrua, *Sulla questione del vaso di sangue: Memoria inedita con introduzione storica e appendici di documenti* (Vatican City: Pontificia Istituto di Archeologia Cristiana, 1944).

past it inspired at all uncommon. Similarly inventive elaborations on the scant remains of Christian antiquity could be found in regional and municipal histories throughout early modern Europe, where patriotic local historians and "Sunday archeologists" ransacked the archives and scrabbled in the soil seeking evidence with which to write their cities' sacred pasts.[72] Others looked to the invented chronicles "discovered" by Annius of Viterbo and Jerónimo Román de la Higuera or to forgeries like the lead books of Granada for their raw materials.[73] These local antiquarians should not be dismissed as small-town scribblers; many brought to their tasks standards of historical evidence and interpretation as rigorous as those advocated by Holstenius and others.[74] All, however, shared Bonfant's primary emphasis on uncovering or inventing lost pasts for their local or regional communities.[75]

Unlike the unlucky Bonfant, relatively few of these homegrown historians suffered the withering scrutiny of the erudite elite in Rome, the sacred heart of Catholic Christianity and the epicenter of the developing field of sacred archaeology. As the competition between Sassari and Cagliari grew increasingly tense, Bonfant's *Triumpho de los santos del reyno de Cerdeña* came to the attention of critics in Rome and Madrid for its very partisan stance on the vexed questions of which city's archbishop could claim the title of primate and of the sanctity and orthodoxy of the fourth-century bishop Lucifer of Cagliari. The issue of Lucifer, whom Baronio

72 On "archeologi della domenica," see Raimondo Turtas, *Scuola e Università in Sardegna tra '500 e '600: l'organizzazione dell'istruzione durante i decenni formativi dell'Università di Sassari (1562–1635)* (Sassari: Centro Interdisciplinare per la Storia dell'Università di Sassari, 1995), 108.

73 For an entry into the large literature on the forged chronicles of the Dominican friar Giovanni Nanni, better known as Annius, see Anthony Grafton, *Forgers and Critics: Creativity and Duplicity in Western Scholarship* (Princeton, NJ: Princeton University Press, 1990); Walter Stephens, "When Pope Noah Ruled the Etruscans: Annius of Viterbo and His Forged Antiquities," *Modern Language Notes* 119(1) (2004): 201–23. On the *falsos cronicones* of the Jesuit Jerónimo Román de la Higuera and their spread through Iberian local histories, see Olds, *Forging the Past.* On the lead books of Granada, see A. Katie Harris, *From Muslim to Christian Granada: Inventing a City's Past in Early Modern Spain* (Baltimore, MD: Johns Hopkins University Press, 2007), especially ch. 3.

74 See especially Ditchfield, *Liturgy, Sanctity and History*; Miguel Morán Turina, *La memoria de las piedras. Anticuarios, arqueólogos y coleccionistas de antigüedades en la España de los Austrias* (Madrid: Centro de Estudios Europa Hispánica, 2010).

75 The literature on local history writing in early modern Europe is dauntingly large. For Spain, the key entry point remains Richard L. Kagan's indispensable, classic essay, "Clio and the Crown: Writing History in Habsburg Spain," in *Spain, Europe and the Atlantic World*, eds. Richard L. Kagan and Geoffrey Parker (Cambridge: Cambridge University Press, 1995), 73–99.

had condemned as a schismatic for his unwillingness to accept repenting Arian Christians into the Catholic fold, had already brought Bonfant into trouble in 1630, when the inquisitors of the Spanish Inquisition, which was housed in Sassari, intervened to suspend the printing of his *Triumpho*.[76] Bonfant and the judges of the Cagliari-based Sardinian *audiencia* (high court) objected that the book had already received the royal license. The dispute had to be worked out in a formal accord between the Council of Aragon and the *Suprema* (the central governing council of the Spanish Inquisition), and the book finally appeared some five years later, in 1635.[77]

The conflictive climate seems to have worsened in the interim. In 1638, with the issue of primacy still pending in the Rota, the archbishop of Sassari complained to the pope that the archbishop of Cagliari tolerated the "idolatrous" veneration of questionable martyrs and the "schismatic and condemned" Lucifer of Cagliari.[78] The troublesome question of Lucifer's orthodoxy was remanded to Rome for judgment by the Congregation of the Holy Office of the Inquisition, an institution separate and sometimes in friction with the Spanish Inquisition.

At the same time, the inquisitors of the Sassari tribunal received denunciations about certain treatises in favor of Lucifer of Cagliari and decided to move against Bonfant. In July of that year, they decided to prohibit Bonfant's *Triumpho* until the pope pronounced on Lucifer of Cagliari's contested orthodoxy, but postponed publication of the edict until they might receive a response from the Spanish king.[79] In November, the *Suprema* decided to put the prohibition into effect. The edict, which was published April 6, 1639, gave owners two days to hand their copies over to the agents of the Spanish Inquisition.[80] Both the city council and the archbishop of Cagliari strenuously protested, and within a few

[76] There is an extensive body of scholarship on Lucifer of Cagliari. For an overview, see Giuseppe Corti, *Lucifero di Cagliari. Una voce nel conflitto tra chiesa e impero alla metà del IV secolo* (Milan: Vita e Pensiero, 2004). On the seventeenth-century debates over the alleged sanctity of Lucifer, see Giulia Piccaluga, "Le prospettive storico-religiose della 'santificazione' di Lucifero di Cagliari," in *La figura e l'opera di Lucifero di Cagliari. Una revisitazione. Atti del I Convegno Internazionale Cagliari, 5–7 dicembre 1996*, ed. Sonia Laconi (Rome: Institutum Patristicum Augustinianum), 199–222.

[77] Angelo Rundine, *Inquisizione Spagnola, censura e libri proibiti in Sardegna nel '500 e '600* (Sassari: Stampacolor, 1996), 101–03, 277.

[78] Manconi, "Storia di un libro di storia," LXI.

[79] Rundine, *Inquisizione spagnola*, 104.

[80] Archivio per la Congregazione della Dottrina della Fede (ACDF), Sanctum Officium (S.O.), Stanza Storica (St. St.), B 4 e, fasc. 17, fol. 10r-v; Rundine, *Inquisizione spagnola*, 107.

months the prohibition had been suspended, due in part to a letter to the Sassarese inquisitors from Cardinal Francesco Barberini, secretary of the Congregation of the Holy Office of the Inquisition, reminding them that the question of Lucifer of Cagliari was pending before the Congregation of the Holy Office and tacitly asserting the superiority of the Roman institution over the Spanish one.[81]

It is within this context of municipal rivalry, institutional conflict, and deepening official scrutiny that Bonfant and his *Triumpho de los santos del reyno de Cerdeña* wandered into Holstenius' sights. Whether he read the book in connection with Cardinal Barberini's duties for the Congregation remains unclear; certainly, the care with which he examined the text and cross-referenced Bonfant's versions of the inscriptions with those of Archbishop Esquivel suggests a more than casual encounter. During the same few years other members of Rome's scholarly community were engaged in related pursuits. Orazio Giustiniani (1580–1649), a consultor to the Congregation and a fellow recipient of Barberini's patronage, scrutinized Esquivel's account of the finds, while his fellow Oratorians Cesare Becilli (1573–1649) and Tommaso Bozio (1548–1610), and the Augustinian friar Fortunato Scacchi (1573–1643) examined the case of Lucifer of Cagliari.[82] By June 1641, both the Spanish Inquisition and Rome's Congregation of the Holy Office had arrived at their decisions. On June 9, the *Suprema* ordered the expurgation of certain sections of Bonfant's text, but left the politically charged question of Lucifer to be decided by Rome.[83] Eleven days later, the Congregation issued a decree that sidestepped the issue by simply prohibiting the archbishop of

[81] ACDF, S.O, St., B 4 e, fasc. 17, 6r-7v, 9r-v, 18r; Rundine, *Inquisizione spagnola*, 107–10. On Cardinal Francesco Barberini's tenure as secretary for the Congregation of the Holy Office, see Thomas F. Mayer, *The Roman Inquisition: A Papal Bureaucracy and Its Laws in the Age of Galileo* (Philadelphia, PA: University of Pennsylvania Press, 2013), 56–60.

[82] Biblioteca Vallicelliana (BV), H42. BV, R45, fols. 11r-12r lists the Genoese jurist Sigismundo Scaccia as taking part in the investigation, but this appears to be an erroneous identification since BV, H42 includes materials written by Friar Scacchi. For Giustiniani's nearly illegible notes on Esquivel, see BV, H48. fol. 165r-166r. On Giustiniani's links to Francesco Barberini, see Mayer, *The Roman Inquisition*, 148–9.

[83] Archivo Histórico Nacional, *Inq.*, Libro 1159, fol. 247r. Unlike the Barberini volume, Holstenius' own copy of Bonfant's *Triumpho* (Biblioteca Angelica, X.15.7) has been expurgated in accordance with the Inquisition's decree. The excised sections related to St. Januarius, bishop of Cagliari, are found on pages 532–3, 535, 537, and 557 and a note signed by Dr. Sisinnius Martis, J.C. on the frontispiece attests to the book's expurgation. The censor did not complete his work, however, since he did not blot out the section dealing with St. Antiochus, lector and martyr (181–2). See *La biblioteca di Lucas Holstenius*, ed. Alfredo Serrai (Udine: Forum Editrice Universitaria Udinense, 2000), 241.

Sassari and his counterpart in Cagliari from engaging in any further public debate about Lucifer of Cagliari's sanctity and orthodoxy.[84] Similarly, the Rota evaded a decision on the sensitive issue of which archbishop could claim to be the island's primate. In April 1640 it ruled that Cagliari was the oldest of Sardinia's churches, but explicitly declined to make a judgment about primacy.[85]

Thus, the conflict between Holstenius and Bonfant was about more than just historical practice. Considered together as a collection of interrelated and nested issues, Holstenius' critique of Bonfant, the Congregation of the Holy Office's examination of Lucifer of Cagliari, and the Rota's intervention in the Sardinian primacy dispute open a window into the complex relations between center and periphery, metropole and province in early modern Catholicism. Recent scholarship has called into question an older body of historiography that tended to see the implementation of the decrees of the Council of Trent and the concomitant reform of the liturgy and devotional practices as Rome's imposition of uniformity upon local churches, substituting a universally shared and Rome-vetted piety for the motley local devotions that prevailed throughout the Catholic world.[86] In his groundbreaking work on local liturgy and historical scholarship in Italy, Simon Ditchfield has suggested that rather than a centrally controlled culture of conformity that pitted Rome against local churches, the Tridentine reforms are better understood as "... an attempt by Rome to particularize the universal (by staffing their parishes with priests trained to a minimum standard by the new episcopal seminaries); and to universalize the particular (by ensuring that local devotions operated within guidelines policed, after 1588, by the Sacred Congregation of Rites)."[87] Similarly, Erin Rowe's work on national patron saints in Spain reveals the limits of the Holy See's control over local and national community choices of a patron saint.[88] The Tridentine Church's reforms of the cult of saints – the trimming of the liturgical calendar, the elimination of doubtful cults, the increased inspection of candidates for sainthood – were not simply unidirectional, top-down impositions, but instead need

[84] BV, O.89, fol. 308r-v, 309r-v; and the cover to ACDF, S.O, St. St., B 4 e, fasc. 17.

[85] Filia, *La Sardegna cristiana*, 2, 284.

[86] For example, the "increase in the central control of the sacred, at the expense of local, unofficial and – one might say – 'wild cat' devotions" posited by Peter Burke, "How to be a Counter-Reformation Saint," in *The Historical Anthropology of Early Modern Italy. Essays on Perception and Communication* (New York: Cambridge University Press, 1987), 48–62.

[87] Ditchfield, *Liturgy, Sanctity and History*, 10.

[88] Erin Kathleen Rowe, *Saint and Nation: Santiago, Teresa of Avila, and Plural Identities in Early Modern Spain* (University Park, PA: Penn State Press, 2011).

to be understood as parts of a flexible system that combined cultic controls from Rome with local and even individual devotional initiatives.[89]

Rome's involvement in the Sardinian debates, all three of which rested upon questions of historical evidence and interpretation, appears both as an attempt to assert the authority of the center over peripheral churches and as efforts by partisans in Cagliari and Sassari to wield Rome's authority as a cudgel against each other. From the perspective of officials in Rome, the investigation and interpretation of the ancient Christian past did not serve scholarship or local patriotism alone; rather, the study of Christian antiquity was fundamental to the Counter-Reformation Church's claims to correct belief and practice derived from unbroken apostolic succession. Erudite investigators of ecclesiastical history did not merely uncover the distant Christian past; they confirmed Catholic doctrine, guided liturgy and cultic practices, and provided precedents and guidelines for ecclesiastical governance.[90] The veneration of the saints, a practice much contested by Protestant critics, found particular support from sacred history in all its genres as "the guide to and guarantor of sanctity and saints' cults."[91] While Holstenius' outraged response to Bonfant's methodological missteps often reads like fussy pedantry, his caustic commentary occasionally revealed that high stakes were at play and that Bonfant's pious credulity had real consequences. Sts. Suinia and Ginia, for example, two martyrs that, together with St. Cecilia, he disgustedly described as "stage saints," found their origin in Bonfant's misreading of *quinquaginta* (fifty).[92] Holstenius exclaimed, "And they inserted these monstrous fictitious names into the public litanies. Who can marvel enough at so much stupidity?"[93] Similarly, the translation of the relics of

[89] Cornelius Conover, "Catholic Saints in Spain's Atlantic Empire," in *Empires of God: Religious Encounters in the Early Modern Atlantic*, eds. Linda Gregerson and Susan Juster (Philadelphia, PA: University of Pennsylvania Press, 2011), 87–105.

[90] See Simon Ditchfield, "What Was Sacred History? (Mostly Roman) Catholic Uses of the Christian Past after Trent," in *Sacred History: Uses of the Christian Past in the Renaissance World*, eds. Katherine van Liere, Simon Ditchfield, and Howard Louthan (Oxford: Oxford University Press, 2012), 72–97; Giuseppe Antonio Guazzelli, "Cesare Baronio and the Roman Catholic Vision of the Early Church," in *Sacred History: Uses of the Christian Past in the Renaissance World*, eds. Katherine van Liere, Simon Ditchfield, and Howard Louthan (Oxford: Oxford University Press, 2012), 52–71.

[91] Simon Ditchfield, "Thinking with Saints: Sanctity and Society in the Early Modern World," *Critical Inquiry* 35(3) (2009): 571.

[92] "sante di comedia." This comment is the first of only two in Italian. See also BAV, Barberini, U.IV.72, 411.

[93] "et haec monstra nominum ficticionum litaniis publicis inseruerunt. quis satis miretur tantum stuporem?" BAV, Barberini, U.IV.72, 232.

St. Asuma to Cagliari's cathedral was "baseless worship."[94] In the view of Rome-based scholars like Holstenius, Bonfant's inventive approach to Cagliari's Christian antiquity was not mere shoddy scholarship but a grave twisting of the historical truth. Unless properly documented with proofs of their historical authenticity and uninterrupted observance, local and regional traditions like those trumpeted (or even invented) by enthusiastic boosters like Bonfant were too dangerous to be left uncorrected.

In Sardinia, by contrast, the situation looked quite different. The archbishops of Cagliari and Sassari strategically deployed the institutional authority of Rome and Madrid as weapons in their contention over preeminence and primacy, while Bonfant turned his historian's pen to support not only the universal Church but the devotional and propagandistic needs of his local community. In his dedicatory letter to King Philip IV of Spain, Bonfant described his book as a "brief sacred history" written in order "to restore the truth of the existence and ancient splendors that this kingdom of Sardinia and this, my homeland of Cagliari, possessed" and "to remove from the dark dungeons of oblivion in which they were cruelly buried the divine heroes that were the many saints who, at the cost of their blood, lives, and holy deeds, defended the Catholic faith."[95] His primary commitment was to the restoration and celebration of Cagliari's sacred past and his methodological choices, his selection of sources, and his interpretive strategy of pious affection served that end. While Bonfant's approach failed to meet Holstenius' high scholarly standards, his efforts to adorn his native city's Christian past with as many martyrs as possible are just one part of a larger Cagliarese initiative to shape the veneration of the saints to local needs. While few would argue that he was a skillful scholar, the real difference between Bonfant and his critic is as much a matter of talent or erudition as of the divergent perspectives and concerns of a local island patriot and an urbane, cosmopolitan member of Rome's intellectual elite.

Despite his many shortcomings as a historian, Dionisio Bonfant should not be laughed off as merely inept and ignorant, a provincial patriot

[94] "adoratio sine fundamento," BAV, Barberini, U.IV.72, 122.

[95] "A fin de restituyr a la verdad de su ser, y antiguos resplandores, que tuuo este reyno de Cerdeña, y ésta mi patria Caller tan leal, y fiel a la diuina, y a V. Magestad quanto dello tiene cabal satisfación; restaurar en el modo possibile a la pluma antigüedades de numerosas poblaciones, y sumptuosos edificios mal logrados a fuerça del disfauor del tiempo, y al pesar destó sacar de las obscuras masmorras del oluido, en que cruel tenía sepultados los diuinos heroes de tantos santos, que a precio de su sangre, vidas, y obras santas defendieron la fe Cathólica, y gozan felices de los eternos bienes; tirando agajes de immortalidad, escriuí esta breue Historia sacra ..." Bonfant, *Triumpho*, unp.

who "lacked the light of healthy criticism," but instead "burned with an extreme zeal [for his] city."[96] Rather, we should seek out the local "social logic" of Bonfant's text.[97] It is not enough to read *Triumpho de los santos del reyno de Cerdeña* from the perspective of his contemporary critics in Rome or to measure its faults according to modern historical metrics; rather, we must also read it within its Sardinian and Cagliarese contexts, as a product of a particular social, political, and cultural climate. Approached in this way, Bonfant appears not merely as an incompetent interpreter of the fragmentary remains of the ancient past, but instead as one of many "local Baronios" whose efforts to defend and advance the sacred histories of their beloved hometowns helped found modern historical method.[98] It is true that Bonfant's positive contributions to historical method are few; perhaps his lasting achievement is standing as an excellent example of what respectable historians ought not to do. Still, careful attention to Bonfant and to his first critic, Holstenius, reminds us that even an "immense structure of errors" can reveal much about the multiple rationales and the contexts that helped shape the development of modern historical practice.

[96] Pietro Martini, *Biografia sarda*, vol. 1 (Cagliari, 1837), 159: "Da questa voluminosa scrittura, divisa in tredici libri, si raccoglie a prima giunta, che lo storico, quantunque non mancasse d'ingegno e di dottrina, pure difettava dei lumi della sana critica, e, ciò che più monta, ardeva d'uno zelo assai esaltato di municipio."

[97] Gabrielle M. Spiegel, "History, Historicism, and the Social Logic of the Text in the Middle Ages," *Speculum* 65(1) (1990): 59–86.

[98] Ditchfield, *Liturgy, Sanctity and History*.

11

The Forces of the King: The Generation that Read Botero in Spain*

Xavier Gil

I

In his instructions to help the king choose good councilors, Fadrique Furió Ceriol detailed the various qualities these advisors should have and suggested the king subject potential advisors to a practical exam with several questions. An essential quality was that the advisor be experienced in the art of war and peace and thus Furió wrote: "The king should inquire him on matters concerning government," one of which was the following: "Is the power of the king based on wealth or good soldiers?"[1]

The question alluded to one of the debated issues of the period. Machiavelli had warned that the topic *pecunia nervum belli* was deceptive since, according to him, it was effective soldiers that won victories which, in turn, procured gold.[2] Another perspective was that a complete and total confidence in Divine Providence preceded any human considerations, a perspective that was perhaps best articulated by the Jesuit Pedro de Ribadeneyra:

Only God concedes kingdoms and rules kings and enlightens and councils them so that they know what to undertake and [gives them] the resolve to actually

* This essay is a shorter, adapted version of that published in *Le forze del principe. Recursos, instrumentos y límites en la práctica del poder soberano en los territorios de la Monarquía Hispánica*, eds. Mario Rizzo, José Javier Ruiz Ibáñez and Gaetano Sabatini (Murcia: Universidad de Murcia, 2004), vol. II, pp. 969–1022. I thank the editors for their permission for this new publication.

[1] Fadrique Furió Ceriol, *El concejo y consejeros del Príncipe* (1559), ed. Henri Méchoulan (Madrid: Tecnos, 1993), 39. The translation of this and the remaining quotes is mine, unless otherwise noted.

[2] Niccolò Machiavelli, *Discorsi sopra la prima deca di Tito Livio*, b. 2, ch. 10, eds. Genaro Sasso and Giorgio Inglese (Milan: Rizzoli, 2011), 316–19.

undertake it as well as the forces and industry to successfully accomplish it and all enterprises undertaken in His name.[3]

While still sharing this superior principle, other authors sought more terrestrial answers. Jean Chokier, for example, one of Lipsius' best disciples, argued in his *Thesaurus aphorismorum politicorum* (Rome, 1611) that the prince's *fama* and *auctoritas* and the *benevolentia* of his subjects were more important than his means (*vires*). The German Jesuit Adam Contzen, on his part, expounded this perspective in further detail in his influential *Politicorum libri decem* (Mainz, 1621):

The resources (*vires*) of a state consist of piety and the virtues, then in those things which follow from the virtues in a principality, the first of which is authority (*auctoritas*), or reputation (*fama*) or majesty (*majestas*), then love (*amor*) of the citizens; thirdly [there comes] wealth (*opes*), arms (*arma*), men (*viri*), fortresses (*arces*), cities (*urbes*).[4]

This careful classification of symbolic and material resources could be simply and clearly established in an author's study. Rulers, on the other hand, usually hard-pressed for money, looked ceaselessly for ways to augment the royal treasury. This concern was nigh universal. Francis Bacon addressed it in his unfinished *Of the true greatness of the kingdom of Britain* (1603–08), which included fragments of a prior speech in the Commons. It was simple enough to measure the territorial dimensions of a kingdom, he argued, but "the just measure and estimate of the forces and power of an estate is an exceedingly difficult task." Antoine de Montchrétien also recognized this difficulty at the very beginning of his *Traicté de l'oeconomie politique* (1615): "In only one aspect are you lacking, oh magnificent state: the knowledge of yourself and consequently the use of your force."[5] Nevertheless, difficult as it was, this was what ultimately mattered and the Junta de Reformación, the famous committee established in 1618 by Philip III to channel the many projects

[3] Pedro de Ribadeneyra, *Tratado de la religión y virtudes que debe tener el príncipe cristiano* (1595), in *Obras escogidas*, ed. Vicente de la Fuente, 60 (Madrid: Hernando, Biblioteca de Autores Españoles, 1899), b. I, ch. 8, 467b–8a.

[4] Both quoted by Robert Bireley, *The Counter-Reformation Prince. Anti-Machiavellianism or Catholic Statecraft in Early Modern Europe* (Chapell Hill, CA: University of North Carolina Press, 1990), 136, 142–3.

[5] Bacon, quoted by Markku Peltonen, "Politics and Science: Francis Bacon and the True Greatness of States," *Historical Journal* 35 (1992): 284; Montchrétien, by Julio Pardos, "Le dénombrement du peuple (para leer un censo)," *Índice. Revista de Estadística y Sociedad* 60 (July 2014): 2.

for economic reform being propounded, admitted as much: "The world does not pay more attention nor render more homage to emperors, kings and princes than that which each of them is able to impose thanks to his forces and power."[6]

The issue of a state's resources was relevant at the end of the sixteenth and the beginnings of the seventeenth centuries in the context of monarchical institutional and territorial development, financial difficulties, augmenting colonial rivalries, and wars of religion. And the Counter-Reformation camp included a resolute response to Machiavelli's challenges: it was possible to become involved in politics while not succumbing necessarily to immorality; at the same time, age-old questions of "necessity" and its repercussions with respect to the practice of governing came to the fore.[7]

In this atmosphere, Giovanni Botero (1544–1617), a churchman, successfully compiled and thoroughly addressed many of these concerns. His noted *Raggion di stato* (1589) went far beyond his stated objective to redirect the uncomfortable teachings of Machiavelli along the path of Tridentine orthodoxy. Fully accepting the Florentine's observation that conserving the state was a task no less difficult and no less worthy of praise than expanding it, the book also implied a far-ranging analysis of the means to strengthen the state and efficiently achieve its goals. Botero did not base this strengthening on an expansion of the crown's legislative or judicial faculties – a topic his contemporaries heatedly discussed and he rarely mentioned – but on ameliorating the prince's human and material resources. And although it was true that this amelioration would benefit the prince first and foremost, as head of the body politic, Botero's global perspective was predicated on the assumption that this self-preservation and amelioration pertained and would be of benefit to both the prince and his subjects.[8]

If at the beginning of his treatise he affirmed that "the self-preservation (*conservazione*) of a state consists in the subjects' quietude and peace," in

[6] Ángel González Palencia, *La Junta de Reformación, 1618–1625*, Archivo Histórico Español, Valladolid, 1932, doc. 42, "Discurso breve y sumario de las causas porque se an disminuido y despoblado muchas villas y lugares en estos reynos de Castilla y León y venido a tanta pobreça los vasallos dellos," anonymous, *c.* 1621, p. 262.

[7] Bireley, *The Counter-Reformation Prince*, ch. 2; Ronald W. Truman, *Spanish Treatises on Government, Society and Religion in the Time of Philip II. The 'De Regimine Principum' and Associated Traditions* (Leiden: Brill, 1999), 366–7.

[8] In general, see Luigi Firpo's entry "Botero," in *Dizionario Biografico degli Italiani*, vol. XIII (Rome: Istituto della Enciclopedia Italiana, 1971), 352–62; and Bireley, *The Counter-Reformation Prince*, ch. 3.

other sections he expressly referred to the prince's forces – those enabling him to enlarge his state, "which I tend to call instruments of prudence and courage." And after confessing that "it would be a long matter discussing in detail all those things which can be called a prince's forces (*forze*)," he summarized these forces as "people and money," so that "I am to use indistinctly the words people and force." He then established a distinction between extensive and intensive means, the latter being those by which courage was increased.[9]

Botero's premises were in fact hardly original. Cartographers, for example, would often accompany their maps and charts with descriptions of the economic resources of the depicted lands and countries, and Venetian ambassadors, above all, would pay close attention in their missives to relaying information on the population, armies, natural resources, judicial courts and the characteristics of the upper nobility and the courtesans surrounding the king in whose presence they were accredited. And although Botero included much economic information, Jean Bodin had previously dedicated the last of his *Six livres de la République* (1576) to finances, taxes, and other public burdens.[10] Although the *Ragion*'s doctrinal contents were rather slim, the breadth of his writing on the practical necessities of the state was remarkable and evinces a refined grasp of the issues, both old and new, confronting rulers and essayists. As a result, Botero's work encapsulated and effectively established an entire field of reflection – to the accolades of many of his contemporaries. In distinction to doctrinaires eager to proclaim broad religious truths but unpreoccupied with or disdainful to the inherent limitations of actual governing practices, Botero and many others always adopted an empirical and realistic point of view.[11] In fact, in his immediately preceding opuscule *Delle cause della grandezza e magnificenza delle città* (1588) he had already elaborated on the interrelationship between natural setting, economic resources, and demographic development – concerns that reappear, with a plethora of facts and on many pages, in his encyclopedic

[9] Giovanni Botero, *Della ragion di Stato* (1589), ed. Chiara Continisio (Rome: Donzelli, 1997), 18, 139–40, 149, 175.

[10] Jean Bodin, *Les six livres de la République* (1576), VI, 1, eds. Ch. Frémont, M.-D. Couzinet, and H. Rochais (Paris: Hachette, 1986), 7.

[11] These two approaches have been labeled in different ways: Bireley considers two pragmaticisms: one, providentialist, the other (Botero's), intrinsic or immanent: *The Counter-Reformation Prince*, 30–1, 51–2, 68; José A. Fernández-Santamaría, dealing with Spanish authors only, divides them between ethicists and realistics: *Razón de estado y política en el pensamiento español del Barroco (1595–1640)* (Madrid: Centro de Estudios Constitucionales, 1986), 14–17.

Relationi universali (1591–5), even though his stated intent in this work was to trace the diffusion of the Gospel in the newly discovered lands.

Botero's works achieved a wide circulation. The *Ragion* (which was usually published in conjunction with the previously mentioned opuscule on cities) had ten Italian editions before his death in 1617. The first foreign translation was to Spanish in 1593 and it was a great success with five new editions by 1606: Barcelona in 1599 and 1605, and Burgos in 1602, 1603, and 1606. There also were translations to German (1596), French (1602), English (1606) and Polish (1609). As for the *Relationi*, it was translated either totally or in part to Latin (1596), German (1596), English (1601), Spanish (1603), and Polish (1609) but not to French, although Pierre d'Avity did apparently begin a translation. In Germany, the *Relationi* as well as the *Ragion* were widely read not only in Catholic but also reformed circles where, Tridentine orthodoxy aside, the teachings on *prudentia regnativa*, useful to all rulers, were welcomed. In England, on the other hand, *Relationi* seem to have had no major impact.[12]

In Spain, Philip II himself ordered the translation of the *Ragion*, confiding it to his royal chronicler Antonio de Herrera Tordesillas with the title *Diez libros de la Razon de Estado, con tres libros de las causas de la grandeza y magnificencia de las ciudades de Juan Botero* (Luis Sánchez, Madrid, 1593). The *Relationi* had two independent translations, both in 1603: one was undertaken by the *licienciado* Diego de Aguiar, *Relaciones universales de mundo de Juan Botero, primera y segunda parte* (Herederos de Diego Fernández de Córdova, Valladolid, folio); the other by the Dominican Fray Jaime Rebullosa, *Descripción de todas las provincias y reynos del mundo, sacada de las Relaciones Toscanas de Juan Botero* (Gabriel Graells y Giraldo Dotil, Barcelona, octavo).

The Spanish edition of *Razón de estado* did not include Botero's dedication to the archbishop of Salzburg that detailed his objectives. But Antonio Herrera parsed them out in his own dedication to the king in his capacity as translator. Botero, as Herrera informed, wanted to "educate a religious and prudent prince who could govern and maintain peace and justice in his state, thereby proving that Niccolò Machiavelli and Cornelius Tacitus' methods were unnecessary." As a result, Herrera was

[12] These data, incomplete, come from Bireley, *Counter-Reformation Prince*, 53; Michael Stolleis, "La scoperta della Ragion di Stato: Giovanni Botero in Germania," *Pensiero Politico* 26 (1993): 185–6; and John M. Headley, "Geography and Empire in Late Renaissance: Botero's Assignment, Western Universalism and the Civilizing Process," *Renaissance Quarterly* 53 (2000): 1134n.

confident that it would prove to be very useful for his son the crown prince.[13] In his dedication to the new edition (Burgos, 1603), Herrera Tordesillas envisaged a limited audience for this work in accordance to the principle of the *arcana imperii*, reserved for princes and rulers. On the other hand, Aguiar and Rebullosa contended that the *Relaciones* would be useful for a much broader audience: "all those who govern" including heads of households, according to the former; and "theologians, preachers, philosophers, doctors, lawyers, orators, poets, princes, governors, captains, pilots and travelers," according to the latter. In addition, Rebullosa emphasized that the books showed that "the Monarchy of the Holy Church" was not at all sidelined but was rising throughout the world. It is hardly surprising that the second edition of his translation was entitled *Historia eclesiástica y estado presente de la religión en todos los reynos de Europa, Asia y África, sacada de la Relaciones Toscanas de Iuan Botero* (Barcelona, 1610).

Once translated into Spanish, both volumes achieved a wide distribution. The expression "the generation that read Botero," used here, should be understood in an approximative way, for the Spanish reception of Botero requires a more detailed study. Moreover, Botero's reception was more or less coeval with Bodin's, Lipsius', Tacitus', and Campanella's and there were conceptual and lexical coincidences among them, if not actual influences and textual interpolations. And in this respect, it is not insignificant that Herrera y Tordesillas translated Tacitus as well. The subject, then, is broad and complex.[14] Nevertheless, it is worth noting that reception was not unidirectional given that Botero, in his diverse collection of

[13] Herrera in Botero, *Diez libros de la razón de estado, con tres libros ...* (Barcelona: Jayme Cendrad, 1599), 3–5. From now on, I quote from this second Spanish edition.

[14] J. H. Elliott spoke some years ago of Botero's presence in Spain and of the need to study his reception: *The Count-Duke of Olivares. The Statesman in an Age of Decline* (New Haven, CT: Yale University Press, 1986), 93, n. 33, 192. See now also Richard L. Kagan, *Clio and the Crown. The Politics of History in Medieval and Early Modern Spain* (Baltimore, MD: Johns Hopkins University Press, 2009), 126. On the reception of the other mentioned authors, see Francisco Sanmartí Boncompte, *Tácito en España* (Barcelona: CSIC, 1951); Beatriz Antón Martínez, *El tacitismo en el siglo XVII en España. El proceso de 'receptio'* (Valladolid: Universidad de Valladolid, 1992); Saúl Martínez Bermejo, *Translating Tacitus. The Reception of Tacitus's Works in the Vernacular Languages in Europe, 16th–17th centuries* (Pisa: Pisa University Press, 2010); Martin Albuquerque, *Jean Bodin na Peninsula Ibérica. Ensaio de historia das ideas políticas e de dirito público* (Paris: Fundacion Gubelkian, 1978); Harald E. Braun, "Making the Canon? The Early Reception of the *République* in Castilian Political Thought," in *The Reception of Bodin*, ed. H. A. Lloyd (Leiden: Brill, 2013), ch.11; John M. Headley, *Tomasso Campanella and the Transformation of the World* (Princeton, NJ: Princeton University Press, 1997), 203–12; Headley, "Geography and Empire," 1142.

information for the *Relationi*, made abundant use of José de Acosta's treatises on the Indies.[15] Moreover, various Spanish essayists, such as Tomás Cerdán de Tallada, from Valencia, and Luis Valle de la Cerda, from Flanders, had already indicated at the beginning of the 1580s some of the characteristic arrows in the Piedmontese's argumentative quiver.[16] And even among *arbitristas*, who generally evince familiarity with Botero, the influences were various: Martín González de Cellorigo was steeped also in Machiavelli and Bodin; and Valle de la Cerda himself could have based his later commentaries on the higher value of merchandise vis-à-vis raw materials in Luis Ortiz's celebrated tract (1588) and in the lively debates that had been taking place in the Cortes of Castile regarding a broad panoply of economic reform proposals.[17]

In any case, Spanish readers seemed particularly sensitive to two of the broad issues formulated in *Razón de estado*. The first consisted of Botero's reflections on which empires and states were more lasting, a question that he divided in two, that is, whether those large, small, or medium sized, and whether those that were geographically united or separated (in the first book); and the second issue consisted of his commentary on Spain's sterility and sparse population (eighth and ninth books). The second of the two issues garnered specific reactions that led to repeated and quite varied references to him.

II

In his treatise on the greatness of cities and with respect to the question "What constitutes a great city?," Botero offered a clear response:

> A city consists of many men gathered together in one place to live in happiness: and the city's greatness is not its setting nor what its walls surround but the conglomeration of its inhabitants and its power.[18]

[15] Headley, "Geography and Empire," 1142.

[16] José Mª Iñurritegui, *La gracia y la república. El lenguaje político de la teología católica y el 'Príncipe Cristiano' de Pedro de Ribadeneyra* (Madrid: UNED, 1998), 137.

[17] Jesús Villanueva, "El reformismo de González de Cellorigo y sus fuentes: Maquiavelo y Bodin," *Hispania* 57 (1997), 63–92; Anne Dubet, *Réformer les finances espagnoles au Siècle d'Or. Le projet Valle de la Cerda* (Aubenas d'Ardèche: Presses Universitaires Blaise-Pascal, 2000), 173 (and with respect to Botero's influence on Valle, pp. 134, 165, 196–8). On the debates at the Cortes in 1580 and 1595, see José Ignacio Fortea, *Monarquía y Cortes en la Corona de Castilla. Las ciudades ante la política fiscal de Felipe II* (Salamanca: Cortes de Castilla y León, 1990), 416–20.

[18] Giovanni Botero, *Dalla ragione di stato libri dieci ... Tre libri della grandezza delle città*, enlarged ed. (Rome: Vicencio Pellagallo, 1590), 325.

The definition was based on the difference and complementariness between *urbs* and *civitas* characteristic of Renaissance culture. Throughout his long and storied career, Richard L. Kagan has written revelatory pages on this topic in significant works on urban views, chorographies, and histories of cities, Latin American urbanism (with Fernando Marías) as well as other aspects of life and culture of urban communities. This essay on Botero's reflections on cities and states and his presence among Spanish authors of the period is then, probably, a suitable way to contribute to his Festschrift.

In the above-mentioned quotation, Botero endowed *civitas* with a much stronger demographic as distinct to juridical meaning and he was to show the same sensibility in *Ragion di stato* when affirming that the prince's real forces lay in the great number of people, since "he who abounds in people is also copious in money, because a great number of people brings larger taxes with which the treasury becomes richer" – an affirmation that retained its full force in Herrera's translation. It was the conclusion to a series of observations. Botero saw the confirmation of his assertion in France and Italy, countries that despite not having gold mines were very populated and as a result had money in abundance due to commerce. The ability to raise great armies was taken as a clear proof of demographic potential. In this sense, Botero indicated that the Iberian Peninsula had maintained many armies in Carthaginian and Roman times as well as during the 700 years of the Reconquista and the last war in Granada and detailed the reason for the contrast with the present: "Not because the nature or quality of its land or air has changed but because the number of its men has diminished." Wars, he went on, had continued in Africa, Naples, the Indies, and Flanders, where many Spaniards went "to trade or to fight wars." Moreover, the expulsion of the Jews from both Spain and Portugal had taken place, about which Botero cited Sultan Beyacid's ironic comment of professed admiration for King Fernando's prudence, "since he deprived himself of what aggrandized and enriched states, that being people." All this led him to conclude that the population scarcity Spain suffered was the cause of its agricultural weaknesses in addition to the natural inclination of its inhabitants for the military and the consequent detachment from manual arts "because there is no province that has less artisans and industry and as a result wool, silk and other things mostly leave the country."[19]

[19] Botero, *Ragion di stato*, 150–2; *Diez libros de la razón de estado*, ff. 97–8.

Botero did not know Spain first hand and it was not until 1603 that he traveled there accompanying the duke of Savoy's sons on a state visit that lasted until 1606. Even so, his observations fell on fertile ground since Spanish society, and in particular the Castilian, was buffeted by an increasing number of economic and financial difficulties in addition to the horrific death toll of the 1598 plague. The sensation of decadence was in the air and was becoming ever more palpable.[20] This sensation was due in large part to the contrast between reality and the rooted ideal of self-sufficiency. In the Aristotelian mold, self-sufficiency was one of the defining characteristics of a civilized and well-ruled community. In Pedro Simón Abril's translation, a city was, according to Aristotle "the multitude of its citizens [that is to say, those who participate in its government] that suffices, generally speaking, for life's needs (...) and to provide for its own security so as to not have to seek it elsewhere and maintain on its own."[21] Self-sufficiency was a requisite for perfection and a city as well as, by extension, a kingdom, claimed to be perfect communities. Hence, urban chronicles would often begin with an ode to the material and human abundance of the city in question and of its hinterland while many authors repeated the idea, each one according to his own objectives, as illustrated by Bartolomé de las Casas and Gregorio López Madera.[22]

Botero himself developed this argument when reflecting on independence and distinguishing between a jurisdictional, "accidental and external," independence and another, effective one, "substantial and intrinsic," that was, consequently, more important. As he explained, political independence "enables me to be an absolute and sovereign lord," while the other one, in contrast, "makes me powerful and gives me enough forces for the self-preservation of my state, and thus I am really a great prince." As he then concluded, independence in this second sense was impossible "without forces of my own," phrases that Herrera translated faithfully

[20] See above all J. H. Elliott, "Self-perception and Decline in Early Seventeenth-Century Spain (1977)", in his *Spain and Its World, 1500–1700* (New Haven, CT: Yale University Press, 1989), ch. 11.

[21] Aristóteles, *Política*, eds. A. Alegre, J. Santolaria, and M. Lavado (Barcelona: Orbis, 1985), according to Pedro Simón Abril's translation, *Los ochos libros de república* (Zaragoza, 1584), b. 2, ch. 1; b. 4, ch. 9: 1275b, 1294b, 121, 194.

[22] Bartolomé de las Casas, *De regia potestate o derecho de autodeterminación*, bilingual ed. by L. Pereña, J. M. Pérez Prendes, V. Abril, and J. Azcárraga (Madrid: CSIC, 1969), fourth *quaestio*, 40, 43; Gregorio López Madera, *Excelencias de la Monarquía y reino de España* (Valladolid, 1597), ed. J. L. Bermejo (Madrid: Centro de Estudios Políticos y Constitucionales, 1999), ch. 9, 123–4.

as was his wont.[23] While for Botero such forces were above all military, Sancho de Moncada, one of the most articulate Spanish economic reformers, rather thought of an agenda for commercial and agricultural development, thanks to which, as he argued to Baltasar de Zúñiga, the man in power at that time, "Spain would be more perpetual, being independent from anyone." Regardless of the emphasis, the time-honored ideal of self-sufficiency seemed to allow for a condition of independence and the great Tacitist writer Baltasar Alamos de Barrientos said as much in an aphorism: "A principality can be happy if it depends strictly on itself and does not need to have recourse to others for its conservation."[24]

In his *Relaciones universales* more than in *Razón de estado*, Botero expounded in detail on the resources and strengths of various countries. This is reflected in the title of *Iudicium Io. Boteri de potentia regis Hispaniarum*, a translation of the sections devoted to the Spanish monarchy in the *Relationi* published in a compilation, *Hispania*, prepared by Johannes de Laet (Leiden, 1629). This partial translation, similar to others that compiled, also in Latin, the chapters in the *Relationi* dedicated to England, France, Poland, Persia, or other countries, must have enjoyed many readers in Europe. But the case is that in Spain, even if some individual authors were based on the *Grandeza y magnificencia de las ciudades*, such as Joan de Xerez and Lope de Deza in their *Razón de corte* (*c.* 1601), a plea for the court to remain in Madrid; or on the *Relationi*, such as Antonio de Herrera in the *Historia general del mundo* (1601) or Cristóbal Suárez de Figueroa in *El Pasagero* (1617) for some aspects of their respective works,[25] it was in fact the *Razón de estado* that was the book that occasioned the most wide-ranging response.

The afore-mentioned contrast between the ideal of self sufficiency and reality was confirmed by another comparison in historical terms: between a better past and a present in decline. By 1600, it was common practice

[23] Botero, *Ragion di stato*, 176, 171; quotes of Herrera's translation, *Diez libros de la razón de estado*, ff. 112vi–113r, 110r–110v.

[24] Sancho de Moncada, *Restauración política de España* (1619), ed. J. Vilar (Madrid: Instituto de Estudios Fiscales, 1974), 122–3; Baltasar Álamos de Barrientos, *Aforismos al Tácito español* (1614), ed. J. A. Fernández-Santamaría (Madrid: Centro de Estudios Constitucionales, 1987), vol. I, 253, aphorism n. 303.

[25] Juan de Xerez y Lope de Deça, *Razón de corte* (*c.* 1601), ed. A. T. Reguera (León: Universidad de León, 2001), quoting Botero only on a lesser passage, pp. 157–8 (I owe this reference to Richard Kagan); Iñurritegui, *La gracia y la república*, 174–5; Jean-Marc Pelorson, *Letrados. Les letrados, juristes castillans sous Philippe III. Recherches sur leur place dans la société, la culture et l'état* (Le Puy-en-Velay: Université de Poitiers, 1980), 356.

to consider the period of the Catholic Monarchs as the culminating period and no one did so more assertively than the great *arbitrista* Martín Gonazález de Cellorigo, who conducted a brief historical review going back to the Goths: "Never did our Spain have in all aspects a higher degree of perfection, its growth augmented and its well-being flourished (...) They situated the monarchy at the apex of perfection and (...) elevated Spain to the highest state of contentment and greatness that it had ever had." But this contentment did not last. "Afterwards it began to decline (...) arriving at the state that we all judge to be the worst of all preceding ones."[26] Juan de Mariana in his *Historia general de España* also combined the topical praise of abundance with the somber panorama of the present marked by a lack of moral fiber and economic vulnerability. And in another section, he recognized that Spain was "subject to a higher degree than other lands to the dryness of the air and lack of rain."[27]

Climate and the fertility of the soil centered part of the debate. While the restoration of "abundance and fertility" was one of the objectives invoked in the discussions in the Cortes of Castile during the decade of 1590, Spanish agricultural fertility and procreative fecundity became a subject widely debated by writers, in part due to Botero's observations. In his first short book on cities, he spoke of man's "generative virtue" and the cities' "nutritive virtue" and subsequently in the *Relationi* again insisted on the importance of the quantity and valor of the population, on supply, and particularly on the favorable balance of trade by exporting manufactured goods as a font of true wealth.[28]

Some writers, such as the Valencian Gaspar Escolano, in his *Décadas* (1610–11), rejected out of hand Botero's observations regarding the weakness of the Spanish economy.[29] But the best *arbitristas* dedicated serious analyses to the issue. The previously mentioned González de

[26] Martín González de Cellorigo, *Memorial de la política necesaria y útil restauración a la república de España y estados de ella y del desempeño universal de estos reinos* (1600), ed. J. L. Pérez de Ayala (Madrid: Instituto de Estudios Fiscales, 1991), 94–5.

[27] Juan de Mariana, *Historia General de España* (Latin ed., 1592; Spanish translation, 1601), in his *Obras*. ed. Francisca Pí y Margall, Biblioteca de Autores Españoles, 20–31 (Madrid: Atlas, 1950), XXX, 7a, 16b.

[28] Fortea, *Monarquía y Cortes*, 437; Botero, *Tre libri della grandezza*, 395–6; Botero, *Le Relationi Universali*, 2nd part (Venice: Giacomo Vincenti, 1598), "Proemio" (no pagination) and 6, 28, 152.

[29] James Casey, "Patriotism in Early Modern Valencia," in *Spain, Europe and the Atlantic World. Essays in Honour of John H. Elliott*, eds. R. L. Kagan and G. Parker (Cambridge: Cambridge University Press, 1995), 194.

Cellorigo is a remarkable case of receptivity to Botero's formulations. He cited Botero in the first paragraph of his well-known *Memorial* and shared his judgment that sterility was the result of lack of people, literally copying the corresponding fragment of Herrera's translation, in sentences that, significantly enough, reappeared a few years later verbatim in one of the reports studied by the Junta de Reformación.[30] Lope de Deza and Sancho de Moncada also rejected in those years the idea that the root of the problem was, in the words of the former, "the tiredness and weakness of the land."[31]

The populationist argument was widely shared in Spain throughout the seventeenth century. Cellorigo, one of the first to openly defend it, invoked the authority of Alfonso X the Learned to affirm that "the dignity of the king is in the number of people and his ignominy in the decrease of population." Sancho de Moncada for his part, quoted Ulpianus and Botero proclaiming: "Without people there is no kingdom because the people are the kingdom, just as a city is its citizens and a school is its students, and (...) the conservation of a kingdom is impossible without people (...) for a kingdom can only do according to the number of its people" so that "a kingdom is considered to be richer because of its people than because of its gold and silver." Like-minded were Gerónimo de Zeballos, who, citing Botero and Moncada, linked the decrease in population with the absence of a protectionist manufacture policy; and Ribadeneyra and Pedro Fernández Navarrete, who based their work on the *Partidas* (Alfonso X's great Castilian legal code) and Justinian and Pompeius Trogus: "Cities are not made by their walls but by their dwellers."[32]

This often heartfelt point of view did not stymie the discussion of other perspectives. Some authors considered the base or the nerve of a kingdom to be money, credit, trade, arms, the love between a prince and his subjects, or the law. These and other opinions reflected both clichés

30 Cellorigo, *Memorial*, 11, 57, 58; González Palencia, *Junta de Reformación*, doc. 42: "Discurso breve y sumario" (see note 6), 236.

31 Lope de Deza, *Gobierno político de agricultura* (1618), ed. A. García Sanz (Madrid: Instituto de Estudios Fiscales, 1991), 41–3; Moncada, *Restauración* (1619), 99–100.

32 Cellorigo, *Memorial*, 7; Moncada, *Restauración*, 134; Gerónimo de Zeballos, *Arte real para el buen govierno de los reyes y príncipes y de sus vasallos* (1623), ed. S. de Dios (Madrid: Centro de Estudios Políticos y Constitucionales, 2003), f.151; Ribadeneyra, *Tratado*, 534b; Pedro Fernández Navarrete, *Conservación de monarquías y discursos políticos* (1626), ed. Michael D. Gordon (Madrid: Instituto de Estudios Fiscales, 1982), 45–6, 65.

and sharp analyses of the present. If land was not sterile, as was often repeated, depopulation was consequently due to other causes and their study was the subject of much discussion. The expulsion of the Jews and Moriscos; the number of religious callings; the abandonment of arts and crafts and the channeling of resources to *juros* and other forms of public debt; idleness and laziness; the presence of foreign traders, a habitual target of criticism; the movement of cash and raw materials to other countries that prospered at the cost of Spanish scarcity; the role of the Indies in the general situation; and the subsequent need to achieve a favorable balance of trade were some of the issues proposed, time and again, by *arbitristas*, rulers, and writers throughout the first decades of the seventeenth century and constitutes, as is well known, an important chapter in European economic thought. And by dint of these debates, the old Scholastic notion of sufficiency understood narrowly as autarchy began to evolve in Spain to an idea of sufficiency achieved by foreign trade.[33]

Botero's commentaries underlay many of these subjects and as a general rule the analyses of the Spanish authors coincided either expressly or implicitly with his. There were two notable exceptions though. The first was with respect to the expulsion of the Jews and the Moriscos (the latter posterior to the publication of Botero's books) – the consequences were mitigated or denied. Fernández Navarrete, for example, openly admitted the demographic losses entailed and even included Beyacid's opinion (seen previously in Botero) but justified them in accordance to the superior criteria of religious uniformity and with the opinion, once again, of Alfonso the Learned that kings should populate their kingdoms with "good people." In contrast, Sancho de Moncada, probably the *arbitrista* who quoted more often and with more precision Botero's three treatises, almost always to support his own arguments, followed a strictly demographic logic in 1619 to dismiss the claim that the expulsion of the Moriscos, begun ten years earlier, had occasioned any negative demographic consequences and he did so by producing some empirical data from parochial registers and other observations.[34]

[33] Marjorie Grice-Hutchinson, *Early Economic Thought in Spain, 1177–1740* (London: Allen and Unwin, 1978); Enrique Fuentes Quintana, dir., *Economía y economistas españoles*, vol. 2: *De los orígenes al mercantilismo* (Barcelona: Galaxia Gutenberg-Círculo de Lectores, 1999). On the development of the notion of sufficiency, see Alain Milhou, "Sufficientia (Les notions d'autosuffisance et de dépendance dans la pensée politique espagnole au XVIe s.: de la Castille des *comuneros* au Pérou colonial)," *Mélanges de la Casa de Velázquez* 17 (1981): 105–45, esp. 117–21.

[34] Fernández Navarrete, *Conservación*, 67, 71–2; Moncada, *Restauración*, 135, 159.

The second exception was with respect to the suitability first of covering the demographic lacuna with the arrival of new populations and second, the suitability of attracting foreign artisans and mechanics. Botero noted that both Portuguese and Spaniards, suffering in a similar pattern from a scarcity of demographic resources, resorted to the slave trade to obtain a labor force in their overseas possessions; and approved of the fact that among the conditions specified by the Polish in electing Henry of Anjou as their king was that a hundred artisan families be sent.[35] The issue of slavery was hardly dealt with by the Spanish authors discussing such matters. And the idea of promoting settlers (which was not original to Botero) did not have a particularly favorable reception among writers. Fernández Navarrete, realizing that it was "necessary" for "kingdoms aspiring to undertake great projects and extend their rule to place the highest hope in the multitude of their people," censured, as the Piedmontese already had, Sparta's practice of closing its doors to foreigners, blaming it for the Spartan military and political ruin, and praising instead the Roman practice of "admitting to their city all those who desired entry." Nevertheless, his view was rather general. Cellorigo, on the other hand, was much more precise, maintaining that the authentic solution was none other than preventing that people feel the need to abandon the kingdom. With different emphases, Zeballos, Lope de Deza, and Moncada shared this reserve against foreigners even if they were to reside permanently and were to become naturalized in Spain. Moncada openly stated: "The authentic population has a true Spanish bloodline."[36]

The issue of population settlements was linked to another very typical issue of the period: the colonies. As he had with many other things Roman, Machiavelli spoke in favorable terms considering colonies useful for consolidating new dominions and defending the boundaries in a much cheaper and more effective manner than military garrisons, in addition to other useful attributes.[37] But the word *colonia* used to incorporate

[35] Botero, *Tre libri della grandezza*, 329; *Ragion di stato*, 172, 156; *Relationi*, second part, 142.

[36] Fernández Navarrete, *Conservación*, 64; Cellorigo, *Memorial*, 57; Zeballos, *Arte real*, ff. 152–3; Deza, *Govierno*, 47; Moncada, *Restauración*, 136–7. The Olivares government took some measures to bring forward skilled people in the 1620s, with limited success: Elliott, *The Count-Duke*, 296, 414.

[37] Nicolás Maquiavelo, *El príncipe. De principatibus*, bilingual ed. by G. Inglese and H. Puigdomènech (Madrid: Tecnos, 2011), ch. 3, 18–21; *Discorsi*, I, ch. 1; II, chs. 6 and 19, 61, 310, 345.

another important meaning even in colloquial speech: enjoying the privileges of the Roman citizenship.[38]

Giovanni Botero dedicated special attention to the colonies. In *Grandezza delle città* he had already presented his essential ideas (colonies "are as one body with the mother" and "they have to add to the mother") and he developed these ideas in *Ragion di stato*, referring to the Romans, in this respect, as "very knowledgeable regarding reason of state," and more extensively in *Relationi*, where he defined colonies as "plants of Rome and seminaries of the Roman people." He mentioned the colonies of the Phoenicians, Greeks, and Romans as well as those established by the Germans in Hungary and Transylvania in more recent times, pointing out that in order to be efficient, colonies should not be too distant from the "patria." As a result, he considered Calais, Ceuta, Tangiers, Melilla, Madeira, Cabo Verde, and other fortifications to be beneficial while other overseas Iberian possessions were overall judged to be prejudicial, with the exception of Goa. He indicated that Portugal and Castile had weakened and bled of people so that that they had practiced "a reason of state mostly opposed to that bringing out Roman greatness and power." The Romans, he explained, enhanced their procreation by means of marriages and other procedures such as conferring citizenship on vanquished peoples and cities. The Portuguese and the Castilians, on the other hand, only made use of their own nationals despite not being very populated and despite, in the Spanish case, that the Milanese, Neapolitans, Sicilians, and Sardinians, by dint of their loyalty and longstanding stability under the Spanish crown, could certainly have been considered natural and trustworthy for distant enterprises. The Portuguese, lacking in conquered subjects, could make use of – he suggested – people brought over on stipends, as long as they were Catholic.[39]

The political and cultural Iberian world was familiar with the term and the world of the colonies. Suárez de Figueroa copied word for word one of Botero's sections on this subject in *Relationi*. Prior to him, Sebastián Fox Morcillo, a humanist and royal advisor, for example, had suggested to Charles V in 1536 that he send to the Indies "colonies of his citizens" in a deliberate imitation of the Romans in the *Mare Nostrum*; and the viceroy Cañete, upon arriving to Peru in 1555, wanted to establish agricultural

[38] Sebastián de Covarrubias, *Tesoro de la lengua castellana o española* (1611), ed. M. de Riquer (Barcelona: Altafulla, 1993), "Colonia," p. 338.

[39] Botero, *Tre libri della grandezza*, 352; *Ragion di stato*, 129, 153, 162–4, 225; *Relationi*, first part, 89, 17–18; second part, 139. The translated fragments, in *Relaciones*, ff. 15v–16r.

colonies to settle the immigrant population.[40] Even so, the word *colonia* was not the dominant one in Spain's political-administrative language since the American possessions were understood mainly as viceroyalties whereas in the English world it was a more common word, together with its vernacular adaptation as "settlement" and "plantation," as early as the occupation of Ireland.[41]

In Spain, the *arbitristas* were the ones who most often employed the word and Moncada and Deza in particular were keen on its use and elaboration. Viewing the American dominions as colonies, Sancho de Moncada judged their function to be positive, since, as he argued, they were capable of providing aid to Spain and consequently were an important factor in the "security of the state." He nevertheless indicated that "the colonies, garrisons and settlements" that Spain had throughout the world were a factor in its depopulation, a view later held by Navarrete.[42] Lope de Deza was more personal. Without citing the Piedmontese, he conceded that having sent people to conquer and settle far-off lands was the primary reason for the abandoned state of agriculture, the ill among those afflicting Castile attracting his attention the most. After drawing an explicit parallel between Rome and the Spanish monarchy, "so similar in greatness and in diversity of foreign provinces, conquered by its military and maintained by its laws," and indicating that consequently "what suited one has inevitably to suit the other with respect to civil matters," Deza proposed a "government of the colonies" according to the Roman practice. Using passages from Titus Livy and Cicero, he explained that practice: whenever there were many poor in Rome, the Senate sent triumviri who ascertained the places in the empire requiring new towns and garrisons where the poor and veteran soldiers were sent and given lands and occupations with optimum results. One only had to adapt this system to the Spanish monarchy "where there is most

40 Suárez Figueroa, quoted by Pelorson, *Letrados*, 435; Fox Morcillo, by Anthony Pagden, *Lords of All the World. Ideologies of Empire in Spain, Britain and France, c. 1500–c. 1800* (New Haven, CT: Yale University Press, 1995), 96; Cañete, by Milhou, "Sufficientia," 122.

41 Nicholas Canny, "The Origins of Empire: An Introduction"; and David Armitage, "Literature and Empire," both in *The Origins of Empire*, vol. 1 of *The Oxford History of the British Empire*, ed. N. Canny (Oxford: Oxford University Press, 1998), 8–10, 108–09; David Armitage, *The Ideological Origins of the British Empire* (Cambridge: Cambridge University Press, 2000), 45, 70–5; J. H. Elliott, *Empires of the Atlantic World. Britain and Spain in America, 1492–1830* (New Haven, CT: Yale University Press, 2006), 9, 106, 318.

42 Moncada, *Restauración*, 99, 133, 137; Fernández Navarrete, *Conservación*, 75.

necessity and opportunity," with the viceroys serving as the triumviri and industrious Castilian farmers in place of the colonists, thereby obtaining the added consequence of assuring the Catholic religion there.[43] On the other hand, the naturalization of the Catholic king's non-Spanish subjects in the Indies, a subject that derived from Botero's approach, was viewed unfavorably. The suspicion evinced in settling foreigners in Castile and the other peninsular kingdoms came to the fore in considering the Indies. Sancho de Moncada spoke for all when he clarified that the colonies were useful to Spain precisely "because they are people who descend from Spaniards."[44] The reserve, if not mistrust, with which foreigners were held was not going to help solve the problems of depopulation that otherwise were so clearly perceived.

III

The degree of territorial and overseas expansion the Spanish monarchy achieved by that time underlay another of the topics Botero studied, mentioned above, namely the degree to which states and empires were likely to survive in relation to size and degree of cohesion. This was the second of Botero's issues that stirred answers from Spaniards. Nevertheless, as occurred with depopulation, rulers, ruling groups, and writers had gathered a rich first-hand experience of their own on these matters, Botero aside.

In *Ragion di stato* Botero opined that small states were very vulnerable and that large ones, despite obvious advantages, easily developed internal vices and external hostilities. Consequently, he preferred mid-size states. He warned that princes, once they succeeded in overcoming the ambition to dominate, should be content with "the terms of mediocrity [*mediocrità*]" and thereby "their empire would become so long-lasting [*durabilissimo*]," with Herrera translating *mediocrità* as *medianía*, as he did in the majority of times Botero used this word. With respect to the physical unity or disunity of the territories integrating a state, Botero formulated various carefully nuanced affirmations, from which it was understood that a noncontinuous state, even though afflicted with certain

[43] Deza, *Govierno*, 45, 117–21, 124.

[44] Moncada, *Restauración*, 137. Other similar opinions, Deza, *Govierno*, 120; Fernández Navarrete, *Conservación*, 75. I deal with Spanish discussions on granting *naturaleza* after the Roman policy on citizenship in "City, Communication and Concord in Renaissance Spain and Spanish America," in *Athenian Legacies. European Debates on Citizenship*, ed. P. Kitromilides (Florence: Leo Olschki, 2014), 195–221, esp. 211–19.

weaknesses, could easily overcome them if its parts were able to help each other (and in such a case "it can not be considered as really lacking unity") and if it was able to take advantage of its dispersion, above all in two situations: the difficulty for factions to spread from one part to another; and the possibility, in situations of revolts, of "punishing the rebel part with the loyal one." He concluded that in these conditions he did not consider disunited territories to be any less sure than united ones, "and this is the case of the kingdom of Spain." Moreover, the Spanish unusual access to money and the naval expertise of Iberian coastal people were reasons to conclude that even though "the empire" seemed to be "divided and dismembered" it needed to be considered "united and almost continuous."[45]

Botero seemed to consider the issue resolved and thus one has to point out that he did not consider a closer politico-administrative cohesion for the Spanish monarchy. He limited himself to recommending, in *Relationi*, a program to strengthen naval forces in order to ensure inter-territorial communication and, as a result, tightening the bonds among the various dominions. On the other hand, by evoking the old Roman idea of war as a purge that liberated the social corpus from bad humors and yielded tranquility, he underlined the contrast between a Spain that was belligerent in the confines of its empire yet peaceful in its interior and a France mired in civil wars.[46]

The fact that these ideas from Botero ultimately favored the Spanish monarchy, as distinct to those dedicated to its depopulation and production capacity, may explain the lack of a direct, explicit response from Spanish writers. There actually was an open response – an endorsement. It came from Fray Juan de Salazar who cited Botero, as well as Abraham Ortelius, at the beginning of his *Política española* (1619), a book with strong religious motivations, to boast of the multitude of kingdoms and provinces spread throughout the four parts of the world over which the Catholic king reigned – a multitude of possessions sufficiently united by Spanish naval power, just as Botero thought. Moncada also quoted Botero in pointing out the Spanish ability to equip its fleet and make Philip II "master of the sea," in a passage that additionally rejected Grocio's *mare*

[45] Botero, *Ragion di stato*, 11–17. Translations, in *Diez libros*, ff. 5v–8v. I deal with Botero's reasonings, as well as with those by Campanella, below, in "Visión europea de la Monarquía española como monarquía compuesta," in *Las monarquías del Antiguo Régimen, ¿monarquías compuestas?*, eds. Conrad Russell and José Andrés-Gallego (Madrid: Editorial Complutense, 1996), 65–95, esp. 76–83.

[46] Botero, *Relationi*, second part, 134–6; *Ragion di stato*, pp. 87–8, 125, 223–5.

liberum. And Fernández Navarrete embraced the same opinion with respect to kingdoms and naval forces while Valle de la Cerda considered the cohesive bonds to be ships and the web of *erarios* of public banks he famously envisaged.[47]

At that time, Tomasso Campanella also partook, from his cell, of this sensibility as seen in *De Monarchia Hispanica* (*c.* 1600). Maritime ways as a nexus of union, a stimulus of both population and artisanal production, and closer internal cohesion through marriages and appointments of subjects from one kingdom to offices of others were the main contents of the treatise in these regards.[48] This suggests that in conjunction with the direct influence among authors there was a wide series of shared ideas and assumptions in circulation. In such a context it only made sense that Botero's reflections, and perhaps Campanella's also, gravitated around the initial plans of the count duke of Olivares to strengthen the Spanish monarchy.[49]

Botero added a more troubling warning, perhaps not originating with him: "No prince will continue in conserving that state that harms him more than it is useful to him."[50] Even though he explained this admonition with past Italian wars, he could well do so with current examples, possibly the Low Countries, which shortly, in 1589, would pass to the sovereignty of the Archdukes. More broadly, the idea that it was advisable to consider a withdrawal or even some territorial concession was in the air. But it was not easy to translate this disquiet to concrete measures, since, as recognized by Fernández Navarrete, two supreme values were being put at risk: religion and reputation.[51]

It was a sharp and pressing dilemma since a principle that was considered absolute was in play: the defense of the Catholic religion. And with the outbreak in 1618 of what would become the Thirty Years' War the dilemma simply became harrowing. The commitment to the Catholic

[47] Juan de Salazar, *Política española* (1619), ed. M. Herrero García (Madrid: Centro de Estudios Políticos y Constitucionales, 1997), 24–5, 34–5; Moncada, *Restauración*, 119, 124–5; Fernández Navarrete, *Conservación*, 83. On Valle, see Dubet, *Réformer les finances*, 259.

[48] Tomás Campanella, *La Monarquía hispánica*, chs. 32, 19, 20, ed. P. Mariño (Madrid: Centro de Estudios Constitucionales, 1982), 266–7, 152–8.

[49] Elliott, *The Count-Duke*, 192, 246, 199, where he points that there is no evidence that Olivares could have known Campanella's work. Conversely, Headley accepts the feasibility that he knew him: *Campanella*, 203, n. 23, 207, 222.

[50] Botero, *Ragion di stato*, 137.

[51] Fernández Navarrete, *Conservación*, 76–81. More generally, Anthony Pagden, "Heeding Heraclides: Empire and Its Discontents, 1619–1812," in Kagan and Parker, eds., *Spain, Europe and the Atlantic World*, 318–26.

religion facing the double challenge of the Turks and the Protestants – a commitment that was widely shared in Spanish society and by Catholics in other countries as well – was a quite considerable means and strength of the monarchy given that it proffered a sense of mission and internal cohesion with evident signs of providentialism. Nevertheless, Philip II himself ended up by experiencing an extreme messianism that led him to an incorrect assessment of reality that affected his decision-making.[52] It was not easy, however, to reach the necessary balance between confidence in Providence and realism in designing and implementing government policies – a hard issue everywhere. Thus, the intense providentialism of the previously mentioned Jesuit Adam Cotzen exercised a strong influence on Emperor Ferdinand II's court, in contrast to the more realistic position of another Jesuit, Johannes Vervaux, that would prevail a little later.[53]

In any case, Providence's help had to be supplemented with the help of allies. Contemporaries, among them Bodin and Botero himself, warned that the states' interest and advantageousness rather than friendship were the decisive factor in making or breaking alliances. In more detail, Botero affirmed in *Ragion* that the Spaniards and the Portuguese with their control over the Philippines could find welcoming islands and ports around the world in their lands or in those of friendly, confederate, or client princes.[54] So positive an assessment was countered in Álamos de Barrientos' somber panorama on the international situation in 1598 – dominated by declared or hidden enemies of the Catholic king. Francis Bacon agreed to a great extent with Álamos in pointing out that one of the weaknesses of the Spanish monarchy was its inability to form lasting alliances, noting that he saw in other states "little of amity and trust" vis-à-vis Spain.[55]

[52] Geoffrey Parker, *The Grand Strategy of Philip II* (New Haven, CT: Yale University Press, 1998), 104, 268, 286–7, 292, 295–6; and *Felipe II. La biografía definitiva* (Barcelona: Planeta, 2010), 240–52. On the weight of Catholic religion on Spanish political thought, see Pablo Fernández Albaladejo, "Católicos antes que ciudadanos: gestación de un apolítica española en los comienzos de la Edad Moderna," in *Imágenes de la diversidad*, ed. J. I. Fortea (Santander: Universidad de Cantabria, 1997), 103–27; and "El pensamiento político: perfil de una 'política' propia," in his *Materia de España* (Madrid: Marcial Pons, 2007), ch. 4.

[53] Bireley, *The Counter-Reformation Prince*, 159–61.

[54] Bodin, *République*, V, 6, 165–6; Botero, *Ragion di stato*, 17, 169–70, 23.

[55] Baltasar Álamos de Barrientos, *Discurso político al rey Felipe III al comienzo de su reinado*, ed. M. Santos (Barcelona: Anthropos, 1990), 10ff, 37, 48, 53; Bacon, quoted by Peltonen, "Politics and Science," 293.

While Botero was clearly aware of the capital importance of maritime communication in governing the Iberian dominions, he also did not fail to mention written communication. Government reports, dispatches, and *consultas* were without a doubt forces of the king, as Cotzen recognized. Nevertheless, Botero mentioned the *consultas* of the Spanish Councils rather as an efficient means to bring about thoroughly studied resolutions.[56] Collecting information of all sorts was a sign of political development and Philip II promoted it by ordering extensive questionnaries; particularly, the *Relaciones topográficas* in Castile and the *Relaciones geográficas* in the Indies, in addition to establishing archives. At the same time, however, the sheer number of letters, *consultas*, and requests that arrived at the Escorial from all corners of the empire could provoke an excess of information and consequently be an obstacle to sound, prompt decision-making.[57] It is not surprising, then, that at the end of the sixteenth century and the beginning of the seventeenth criticisms began to arise regarding what many considered to be excessive bureaucratic and legislative development. Zevallos, for example, argued that the Romans, "having as many provinces to govern as they did," managed with the Laws of the Twelve Tables and that "God governs the world with the Ten Commandments," while Quevedo recalled that "Christ did not send reports [*memoriales*]." Some voices went as far as to lament that the increase of both study and students would deprive the fields of workers in what has been called *biblioclast arbitrismo*.[58] Botero also spoke about the advantages incurred by princes who made sure that the chronicles and histories of their kingdoms were written down – something, he noted, done by the Portuguese but not by the Castilians. In effect, Philip II was not particularly diligent in this regard, an attitude that Richard Kagan has studied as one of his characteristic fields of research.[59]

[56] Botero, "Dela reputación" and "De la neutralidad," both included in the Spanish partial translation, *La razón de estado y otros escritos*, ed. M. García Pelayo (Caracas: Universidad Central de Venezuela, 1962), 207–08, 220 (but not in Continisio's); *Relationi*, second part, 134, 137–8. On Cotzen, see Bireley, *The Counter-Reformation Prince*, 154.

[57] Kagan, *Clio and the Crown*, 97, 167; Parker, *Grand Strategy*, 65–71; Parker, *Felipe II*, 197–204.

[58] Zeballos, *Arte real*, ff. 165; Francisco de Quevedo, *Política de Dios, govierno de Christo* (1626), ed. J. O. Crosby (Madrid: Castalia, 1966), 64. On that *arbitrismo*, see Fernando Bouza, *Comunicación, conocimiento y memoria en la España de los siglos XVI y XVII* (Salamanca: SEMYR, 1999), 113–14.

[59] Botero, *Ragion di stato*, 198–9; Kagan, *Clio and the Crown*, chs. 3, 4.

IV

The generation that read Botero in Spain did so when it already carried on their shoulders a remarkable experience in the matters the Piedmontese wrote about, an experience acquired the hard way and that was to increase as the seventeenth century advanced. This explains the wide confluence of interests, much broader than that resulting from a direct influence by reading his books. An underlying contractual sensibility was to be found in that confluence. It was not an overtly parliamentarian contractualism, but one evinced in an essential sense of reciprocity between king and subjects insofar as the self-preservation of them all – that supreme goal – could only be conceived of and reached through mutual help.

The challenge of self-preservation of both king and the monarchy and how this was faced make this generation pivotal in Spanish political and intellectual history. In the course of the following years, amid the scourge of a war waged on all fronts, much energy would be expended on reaching this goal with the conceptual baggage that had been defined in the course of the last years of the sixteenth century and the beginning of the seventeenth. Eventually, around 1640, when so many attempts had been tried and so many resounding failures began to be accounted for, the concerns do not appear to have changed in any substantial form. From his desk, Baltasar Gracián addressed in *El político* (1640), an archetypical portrait of the prudent king, the old question of "the forces in his kingdom" and gave a standard answer in pointing out that "gold and people are the principal and foremost source of wealth." For their part, the close collaborators of the dismissed Olivares, carrying their bitter experience of rulers, stated in the *Nicandro* (1643), a vindication of the Olivares ministry, that "all great monarchies are based on their people, monies and an abundance of all things crucial to conducting great armies that defeat enemies" – thereby adducing that neither the fallen favorite nor Philip IV's monarchy had counted on these resources. And Diego de Saavedra Fajardo, a man of action and reflection, reiterated in his *Empresas políticas* (1640–2) a series of well-known maxims, in brilliant style: "A kingdom's strength consists of the number of its vassals. The greatest prince is the one who has more [vassals], not the one who has more states because these do not offend, nor are they defended by themselves but by their inhabitants, who provide the states with a most firm ornament."[60] An increasingly acute mindset of this kind explains that

[60] Baltasar Gracián, *El político don Fernando el Catholico* (Zaragoza, 1640), facsilmile, ed. Aurora Egido (Zaragoza: Institución Fernando el Católico, 1985), 49, 89; "El Nicandro",

proposals – finally concrete proposals – were put forward by politicians and writers alike, arguing that one of the Crown's dominions be abandoned – the Netherlands.[61]

In effect, Botero's concerns, and along with them the preoccupations of his contemporary Spanish writers and reformers, were still alive in 1640, if anything even more pressing. Saavedra Fajardo's *Empresas* are the epitome of all this. With Prince Baltasar Carlos as his dedicatee, the great diplomat and writer explained that observing the past and comparing it with the present provided a fruitful lesson in political prudence. He thus invited the young prince to examine a large number of issues in this fashion, issues that were none other than the great worries of the last decades, always concerning the forces of the king:

Your Highness, let your gaze rest on the past, on the time from King Ferdinand the Catholic to Philip the Second. And in parallel with the current times, let your Highness ask if Spain is as populated, as rich, as full of abundance as it was then; if arts and arms are as flourishing, if commerce and culture are not lacking. And if you were to find one of these diminished, let your Highness examine the anatomy of this body, study its arteries and sections, verify which are healthy which are not and ascertain the causes of its sickness. Let your Highness consider if the sickness comes from one of these causes, as it usually does: a loss of so much population; negligence in reproduction; a multiplicity of religious orders; the large number of holidays; the abundance of universities and studies; the discovery of the Indies; expensive peace; wars undertaken on whim or slowly waged; the disappearance of the headships of the military orders; the paucity of prizes; the burden of exchange rates and usury; the export of money; the disproportion of currency; or other similar causes. Because if your Highness ascertained from which one of these causes your kingdom suffered, the remedy would not be difficult.[62]

in *Memoriales y cartas del Conde Duque de Olivares*, eds. J. H. Elliott, J. F. de la Peña, and F. Negredo (Madrid: Marcial Pons, 2013), 2nd part, doc. XXb, 417; Diego Saavedra Fajardo, *Empresas políticas*, ed. F. Javier Díez de Revenga (Barcelona: Planeta, 1988), empresa 66, 461.

[61] Elliott, *The Count-Duke*, 461, 191–492; Xavier Gil, "L'engany de Flandes. Les anàlisis de l'aragonès Juan Vitrián sobre la Monarquia espanyola i la seva proposta d'abandonar Flandes (1643)," in *Miscellània Ernest Lluch i Martín* (Vilassar de Mar: Fundació Ernest Lluch, 2006), I, 411–29.

[62] Saavedra Fajardo, *Empresas*, empresa 28, 188.

PART IV

PERFORMANCE AND PLACE

12

Censuring Public Images: A Woodcut in the Inquisition Trial of Esteban Jamete[1]

Fernando Marías

Translated by Consuelo López-Morillas

In 1557–8, in the course of his trial before the Inquisition in Cuenca, the artisan Esteban Jamete described some of the acts that might have led to the charges against him. Beside speaking inappropriate words and keeping bad company, he had read three books and meditated on an image on paper. The first book was "a little one in the French language ... of no more than two sheets of paper ... of less than octavo size" which two French stonemasons, Santos Picardo and a Gascon named Juan,[2] had shown him in about 1545, soon after he had arrived in Cuenca. The book must have struck him as suspicious, because "it had no [name of an] author nor place of publication, although it was printed"; he read it and "thought that it was bad," because it contained some "little songs in the French language ... against the Pope and the Mass and the Sacrament of the Mass." This "nameless" Lutheran pamphlet might also have contained different engravings that showed attacks on the papacy.[3] The "German" artist Giralte (or Giraldo) de Flugo (1512–91)[4] – actually a

[1] This research has been done thanks to the Proyecto de investigación "El Greco y la pintura religiosa hispánica," HAR2012-34099/ARTE from the Ministerio de Ciencia e Innovación (MICINN).

[2] Picardo, who came from the town of Lan, was in Cuenca with a fellow countryman named Pierre between 1553 (a date that conflicts with Jamete's statement) and 1571: see María Luz Rokiski Lázaro, *Arquitectura del siglo XVI en Cuenca. Arquitectos, canteros y carpinteros* (Cuenca: Diputación, 1989), 230–1.

[3] We cannot identify this text, since no copy has survived to the present day. See Francis M. Higman, *Censorship and the Sorbonne: A Bibliographical Study of Books in French Censured by the Faculty of Theology of the University of Paris, 1520–1551* (Geneva: Droz, 1979).

[4] Pedro Miguel Ibáñez Martínez, "El escultor Giraldo de Flugo," *Boletín del Museo e Instituto Camón Aznar* 42 (1990): 101–12.

Frisian who had lived in Cuenca since 1539 – and the Dutch glassmaker Esteban Giraldo both identified it as a text written by Martin Luther himself. Jamete burned his copy, but when his wife advised him to go to confession, the priest upbraided him for not having shown the booklet to the Inquisition or denounced its owners – especially since, as we know, Jamete had been a familiar of the Holy Office at least since 1548.[5] As Jamete testified in 1558, Santos Picardo and Juan had brought the booklet from Navarre and took it to be "el Vitruvio," Vitruvius' basic textbook for architects, stonemasons, and artisans who sought to work in the new Roman style.

The second book in Esteban Jamete's possession (unfortunately we do not know the contents of his whole library)[6] was one in French by Clément Marot, chamberlain to the king of France. Jamete declared that because he was not educated, "I cannot tell if it is good or bad"; other witnesses testified that it included a "fiction about friars" in which the holy men were portrayed as crows.

The third book was the *Propalladia* by Bartolomé Torres Naharro (*c.* 1485–1524), the dramatist from Extremadura. He had belonged to the household of Cardinal Bernardino López de Carvajal in Rome, where he had produced several plays at the papal court; the *Propalladia* was a collection of his poetic and dramatic works, published in Naples (1517) and later in Seville (1520). His free thinking, Erasmist views, and satires of Roman life earned his volume a place on the Index of Prohibited Books in 1559.[7] In Jamete's trial, two of Torres Naharro's poems in particular drew the Inquisitors' censure. The first, cited by Jamete himself, was the penultimate quatrain of *Canción* V, in a macaronic distortion of the original Latin: *Judas que das jubileo penitentes ginimundo, tu serás según que veo condenatus in profundo* ("Judas, you who give joy to the penitents of this world [*penitentibus in mundo*], you will be, as I can see, with the damned in Hell [*cum dannatis in profundo*]"). The second, quoted by the witness Giraldo of Holland, came from the end of Chapter III: *Es, en fin / nuestra Roma un gran jardín / de muchas frutas poblado: /*

5 The two stonemasons fled Cuenca on hearing that the book had been burned, although Santos Picardo continued to live in the village of Cañaveras. He was brought to trial in 1558, condemned as an "apostate heretic and protector of heretics," and sentenced to a year in prison and to wearing the *sambenito*, along with other lesser punishments.

6 Another work mentioned at the trial was a magical or necromantic text belonging to Jamete's employee Hans of Brabant. It told, among other scandalous things, how to have carnal "access" to a woman even in public without being seen.

7 Jesús Martínez de Bujanda, *Index des livres interdits* (Geneva: Droz, 1985).

son las flores de jazmín / blasfemar por un quatrín, / renegar por un cornado / una esgrima / do ningún tiro lastima / que lo sientan sus conciencias ("In short, our Rome is a great garden filled with many fruits: its jasmine flowers are blasphemies spoken for a penny, oaths for a farthing; a fencing-match that injures no one's conscience").

As for the image on paper in Jamete's trial, we shall see that it was a woodblock print that had circulated publicly. The accused stated that in about 1552 or 1553, in the house of the illiterate German glassmaker Esteban Giraldo or Giralte, he had seen:

> on a wall a painted paper from Flanders, and [in] the painting that I saw there was an owl, and in the claws of one foot it held a pair of spectacles, and below there was a candle in a candlestick. And there was a banner written in German, and I asked ... Giraldo the glassmaker, who was present, what that owl meant, and Giraldo ... replied that it was the Pope, who wished to see and consider nothing. And Giraldo ... said to me that the words written in German or Flemish meant in our Castilian language: "What good are a candle or spectacles to me if I do not wish to see or consider?" And about a year later I entered ... the house of ... Giraldo the glassmaker and looked at the painted paper, and saw that there was another banner in French that said the same thing ... Then I asked him why he had that there, and that he should remove it, and I even threatened him with the Holy Office and ... Giraldo said that it was nothing, that it was a bird, and I think that he still has it at home.[8]

Little attention has been paid until now to the words that Jamete read and recited in French, or to the image that he saw in Giraldo's house – hanging on a wall, like so many Northern paintings and prints at the time, and therefore exposed to the gaze of anyone who entered the home.[9]

Our contribution to this homage volume will be to study these issues from the standpoint of our area of research. Our intention would be to go

[8] "... en una pared un papel pintado de Flandes e [en] la pintura que en él vide estaba un búho, y en las uñas de un pie tenía unos anteojos, e debajo una vela con un candelero, e estaba un rótulo en alemán e pregunté ... a ... Giraldo vidriero que estaba allí presente, qué significaba aquel búho ... e respondió ... Giraldo que era el Papa que no quería ver ni considerar. E me dijo ... Giraldo que lo que decía el rótulo escrito en alemán o flamenco era en nuestra lengua castellana: 'Qué me pueden candela e anteojos aprovechar cuando yo no quiero ver ni considerar ...'. E después, dende a un año, poco más o menos, entré ... en casa de ... Giraldo vidriero e miré el ... papel pintado, e vi que estaba también otro rótulo en francés que decía lo mismo ... Entonces me le dije que para qué tenía allí aquello, que lo quitase, e aun le amenacé con este Santo Oficio e ... Giraldo me dijo que no era nada, que era un pájaro, e creo yo que aún está en su casa ...".

[9] André Turcat, *Étienne Jamet alias Esteban Jamete. Sculpteur français de la Renaissance en Espagne, condamné par l'Inquisition* (Paris: Picard, 1994), 93–5 and 97. Turcat's interpretations of both quotations are generic and erroneous; he does identify the work of Torres Naharro correctly.

further than presenting a case study, with the first identification of a print first suspected and later censored.[10] There has been very little study, in the field of Spanish art, of how artists could deviate from pre-Reformation or Counter-Reformation Catholic norms, or of works of art – dual products in which the client could even be the more responsible party – that can be susceptible to an alterante reading from the traditional Catholic interpretation. The historiography of art has taken a priori censorship for granted, as a posteriori persecution should have left a trace, and there is a lack of even dubious or suspect cases; the watchful presence of the Inquisition and the episcopal councils would have excluded such a possiblity, in a space in which, moreover, the complete orthodoxy of artisans and patrons has been taken as secure, all of them free of all suspicion only by working in the Iberian Peninsula. It seems clear that the analysis of the images of the Golden Age – and not only of the role of images in the activities of the Inquisition or Inquisitors – from this perhaps voluntarily forgotten and barely trodden perspective, as yet delved into by only a scant few historians (as can be deduced from the bibliographical citations), could result in another panorama less ideologically homogenous than that which has been considered to date. The world of images and their makers would benefit from a more coherent relationship to other areas of religious and material life in a society less unitary than some desired.

We are conscious of the difficulties of this task; it means, in most cases, in the absence of an inquisitorial trial, to intuit and attribute intentions to the "plural" creators, in a certain way to assume – without any pretense of condemnation – the role of denouncers, without other evidence than perhaps a few textual references, visible and legible in figurative works only in the form of legends or inscriptions, or some references to previous images of a heterodox nature, always in a dynamic and mined field; that character would depend on their place of publication and their

[10] Sara T. Nalle, *Mad for God. Bartolomé Sánchez, the Secret Messiah of Cardenete* (Charlottesville, VA: University Press of Virginia, 2001) *[Loco por Dios. Bartolomé Sánchez, el Mesías secreto de Cardenete* (Cuenca: Fundación de Cultura "Ciudad de Cuenca"- Ayuntamiento, Cardenete, 2009)] referred to some visions of the carder of Cuenca, Sánchez; however, it identified as the source of his visions an image of the *Coronation of the Virgin* included in his book of hours; it would be as likely a miniature as a print included in the book of hours printed for him. The difference was that the image was perfectly orthodox and what deviated was only the interpretation conferred on it by the "Messiah of Cardenete," holding it as a representation of the *true Trinity* – God the Father, the Son, and the Mother – united in intention by the Holy Spirit. It is evident that any vision of "the invisible" would have to base itself in a graphic source.

authorship, a place of publication and an authorship of the printed text which in the majority of cases would be closely linked. And always, in addition, artists and clients could have argued their innocence by ignorance of the meaning and implications of those graphic and textual references. Nevertheless, by being able to establish another zone of dissent, that of the images of *alumbrados*, evangelicals, or Lutherans, the history of art – as social construction more than individual product, decontextualized and stripped of ideological considerations – would contribute from its own field to the history of the religious practices of Europe in the era of confessionalization, and to the global reconstruction of the life and work of some artists both as readers and as consumers conscious of other images. Thus we return to Jamete's reading.

The author of the second book cited in our trial was Clément Marot, a native of Cahors (1496–1544) who was a *valet-de-chambre* to King Francis I and belonged to the circle of his sister Marguerite of Navarre. He translated the Psalms of David, Vergil's *Eclogues*, and Ovid's *Metamorphoses* into French, and composed poetry in the style of François Villon. Accused of heresy in 1526 and 1531, he fled to Ferrara in 1534 and, after abjuring his errors in France in 1538, again went into exile in Geneva in 1543; from there, accused of freethinking, he had to flee in 1544 and take refuge in Turin. In his epistles, *chansons*, *cantiques*, *ballades*, and even his translation of the Psalms,[11] Marot displayed in light-hearted fashion not only nationalist and anti-Spanish sentiments but a clear disdain for the bans on translations of the Bible that issued from the Church and the Sorbonne. He also criticized the pope and members of the clergy, although always insisting that he was not a Lutheran.

Given the complicated editorial history of Marot's works, it is not easy to identify the book in Jamete's Inquisition trial. It might be the fictional "Ballade de Frère Lubin" from *L'Adolescence Clémentine* (1531), or the epigram – eventually found to be apocryphal – titled "D'un gros prieur," which appeared as Marot's in his *Oeuvres* of 1544.[12] A third possibility is the work *L'enfer*, written after his first spell in prison and first published

[11] See Artus Désiré, *Le contrepoison des cinquante-deux chansons de Clément Marot, faulsement intitulées par luy "Psalmes de David," fait et composé de plusieurs bonnes doctrines et sentences préservatives d'hérésie* (Paris: Jehan Ruelle, 1567). Werner Thomas, *Los protestantes y la Inquisición en España en tiempos de Reforma y Contrarreforma* (Louvain: Louvain University Press, 2001), is almost always able to identify Marot's texts with his French translation of the Psalms, which were sung by Jamete's French coreligionists.

[12] The *Oeuvres* were not placed on the Spanish Index until 1583.

in 1540. But the most probable identification is the satirical epigram "Mes beaulx religieulx," directed against minor friars, which was printed in 1539 in *L'Adolescence* and in the small volume *Plusieurs traitez par aucuns nouveaulx poetes de different de Marot, Sagon & la Huetterie, avec le Dieu gard du dict Marot.* A copy of this text was found in Spain less than a decade ago, having been walled up in a house in the village of Barcarrota (Badajoz) in the mid-sixteenth century by the Judaizing physician Francisco de Peñaranda.[13] The fact that the doctor hid a work by this French evangelist (along with a copy of *Lazarillo de Tormes*) shows that Marot's writings were the object of suspicion and self-censorship.

The Flemish "painted paper," said to contain text "written in German or Flemish," can be identified, thanks to Jamete's detailed description – though without the "French words ... that said the same thing" – as a leaflet known as *Das Lichtscheue Eule* ("The Light-Shy Owl") (*c.* 1540)[14] by the German engraver Erhard Schön (*c.* 1491–1542) (Fig. 12.1).[15] One print of this woodcut (36 cm × 25.9 cm), by Georg Wächter of Nurenberg, is found in the Herzogliches Museum (Landsmuseum) in Gotha; another, somewhat larger (44.8 × 32.1 / 40.5 × 28.8 cm), colored at an early date, is in Leipzig's Stadgeschichtliches Museum. No example seems to have survived in the Iberian Peninsula.[16]

If this identification is correct, we would be dealing with one of the few examples of a sixteenth-century image that was confessed to by a victim of the Inquisition. It might, of course, have been confiscated immediately by the Inquisitors, though we do not know if they went to the illiterate glassmaker's house to seize the woodcut and throw it in the fire.

Although Jamete may not have read German, and we know of no similar prints that contained words in his native French, we must consider the texts that appear in this woodcut.

[13] Fernando Serrano Mangas, *El secreto de los Peñaranda. El universo judeoconverso de la Biblioteca de Barcarrota. Siglos XVI y XVII* (Madrid and Huelva: Hebraica, 2004).

[14] Other versions, later than the one we are discussing here, are also known.

[15] Heinrich Röttinger, *Erhard Schön und Niklas Schön, der Pseudo Schön: Zwei Untersuchungen zur Geschichte des alten Nürnberger Holzschnittes* (Strasbourg: J. H. Heitz, 1925); Walter L. Strauss (ed.), *German Masters of the Sixteenth Century: Erhard Schoen, Niklas Stoer*, in *The Illustrated Bartsch* (New York: Abaris Books, 1984).

[16] Enrique Cordero de Ciria, "Arte e Inquisición en la España de los Austrias," *Boletín del Museo e Instituto Camón Aznar* 70 (1997): 29–86, and "Las artes plásticas en Erasmo y el erasmismo español," *Boletín del Museo e Instituto Camón Aznar* 89 (2002): 93–164, 43–4, and Illus. 1. The author refers to two seventeenth-century English engravings on the same subject and reproduces (without identification) a German print of unclear date and uncertain authorship.

Wer arges thut/hasset das liecht/vnd kumpt nit an das liecht/auff das seyne werck nicht gestrafft werden/Joannis.iij.c.

Das ist aber das gericht/das das liecht in die welt kumen ist/vnd die menschen liebten die finsternis mer denn das liecht/denn jre werck waren böse/Joan.iij.c.

Denn dise sinds auch wert/das sie des liechts beraubt/vnd im finsternus/als im kercker/gefangen ligen/Sapien.xviij.c.

¶Ein Ewl bey dem tag nit gesicht
Sie hab gleych Sunnen oder licht
Sie hab gleych fackel oder prillen
Also auß eygnem bösen willen
So seind auch aller menschen kindt
Verstockt/verstarret vnd erblindt
In wůcher/geytz vnd Tyranney
In lůgen/list vnd triegerey
Im Ehbruch/hůrerey vnd fraß
In kriegen/zoren neyd vnd haß
In vngehorsam vnd Gotßlestern
Ey heůt vil gröber noch dann gestern
On alle straff so gar verrucht
On forcht/on scham/on scheuch vnd zucht
Als ob es in der finster gschech
Das weder Got noch menschen sech
Ir vnchristlich leben vnd fell
Vnd leucht doch Gottes wort so hell
Das lockt vnd zeucht/vermant vnd bitt
Verheyst vnd dröwet auch darmit
Gottes zoren vnd grymmig straff
Wie der die sünder all mal traff
Wo sie nit würckten frucht der büß
Vnd das täglich mit vberfluß
Man predigt/schreybt/man singt vnd malt
Noch bleyben beyde jung vnd alt
Wie vor in sünden blind vnd thum̃
Niemand thůt büß das er würd frum
Sonder man wirdt verstockt vnd erger
Vnbůßfertiger böß vnd kerger
Das wort gentzlich kein frucht wil würcken
Als ob wir wern Heyden vnd Türcken
Die doch in jrem leben trůgendt
Gůt ebenbild/sitten vnd tugendt
Gůt burgerliche Policey
Das man dennoch spürt klerlich bey
Iren büchern die sie beschriben
Die zů zeugnus sind bey vns bliben
Das wir vil erger sind dann sie
Weyl wir yetz bey vns haben hie
Die Götlichen hymlischen warheyt
Mit solcher durchleuchtigen klarheyt
Doch wenig beßrung kumpt daruon
Derhalb das Euangelion
Wirdt vor dem endt/wie Christus meldt
Gepredigt durch die gantzen welt
Nur zů zeugnus das wir nicht glaubet
Haben/sonder blind vnd betaubet
Bleyben/in schwerer sünden schlaff
Auff das die hefftig Gottes straff
Vns dester hefftiger müg plagen
Eylend vnd schnell in kurtzen tagen
Mit hunger/krieg vnd pestilentz
Vnd dergeleychen plag onentz
Die allenthalben auff vns deuten
Vns zů vertilgen/auß zů reüten
Derhalb du sündig blinder hauff
Wach auß der nacht der sünden auff
Vnd ermunter deyn blöd gesicht
Weyl du noch hast das helle licht
Christum die Sonn der grechtigkeyt
Sampt seynem wort das diser zeyt
Dich freundtlich locket zů der büß
Auff das du setzest deynen füß
In ein Christlich bůßfertig leben
Wo du nicht wilt verderben eben
In der ewigen finsternus
Vor welcher vns behůt Christus
Der geb das sein wort in vns wachs
Bring frucht der büß wünscht vns Hans Sachs.

¶Gedruckt zů Nürnberg
durch Georg Wachter.

FIG. 12.1. *Light Shy Owl*, Georg Wächter of Nuremberg and Erhard Schön (*c.* 1540), Gotha, Herzogliches Museum (Landsmuseum) www.zeno.org/Kunstwerke/B/Schoen,+Erhard%3A+Lichtscheue+Eule http://images.zeno.org/Kunstwerke/I/big/HL21111a.jpg

Along the top there is a passage from the Gospel of John (3:20): *Wer arges thut / Der hasset das Liecht / und kumpt nicht an das liecht / Auff das seine Werck nicht gestraffet werden [Joannis iii c]* ("For every one that doeth evil hateth the light, neither cometh to the light, lest his deeds should be reproved"). The verse immediately preceding (3:19) appears along the left margin: *Das ist aber das Gerichte / Das das Liecht in die Welt komen ist / Und die Menschen lieb[e]ten die Finsternis me[h]r denn das Liecht / Denn jre werck waren böse* ("And this is the condemnation, that light is come into the world, and men loved darkness rather than light, because their deeds were evil"). The right margin contains a fragment of the Book of the Wisdom of Solomon (18:4): *Dann dise sinds auch wert / das sie des Hechts beraubt / und im finstemus [finsternis] / als im kercker / gefangen ligen / Sapien. xviij. a* ("For they were worthy to be deprived of light and imprisoned in darkness, who had kept thy sons shut up, by whom the uncorrupt light of the law was to be given unto the world"). Finally, on the image itself, a cartouche encloses another quotation: *Was hilfft mich sün / licht oder prill / weyl ich doch selbs nicht sehen will*; we could translate this proverb as "Of what good are candles or lenses, if I do not wish to see?"[17] Jamete's version was *¿Qué me pueden candela e anteojos aprovechar, cuando yo no quiero ver ni considerar?* ("What good are a candle or spectacles to me if I do not wish to see or consider?"), a saying or proverb whose use is not otherwise recorded in sixteenth-century Spain.

These biblical quotations in German would cause suspicion by themselves, but the woodcut also included the German text of a lengthy poem, *Yergleicliüng der blinden weltkinder einer ewlen*, by the prolific poet, shoemaker, and reformer of Nurenberg, Hans Sachs (1494–1576).[18] We are not concerned here with the full text or meaning of the verses, since the Dutch glassmaker, being illiterate, would not have understood them.[19]

[17] See, for example, Brett D. Hirsch, "From Jew to Puritan: the Emblematic Owl in Early English Culture," in *This Earthly Stage: World and Stage in Late Medieval and Early Modern England*, eds. Brett D. Hirsch and Christopher Wortham (Turnhout: Brepols, 2010), 131–71.

[18] *Hans Sachs*, eds. Adelbert von Keller and Edmund Goetze, 26 vols. ([Stuttgart, 1870] Hildesheim: G. Olms Verlagsbuchhandllung, 1964); Winfried Theiss, "Der Bürger und die Politik: Zu den zeitkritischer Dichtungen von Hans Sachs," in *Hans Sachs und Nürnberg*, eds. H. Brunner, G. Hirschmann, and F. Schnelbögl (Nuremberg: Vereins für Geschichte der Stadt Nürnberg; M. Edelmann, 1976), 76–104; *Die Welt des Hans Sachs*, Stadtgeschichtliches Museum, Nurenberg, 1976.

[19] *Ein ewl bey dem tag nit gesiebt,/Sie hab gleich sannen oder liecht,/Sie hab gleich fackel oder prillen;/Also auß eygnem bösen willen/So sind auch aller menschen kind/*

Perhaps they were translated for him by some more educated German or Flemish artisan resident in Cuenca, whose name has not been preserved. In any case, no reference was made to the verses during the trial by either the accused or his judges. This, and the space available, frees us to pursue the case further.

Therefore we can now add this additional detail to the history of Esteban Jamete's Inquisition trial in Cuenca.[20] While scholars have

Verstockt, verstatret und erblind/In Wucher, geilz und tyranney,/In lügen, list und triegerey,/In ehbruch, hurerey und fraß,/In kriegen, zoren, neyd und haß./In ungehorsam und gotslestren,/Ey heut vil gröber noch, denn gestren,/On alle straff so gar verrucht,/ On forcht und schäm, an scheuch und zocfat,/Als ob es in der finster gschech./Das weder Got noch menschen sech/Ir unchristlich leben und feil/Und leucht doch Gottes wort so hell./Das lockt und zeucht, yermant und bitt,/Yerheist und troet auch darmit/ Gottes zoren und grimmich straff,/Wie der die sündter allmal draff,/Wo sie nit würckten frucht der büß,/Und das teglich'mit Überfluß./Man predigt, schreybt, man singt und malt,/Noch bleiben beyde jung unnd alt/Wie vor in Sünden plind und dumb./Niemand thut büß, das er wurt frumb./Sander man wirt verstockt und erger,/Unbüßfertiger, böß and kerger./Das wort gentzlich kein fracht kan warcken,/Als ob wir wem haydn and Türcken,/Die doch inn ihrem leben tragend/Gut ebenbild, sitten and tagend./Out bürgerliche policey,/Das man dennoch spürt klerlich bey/Iren büchren, die sie beschrieben,/Die za zeugnuß seind bey ans blieben,/Das wir viel erger sind, wann sie,/Weyl wir yetz bey ans haben hie/Die gödelichen, himlischen warheit/Mit solcher durchleachtigen klarheyt,/Doch wenig beßrung kumpt darvon./Derhalb das evangelion/Wirt vor dem end, wie Christus melt,/Gepredigt durch die gantzen weit./Nur zu zeugnaß, das wir nicht glaubet/Haben, sunder blind und betäubet/Bleyben in schnOder sflnden schlaff,/ Auff das die heffltig Gottes straff/Uns dester heftiger mfig plagen/Eylend und schnell in kurtzen tagen,/Mit hunger, krieg und pestilentz/Und dergeleichen plag unentz,/Die allenthalben auff uns deuten,/Und zu vertilgen auß zu reuten./Derhalb, du sündig blindter hauff,/Wach auß der nacht der Sünden auff/Und ermunder dein blöd gesiebt,/ Weyl du noch hast das helle Hecht,/ Christum, die Sonn der grechtigkeyt,/Sampt seynem wort, das dieser zeyt/Uns freundlich locket zu der büß,/Auff das du setzest deinen fuß/Inn ein Christlich büßfertig leben,/Wo du nicht wilt verderben eben/In der ewigen finsternuß,/Vor welcher uns behüt Christas!/Der geh, das sein wort in uns wachs,/Bring frücht der baß, wünscht ans Hans Sachs.

20 On the trial, see Jesús Domínguez Bordona, *Proceso inquisitorial contra el escultor Esteban Jamete* (Madrid: Centro de Estudios Históricos, 1933). Thomas, *Los protestantes*, 397–414. Richard L. Kagan and Abigail Dyer, *Inquisitorial Inquiries: Brief Lives of Secret Jews and Other Heretics* (Baltimore, MD: Johns Hopkins University Press, 2004); a life of Jamete is appended to the Spanish translation, *Vidas infames. Herejes y criptojudíos ante la Inquisición* (Hondarribia: Nerea, 2010), 51–77. I am grateful to Felipe Pereda for showing me his unpublished paper, "*Ad modum Hispaniae*: Simulating Devotion in Renaissance Rome," presented to the conference "Hypocrisy and Dissimulation in Judaism, Christianity and Islam," Chicago Center for Jewish Studies, The University of Chicago-Jerusalem, 2012. See also María Luz Rokiski Lázaro, "Proceso de Alexandre Francés y noticias de artistas que trabajaron en Cuenca," *Archivo español de arte* 184 (1973): 440–8. For the context, see also Jean-Louis Flecniakoska, "La propagation des

known of the trial for eighty years (*pace* some who have preferred to ignore it), it has been looked at with fresh eyes in the last decade; we know a good deal about its artisan protagonist, although we lack some data that would be valuable for a truly contextualized history of art.

This artisan's life and times are well known to us, thanks to the *vida* or biography included in his Inquisition trials of 1557 and 1559 and to documentation from cities where he worked.[21] Étienne Jamet (*c.* 1515–65) was born in Orléans (where none other than John Calvin preached in 1530), a son of the master stonemason Jamet Loxin and his wife Nicolasa. He arrived in Spain as an itinerant artisan (the exact nature of whose work is still unclear) in 1535, "the year that His Majesty [Charles V] took Tunis and La Goulette," at the age of "twenty-one years, more or less." He probably went there in search of work, rather than as a *luthériste* or Lutheran fleeing his homeland for religious reasons. He probably kept ties to Orléans, perhaps through a younger sister who went to Spain in the 1550s, for he learned that his family home had been destroyed by fire, apparently causing the deaths of his parents and two of his nine siblings. (By 1557 two more of them had died, or at least he did not know if they were still alive.)

After crossing into Spain at Fuenterrabía, the man now known as Esteban Jamete worked in León, Zamora, Benavente, and Salamanca. In Medina del Campo he helped build the house of Dr. Diego Beltrán, a member of the Council of the Indies; in Valladolid, the home of Francisco de los Cobos, the emperor's secretary; and in Medina de Rioseco, the Convent of San Francisco, for the second admiral of Castile Don Fadrique Enríquez and his wife Ana de Cabrera, countess of Modica.[22]

idées protestantes par les français en Espagne et l'Inquisition de Cuenca, 1554–1578," *Bulletin de la Société de l'Histoire du Protestantisme Français* 120 (1974): 532–54; Miguel Jiménez Monteserín, "Los luteranos ante el tribunal de la Inquisición de Cuenca, 1525–1600," in *La Inquisición española: Nueva visión, nuevos horizontes*, ed. Joaquín Pérez Villanueva (Madrid: Siglo Veintiuno de España, 1980), 689–736; and Sara T. Nalle, *God in La Mancha: Religious Reform and the People of Cuenca 1500–1650* (Baltimore, MD: Johns Hopkins University Press, 1992).

[21] On Jamete's life, see Turcat, *Étienne Jamet alias Esteban Jamete.*

[22] The Admiral Don Fadrique was an adherent of the *alumbrado* or Illuminist movement, which encouraged an interior piety, in the decade of the 1530s. While he naturally protested his orthodoxy in the wills (later made public) that he drafted in 1540 and 1577, he was imprisoned for several months in 1567 by the Inquisition in Valladolid. See Agustín Bustamante García, "El Santo Oficio de Valladolid y los artistas," *Boletín del Seminario de Arte y Arqueología* 61 (1995): 455–66. It would be useful to determine his connections with evangelical circles, which would have exposed him to accusations of heresy. See also Felipe Pereda, "Laughing at the Foot of the Cross: Alternative Imaginaries in 16th-Century Spain," in press.

He ended the year on a pilgrimage to Santiago de Compostela, and stated that it was for "the Jubilee; I stayed there for two days, confessed, and received communion." This declaration would be important in his trial, for it affirmed that he believed in the intercession of the saints[23] and in the forgiveness of sins following confession, thus implying that he accepted the doctrine of Purgatory.

From Santiago, Jamete returned to Castile and continued to work, appearing in the following towns and cities:

Medina de Rioseco

Burgos, in the workshop of a French sculptor from Langres, Philippe Bigarny

Valladolid, for a captain resident in the parish of San Esteban

León, on the convent of San Marcos

Burgos, from August 1536 to the end of that year, as a stonecutter for the cathedral. He spent the last month and a half of the year in the local jail for reasons unknown (he would be imprisoned in Cuenca for blasphemy in 1550, and again in 1555)

León in 1537, for Claudio Fernández Vigil de Quiñones, fourth count of Luna (*c.* 1508–63)

Carrión de los Condes (Palencia), on the Benedictine monastery of San Zoilo or San Gil

Valladolid, on the convent of San Francisco, for the bishop of Mondoñedo Antonio de Guevara

Segovia, on a house in the parish of San Martín

Toledo, from 1537 to 1538, on the convent of San Agustín, the chapel of La Torre or San Juan, and the cathedral's Puerta del Reloj

Madrid in 1539, for the bishop of Calahorra

Toledo once more in 1539–40. In 1540 he moved south, and worked in:

Chinchilla de Montearagón (Albacete) in 1540–1, on the parish of Santa María de Chinchilla

Úbeda (Jaén), for two years between 1541 and 1543,[24] on the chapel of El Salvador, endowed by the emperor's secretary Francisco de los Cobos and his wife María de Mendoza

Seville in 1544, on the town hall, the cathedral, the convent of San Francisco and the Carthusian monastery of Nuestra Señora de las Cuevas

Marchena, on the palace of the duke of Arcos

Seville once more, where he joined his first wife, Mari González, a native of Becerril de Campos (Palencia), whom he had married in 1537.

[23] See Robert Bartlett, *Why Can the Dead Do Such Great Things? Saints and Worshippers from the Martyrs to the Reformation* (Princeton, NJ: Princeton University Press, 2013).

[24] It has been assumed that he also worked on the convent of San Francisco in Baeza (Jaén). The traditional belief that he worked on the sacristy of El Salvador, however, is not supported.

Although Jamete tried to impress his interrogators with his distinguished list of clients, he could not disguise his itinerant history or the fact that, whether willingly or not, he had been unable to establish himself in Valladolid, Toledo, or Seville, all cities that were more important than Cuenca and offered more opportunities. So much moving around in itself proved suspicious.

While in Seville he received an offer of work in Cuenca, and moved there in 1546; it was there that he was widowed in 1554, and in 1555 he married his second wife, the eighteen-year-old Mari Fernández de Castro from Villanueva de los Escuderos (Cuenca). His first commissions in Cuenca suggest that he was employed by Bishop Sebastián Ramírez de Fuenleal (*c.* 1490–1547), who had held the See of León in 1539–42. He was assigned to work under the master builder Francisco de Luna on the hall and arch (today called "Arco de Jamete") that gave entrance to the cloister (1546–50).[25] He might have worked on the chapel of San Honorato (1547–9, but no longer extant in the seventeenth century), endowed by Francisco [Hurtado] de Mendoza y de Silva y Ribera (1481–1547), archdeacon of Toledo and Moya and prior of Aroche (Huelva);[26] this cleric was the brother of the first marquis of Cañete and viceroy of Navarre, Diego Hurtado de Mendoza.[27] Jamete went on to work for the dean of the cathedral and apostolic protonotary Constantino del Castillo (d. Rome, 1565) on the chapel of Santa Elena (1548–51).[28]

In later years his activities were concentrated in the city of Cuenca and its region: in Alarcón (1550–9), perhaps Huete (1552), Cañaveras (1552), Villanueva de los Escuderos (1554), and Castillo de Garcimuñoz (1557), where he would be arrested by the Inquisition after having been denounced by his parents-in-law.

As a result of the tortures he suffered under questioning in 1558, Jamete was too badly injured to resume work as a carver and sculptor.

[25] José Luis Sáez, S.J., *Don Sebastián Ramírez de Fuenleal, obispo y legislador* (Santo Domingo: Banreservas, 1996).

[26] This house depended in part on the Hieronymite monastery of Buenavista on the outskirts of Seville.

[27] Not to be confused with his nephew, the cardinal and bishop of Burgos Francisco de Mendoza y Bobadilla (1508–66), who also became prior of Aroche, in the archbishopric of Seville.

[28] On this figure, see Gonzalo Redín Michaus, "Giulio Mazzoni e Gaspar Becerra a San Giacomo degli Spagnoli. Le cappelle del Castillo e Ramírez de Arellano," *Bollettino d'Arte* 120 (2002): 49–62; "Sobre Gaspar Becerra en Roma. La Capilla de Constantino del Castillo en la iglesia de Santiago de los Españoles," *Archivo Español de Arte* 298 (2002): 129–44; and *Pedro de Rubiales, Gaspar Becerra y los pintores españoles en Roma, 1527–1600* (Madrid: CSIC, 2007).

He made a design for an iron grille in 1561. Always a heavy drinker – some said that he worked better and with more energy when drunk – by 1564 he was a hopeless alcoholic. When he died intestate on May 5, 1565, the priest noted in the parish record, with barely disguised irony, that "his heirs have done what they could" for the soul of this man who had been reconciled to the Church. The Holy Office had sentenced him to excommunication, stripped him of his status as its familiar,[29] confiscated his goods, made him recant in a public *auto da fe* on May 15, 1558, and forced him to wear the *sambenito* for three years (although his father-in-law denounced him again in 1559 for not wearing it). It had also goaded him into a quarrel with another foreign artisan, the Parisian silversmith Tristan "Alexandre" del Vago, who also had to recant in 1567.[30] Jamete, acting as a now-reconciled collaborator with the Inquisition, had accused him of being "a bigger Lutheran than all the other Lutherans put together."

Jamete, arrested in 1557, was a well-documented representative of Protestant Nicodemism in sixteenth-century Castile. He was fully aware of the risks involved in seeking and obtaining the post of familiar of the Holy Office of the Inquisition (one that he held from at least 1548); this network of informants, eager to suspect and quick to denounce, might have protected him from the suspicions of others. But his character flaws denied him protection and, indeed, contributed strongly to his arrest. He drank to excess and had a rash tongue. He was violent not only towards his colleagues (as soon as he arrived in Cuenca he attacked the quarryman Sebastián de Arnani for having called him a *gabacho* or "Frenchman," virtually a synonym of "drunkard") but also towards his young wife María de Castro (born in 1538, by the 1560s she was cuckolding him flagrantly)[31] and towards her own relatives.

Jamete stated in 1557 that before reaching Cuenca in 1545 he had been entirely ignorant of the Protestant heresy. This claim has been accepted by the traditional Catholic historians of the city, including most recently María Luz Rokiski Lázaro. But Werner Thomas has shown convincingly that our sculptor, far from serving as a passive subject, had taken an active role in disseminating his heretical ideas.

[29] Thomas, *Los protestantes*, 341–2.

[30] Rokiski Lázaro, *Documentos*, 338–45. See also Dimas Pérez Ramírez and Francisco Javier Triguero Cordente, *Papeles sueltos de la Inquisición de Cuenca* (Cuenca: Diputación, 1999).

[31] See the case against Tristán del Vago.

According to his own biased account, Jamete slipped gradually from a position of relative indifference and skepticism to a frank anticlericalism with Calvinist elements, a product either of his contacts with other Frenchmen and Flemings or of his own youth in Orléans. He seems to have had decisive encounters with one Santos Picardo (from Huguenot La Rochelle) in 1547 and with the French artisan Francisco in 1553, though we know nothing else about either individual. On the other hand, Jamete had local ties: he was a resident of Cuenca and had married a woman from the city, which must have helped him escape the two expulsions of French citizens from Castile in 1551 and 1555 (his first wife had died in the summer of 1554).[32] These dates opened a period of repression of possible Lutherans in Spain, with foreigners being singled out as the chief suspects. Jamete's situation changed dramatically with his second marriage in 1555, since his physical and emotional abuse of his wife set the Fernández de Castro family against him. In 1548 and 1550 he had eluded the tribunal of the Holy Office, and in 1555 had escaped jail in Cuenca, but in the latter year he began to slide towards his own destruction. Seven of the nine witnesses in his defense were clerics who proclaimed him a good Christian, although of the eighteen favorable witnesses whom the accused had proposed, six did not appear. His accusers included, naturally enough, his wife's family as well as fellow workers, both Spanish and foreign, his own apprentices and employees, and others from different workshops. He called them his "sworn enemies," not surprisingly in light of his history of "annoyances, disagreements," disputes and "quarrels"; besides, denouncing him helped them to protect themselves. Among more than twenty prosecution witnesses were his wife's parents and his two brothers-in-law, six of his laborers, and his own parish priest, together with various fellow workers and artisans.[33]

Aside from his anticlericalism, there were his open criticism of the papacy and of basic doctrines of the Catholic Church, and his attempts to keep his wife from practicing her religion, going regularly to Mass (he could accept only the readings from the Gospels and the consecration of the Host), or attending other celebrations or liturgies. He would argue with friends and relations about the Trinity, the existence of Purgatory, and transubstantiation. As for the placement of images on altars and the reverence due to them, he claimed that they distracted believers from their prayers and from the Holy Sacrament.[34]

[32] Thomas, *Los protestantes*, 209.
[33] Thomas, *Los protestantes*, 126.
[34] Thomas, *Los protestantes*, 397–414.

Few of these opinions can be inferred from his artistic work. He must have been able to separate in his own practice (and in obedience to his clients' iconographic preferences) the ideas of evangelist groups from what has been called (for Spain) a "strong theory of representation."[35] He might have had feelings about what kind of worship was due to certain images (as a function of what was owed to their referents), and about whether they held meaning as signs or merely as representations; but it would be something else entirely to put these feelings into visible form:

> About four or five years ago, as I [Jamete] was going ... with [Bartolomé de] Íñig[o], a painter from Huete, [and] another painter whose name I don't remember, from Huete to Cañaveras to meet the bishop's inspector, this Íñig[o] told me that when he was traveling [with] Giralte, an image-maker from Cuenca, they came across an image. When Íñig[o] took off his cap to it, Giralte said to him, 'I don't take off my cap for images because I make them myself, but only for the Cross.' And on that trip there were Giralte and Íñig[o] and Briones, a resident of Huete ...[36]

This text shows Giralte's inconsistent practice, since he was prepared to worship the cross with *latria* (due only to the Divine) in spite of its nature as a sign rather than an icon (according to defenders of the "strong" theory of representation), but did not offer such worship to other images, whether of Jesus (to whom *latria* was due), the Virgin (the proper object of *hyperdulia*), or the saints (the proper object of *dulia*). The problem is not with these definitions but with how they were put into practice, that is, how one behaved when faced with different types of images.

[35] Felipe Pereda, *Las imágenes de la discordia. Política y poética de la imagen sagrada en la España del cuatrocientos* (Madrid: Marcial Pons, 2007); "Sombras y cuadros: teorías y culturas de la representación en la Europa de la reforma católica," in "*Sacar de la sombra lumbre*". *La teoría de la pintura en el Siglo de Oro (1560–1724)*, ed. José Riello Velasco (Madrid: Abada-Museo Nacional del Prado, 2012), 69–86; "La conversión por la imagen y la imagen de la conversión: notas sobre la cultura figurativa castellana en el umbral de la Edad Moderna," in *Cartografías visuales y arquitectónicas de la modernidad, siglos XV–XVIII*, eds. Silvia Canalda Llobet, Carmen Narváez Casas, and Joan Sureda Pons (Barcelona: Universidad de Barcelona, 2011), 228–41; and "'Miracles Unmasked': El debate de las imágenes sagradas en la controversia del barroco," (Paper presented to the Congress *Teoría y literatura artística en España. Siglos XVI, XVII y XVIII* (Malaga, April 10–12, 2013).

[36] "Puede haber cuatro o cinco años más o menos que yendo ... con [Bartolomé de] Iñig[o], pintor vecino de Huete, [y] otro pintor, que no me acuerdo de su nombre, que íbamos de Huete para Cañaveras a buscar al visitador del obispo, me dijo este Iñig[o] que yendo camino [con] Giralte, imaginario vecino de Cuenca, toparon con una imagen e este Iñig[o] se le quitó el bonete e gorra que llevaba, e este Giralte le dijo: 'No me quito yo el bonete a las imágenes que yo las hago, sino es a la cruz'. E iban el camino este Giralte e Iñig[o] e Briones, vecino de Huete ...".

This supposed divorce between his sculptural[37] and architectural[38] work and his private and public behavior should receive more attention than it has been given so far. It has not been possible to identify any feature of his art or iconography that reveals his personal opinions.[39] It emerged at trial that he might be opposed to representing the Trinity as three equal persons "of the same countenance and likeness, for though they are three persons they are only one God and one eternity"; to showing the Virgin Mary in a swoon at the foot of the Cross ("he found there and read a book named 'Mount Calvary'[40] that said that during the Passion ... [she] did not lose consciousness but was always very much awake and observing the travails of her son"); or to portraying the saints. He would contradict himself the most when he swore that "from then on, he would place as few sculpted images as he could on altarpieces, because those were rats' nests; instead, he would make as many painted images as possible." Nonetheless, no painted work by him has been identified.

Nothing strange appears in his Annunciations, or in the Assumption and the Immaculate Conception of the doorway and retable of the chapel of Santa Elena, in spite of the ideas he expressed while he worked on them and the accusations he made against others:

This image-maker Giralte was in the [cathedral] cloister with some Basques whose names I don't remember, except for one quarryman called Adrián who has died. While they were all working Giralte [remarked that] if a farmer who has

[37] José María de Azcárate Ristori, *Escultura del siglo XVI* (Madrid, 1958); Turcat, *Étienne Jamet alias Esteban Jamete*; María Luz Rokiski Lázaro, *Documentos sobre escultura del siglo XVI en Cuenca* (Cuenca: Diputación, 2001), and "Esteban Jamete en Burgos," *Archivo Español de Arte* 297 (2002): 55–7.

[38] The degree of his involvement in the design and creation of certain architectural projects is still not clear. See José María de Azcárate Ristori, "Sobre el arco de Jamete de la catedral de Cuenca," *Archivo Español de Arte* (1945): 178–80; Fernando Chueca Goitia, *La arquitectura del siglo XVI* (Madrid: Ars Hispaniae XI, 1953); María Luz Rokiski Lázaro, *Arquitectura del siglo XVI en Cuenca* (Cuenca: Diputación, 1985), 145–69, *Arquitectura del siglo XVI en Cuenca. Arquitectos, canteros y carpinteros* (Cuenca: Diputación, 1989), and *Arquitectura de Cuenca. El paisaje urbano del casco antiguo*, 2 vols. (Cuenca: Diputación, 1996).

[39] Some of his master masons and apprentices testified that he used to sing songs like *El papa de Roma que se faze Dios / engaña a los hombres en cada un lugar. / Matad vuestras candelas, ypócritas, ypócritas, / dexad los ýdolos e adoremos a Dios* ("The pope in Rome thinks he is God; he deceives men wherever he goes. Put out your candles, hypocrites, hypocrites; abandon your idols and let us worship God").

[40] This must be the work titled *Libro llamado del Monte Calvario* by Fray Antonio de Guevara, O.F.M. (*c.* 1471–1545), for whom Jamete had worked years before in Valladolid. Part I appeared in Salamanca in 1542 and Part II in Valladolid in 1543. The Virgin is not shown fainting in its depictions of either the burial of Jesus or the Pietà on the Retable of the Apostles in Cuenca Cathedral.

good land is rich and sows it with good wheat and then harvests it, the land afterward remains the same. And he meant it in reference to Our Lady, that after she gave birth to Our Redeemer she remained the woman she had been before. And she had no greater virtue after giving birth than before, just like the earth after the wheat is harvested. And Giralte said that not on his own account but because it was what they said in his native land of Germany.[41]

Nonetheless, there are two examples that we can refer to, although in principle they contradict each other. The first is the altar of San Fabián and San Sebastián (Sts. Fabian and Sebastian, 1551–7), commissioned by the archdeacon of Cuenca Juan Fernández de Heredia (1468–1557) but paid for in part in 1553 by Diego de Tiedra (active 1535–59). It presents the figures of the two titular saints but also, in the upper part, another figure seated on a cloud (in the manner of Michelangelo's Moses) and surrounded by heads of seraphim; it appears to be God the Father, but holds a Resurrection cross (like Michelangelo's Cristo della Minerva in Rome). Although it seems that Jamete did not work alone on this group – the Gascon artisan Joannes or Juanes and the Castilian Miguel Fernández also took part – the image of the "Resurrection" God the Father should have received closer attention from students of iconology.

In contrast, the contemporaneous altar of San Mateo and San Lorenzo (Sts. Matthew and Lawrence, 1551–2), created perhaps for Pedro del Pozo in collaboration with the painter Martín Gómez the Elder (*El Viejo*) (*c.* 1503–62),[42] presents on its lower part a relief of Veronica's Veil or the Holy Face. This image, because of its legendary origins, was one of the favorite objects of John Calvin's invective, for instance in his treatise on relics of 1543. As Felipe Pereda has noted, it may be an iconic witness to a little-studied phenomenon in sixteenth-century Spain: the presence of Nicodemism, especially among foreign artisans and artists – embroiderers, weavers, silversmiths, watchmakers, printers, and so on[43] – who lived under constant scrutiny by the Holy Office.

41 "Este Giralte, imaginario, estando en la claustra [de la catedral], e estaban unos vizcaínos, que no me acuerdo de sus nombres, mas que el uno se decía Adrián, cantero, difunto, estando todos trabajando, [comentó que] un hombre labrador que tiene buenas tierras e es rico e siembra buen trigo en ellas e después lo coge, queda la tierra como de antes. E traía esto a comparación que Nuestra Señora, después que parió a Nuestro Redentor, quedó como una mujer como era antes. E porque esto no tenía más virtud después de parida que antes, como queda la tierra después de haber cogido el trigo de ella. E esto decía el dicho Giralte, no de suyo, sino que lo decían en su tierra de Alemania."

42 Pedro Miguel Ibáñez Martínez, *Pintura conquense del siglo XVI* (Cuenca: Diputación, 1993–6), II (1994), 171–6.

43 Clive Griffin, *Journeymen-Printers, Heresy, and the Inquisition in Sixteenth-Century Spain* (Oxford: Oxford University Press, 2005).

We cannot, however, accuse Jamete of hypocrisy, given the nature of his art. If he was producing figures in which the emotive traits of the religious subjects were the chief element of devotional histories or images, these traits could be recognized even from abroad as a style that was explicitly Spanish and hypocritical. It was based not only on a provincial artistic traditionalism but also on a willed obsolescence or retardation; this would be justified by a wish to appear pious rather than by a sincere religious inclination. Jamete often carved moldings and medallions containing ancient or modern heads, and decorations and ornaments in a style not only antiquarian and "Roman" but mythological. Perhaps they were even, as he claimed during his trial, "invented by his own invention" (*fantaseadas en su fantasía*): he could have been imagining a Purgatory that might exist on the earthly plane or even enclosed in a "hazelnut shell." He also carved three-dimensional statues that represented allegorical figures, virtues, churches, and synagogues; Bible stories, like the Virgin and the Three Wise Men on the Arco de Jamete; saints, battles, or the Last Supper, as in the chapel of Santa Elena in the cathedral. More complex groups were the Epiphany and the Adoration of the Shepherds from the doors of the chapter house or the doorway of the parish church of Santa María de Castejón in Huete (merely attributed to him), or the Resurrection and Ascension on the Retable of the Apostles. In many of these he had palpably absorbed the lessons of masters like Philippe Bigarny and, above all, Alonso Berruguete.

Jamete protested that these visions came to him in waking or sleeping dreams, rather than through his readings – it was the readings that provoked suspicion in his interrogators, who always cared more about the written word than about images like the owl. Still, his ideas clearly came from his visual culture and the memory of images he had seen: his reference to a hazelnut shell reminds us that Jeroen van Aken, alias Hieronymus Bosch (engravings of whose paintings circulated widely), painted figures contained in eggs, mollusc shells, pumpkins, and tree trunks; there were also the contemporary *Kunstkammern* miniatures, carved into walnut shells.

Let us return now to Erhard Schön's owl (Fig. 12.1).[44] Although the words printed next to the image refer generically to human blindness, the glassmaker Giraldo from Holland identified the non-seeing owl with

[44] The difference between the horned owl (*buho*) and the barn owl (*lechuza*) was not clear even to the contemporary naturalist Conrad Gesner, *Icones animalium quadrupedum viviparorum et oviparorum*, III (Zurich: Froschauer, 1553; Lyon, 1560), no. 15 (*Ewl*) and no. 14 (*Herzog*). The former was much less similar to our image of an *Ewl*.

FIG. 12.2. *An owl with spectacles*, Hendrich Bloemaert and Cornelis Bloemaert (*c.* 1625), Amsterdam, Rijksmuseum, http://search.smk.dk/Vark.asp?Type=Udvidet&ObjectId=40915
Copenhagen, Statens Museum for Kunst, Inventarnr. KKSgb4959

the pope; from that moment, this loose sheet of paper took on a greater significance that would accompany it for 200 years. In the process the polysemy attached to the owl increased: its dark side suggested night, the power of nocturnal wanderings, shadows, and death, but its association with Pallas Athena and Minerva also linked it to wisdom and to the mechanical and liberal arts, creating a bifaceted symbol (Fig. 12.2).

In Apuleius' *Golden Ass* III, 21–3,[45] Lucius' maid and lover cast the wrong spell and turned him into an ass instead of an owl. Elsewhere, the

[45] Translated into Spanish as *Metamorfosis de Lucio* (Seville, 1513), and placed on the Inquisition's Index in 1559. See Francisco Javier Escobar Borrego, Samuel Díez Reboso, and Luis Rivero García, eds., *La Metamorfosis de un Inquisidor. El humanista Diego López de Cortegana (1455–1524)* (Huelva: Universidad, 2012).

study of physiognomy saw an image of envy in the bird's large head, since it could look in both directions and stand for both the envier and the envied (Fig. 12.3).[46] We should not forget, in the public artistic context of drawn or printed images, the role played by an owl in Bosch's representation of himself: the bird is vigilant at night and envied by birds of prey.[47] In an incised 1567 etching by Hieronymus Cock (Fig. 12.4), showing a comic scene in a kitchen or tavern, there is an image tacked above the chimney of a pilgrim in the guise of a maimed and limping owl; its author is identified with "Hiero. Bos. Invenit," but it leaves open the idea of Bosch himself as a *Herzog* of 's-Hertogenbosch, a wanderer through life like an owl, limping like a lame watchman.

In the case of our own image we find, almost for the first time, a woodcut that was seen in sixteenth-century Spain by Nicodemist Protestants and probably also by the Inquisitors as contrary to the faith and doctrine of the Catholic Church, whether or not it had been conceived as Lutheran propaganda. We do not know its ultimate fate: it was more likely to have been burned than to have appeared on one of the post-1551 Indexes, which were meant for books rather than loose sheets. One detail of our image may refer to the sale of indulgences and the corrupt wealth of the Church: the owl, like any nocturnal bird, turns away from the sun's light, but the candlestick contains a keyhole that looks just like the lock of a money-box.

The relationship between art and the Holy Office has yet to be explored from diverse and impartial points of view; up to now, all arguments have seemed to come from the Inquisition side.[48] Most research, based on the surviving documents, has explored the trials of artists and artisans who in most cases were tried not for holding the wrong religious opinions but simply to uphold Counter-Reformation morality. Occasionally the cause was having produced a work considered heretical or merely dubious, even if it did not

[46] Giambattista della Porta, *De Humana Physiognomia libri IIII*, II (Naples, 1586), 29.

[47] Matthijs Ilsink, "On Three Drawings by Jheronimus Bosch: The Field has Eyes, the Wood has Ears, The Treeman, The Owl's Nest," in *Jheronimus Bosch and His Sources. Second International Jheronimus Bosch Congress* ('s-Hertogenbosch: Jheronimus Bosch Art Center, 2010), 175–90.

[48] See, for a general view, Michael Scholz-Hänsel, *Inquisition und Kunst. "Convivencia" in Zeiten der Intoleranz* (Berlin: Frank & Timme, 2009). In contrast, see Bustamante García, "El Santo Oficio de Valladolid y los artistas," replying to Scholz-Hänsel's "¿La Inquisición como mecenas? Imágenes al servicio de la disciplina y la propaganda inquisitorial," *Boletín del Seminario de Arte y Arqueología* 60 (1994): 301–17. However, Bustamante García later published "Alboraique. Un dato inconográfico," *Archivo Español de Arte* 280 (1997): 419–26.

LIBER SECUNDUS. 29

De Capite. Cap. I.

NOS cum Aristotele totum id caput vocamus, quod collo substentatur. Primo integrum considerabimus, post eius partes, frontem, ocul[illegible]percilia, & reliqua. Galenus capitis considerat quantitatem, & formam, quant[illegible]tis varietas in magnitudine, & paruitate consistit; loquemur primo de quantitate.

Hac tabella maximum bubonis caput, & Vitellij C[illegible]vastum exprimitur, vt ex marmorea statua, quæ in Musæo Hadriani Spatafor[illegible] doctissimi, & rerum antiquarum studiosissimi asseruatur, excipi curauimus.

Maximum Caput.

POlemon, & Adamantius in eorum Physiognomonicis, caput, quod magnitudine exuperat, obtusi, & indocilis hominis argumentum dicit. Albertus animalium libro, Enorme caput, quod præ sui magnitudine ad humeros vergit, sensu, & virtute carere dixit: & Paulo post. Immensam autem stultum, stolidum, & vehementer indocile. Aristoteles libro de somno. & vigilia, ait. Qui minimo plus in caput extumuerint, somniculosi sunt, nam quo amplius est receptaculum, eo plus item capiat necesse est. In nanis, siue pumilionibus vaporum multitudinis vis adcaput surrigitur, quoniam grandius membrum alimenti amplius requirit, quod verò copiose scandit, copiose remeet consequens est, sed lentius, quia insitus calor vim tantum sublatæ euaporationis breui percoquẽdo nequit conficere. & in suis Physiognomonicis in figura sonnolenti, tribuit ei superiora maiora, idest caput. Capitis magnitudo materiei potius abundantiam, quam virtutis excellentiam indicat. Ego ad asinos referrem, quorum caput grande, & ineptissimis sunt moribus, stolidi, degeneres animo, timidiq;. Inter aues ad nocturnas, vt noctuas, & bubones, qu[illegible] omnes capita enormia magnitudine habent, ex otio prodeuntes, è pedibus exeunt, nimio pond[illegible] capitis detente[illegible]. Ouidius de Ascalapho in bubone mutato.

In ro-

FIG. 12.3. Giambattista della Porta, *De Humana Physiognomia libri IIII* (Naples, 1586), bk. II, p. 29

FIG. 12.4. "Scene in a kitchen or tavern, with an etching with the script 'Hiero. Bos. Invenit'", Pieter van der Heyden and Hieronymus Cock (1567), Madrid, BNE, INVENT/36991

violate Catholic orthodoxy; sometimes the motive was a dispute between religious orders. We know that the Italian sculptor Pompeo Leoni, who had reached Spain in 1556, had been arrested by 1558 and subjected to an *auto da fe* in Valladolid for having said "Lutheran things"; he was sentenced to one year of seclusion in a monastery. In the same city Juan de Juni, a French sculptor and an old acquaintance of Jamete's, was arrested in 1567.[49] The Flemish painter Simon Pereyns was seized in Mexico City in 1568, also for rashly expressing the wrong religious opinions. The Inquisition detained the French sculptor Nicholas de Chanterenne (*c.* 1511–51) in Portugal in 1538; his denouncer, the artisan João Rombo, claimed that he possessed a *mendracolla* (mandragora or mandrake) and a Bible that contained "many heresies," probably illustrated with woodcut images.[50]

[49] Eugène Plon, *Léone Léoni, sculpteur de Charles Quint, et Pompeo Léoni, sculpteur de Philippe II* (Paris, 1887), 131–2, 380–8, and 421–3; Bustamante García, "El Santo Oficio de Valladolid y los artistas," 456–8.

[50] António Baião, *A Inquisição em Portugal e no Brasil. Livro I. A Inquisição no século XVI–XIX. As Denunciações da Inquisição de Lisboa*, VI (Lisbon: Edições do Arquivo

In other (perhaps more interesting) cases it was iconoclasm, of various kinds, that was persecuted.[51] The Italian Pietro Torrigiano landed in the Inquisition's prison in Seville in 1528, after he had destroyed one of his own carvings of the Virgin. Certain cases did not even have to do with the worship of images: especially in the later period the figures that the authorities were pursuing might be lascivious or libertine ones,[52] together with their makers, if they could be found.

Documents and images are always out of phase with each other, when images refer to documents and even more so when documents or literary texts speak of images. There may always have been artists whose works approached the ever-shifting boundaries of what was acceptable at a given place and time, yet who somehow escaped being caught, so that we have no clear and full account of their trials (as we do for Jamete's) or even a laconic notice of their sentences; in some cases we have not yet identified the condemned images, whether made by or merely possessed by the accused. The former category included images by Lucas Cranach, the Elder of the *Passional Christi und Antichristi* of Martin Luther (of which only one example has survived in Spain [Madrid, Biblioteca Nacional, U/3139], but of which there were apparently imported examples in foreign languages in circulation). There were also images of some versions of illustrations by Cranach that appeared in *The Image of the Antichrist* (n.pl. [Geneva: Jean Crispin], n.d. [1556]) (Figs. 12.5–6), attributed to the Italian evangelist Bernardino Ochino of Siena (1487–1564) and translated by Alonso de Peñafuerte (Madrid, Biblioteca Nacional, U/11097[3]).[53]

Histórico Português, 1908), 81; Vergílio Correia, *O imaginário francês Nicolau Chanterene na Inquisição (uma denuncia em 1538)* (Lisbon. Tipografía Comercial, 1922), and in *Obras. Estudos de história da arte, escultura e pintura*, III (Coimbra: Por Ordem da Universidade, 1953), p. 122 ff. See also Pedro Dias, *Nicolau Chanterene, escultor da Renascença* (Lisbon: Ciência e Vida, 1987), 35–44; and Fernando Jorge Grilo, "Nicolau Chanterene e a afirmação da escultura do Renascimento na Península Ibérica (c. 1511–1551)," doctoral dissertation (Universidade de Lisboa, 2001).

51 Virgilio Pinto Crespo, *Inquisición y control ideológico en la España del siglo XVI* (Madrid: Taurus, 1983). Of little value (and too heavily Catholic) is José Sánchez Castro, "La censura de la figuración artística en España (1487–1820)," *Boletín del Museo e Instituto Camón Aznar* 66 (1996): 37–98; see also Cordero de Ciria, "Las artes plásticas en Erasmo y el erasmismo español," 93–164.

52 Enrique Cordero de Ciria, "Mostrar el arte. Consideraciones sobre la cuestión del desnudo en España en los siglos XVI y XVII," *Boletín del Museo e Instituto Camón Aznar* 92 (2003): 29–65; "'Con la ocasión de ponerlos desnudos, y castos': lascivia y castidad en la pintura del Siglo de Oro," in Riello Velasco (ed.), "*Sacar de la sombra lumbre*," 105–34.

53 Also cited by Cordero de Ciria, "Las artes plásticas en Erasmo y el erasmismo español," 36–8, in reference to Luis Usoz y Río, *Reformistas antiguos españoles* (London, 1863), and Ramón Alba, reed., *Del Anticristo* (Madrid: Editora Nacional, 1982).

EL SANCTISSIMO PADRE

De ſu padre el diablo recibe el Antechriſto las leyes,
Con que tiraniza cóſciencias de vaſſallos y Reyes.
1. a Timotheo. 4.

Auiſo neceſsario.

CHriſtiano Lector, el error por ſer antiguo, no es preuilegiado, ni el engano por ſer enuejecido, y tener ya hechos callos, no es exempto de ſer conocido por tal qual es. Porque viſto que Dios ha dado tantos medios de conocello y huyllo, es ſenal que no quiere que eſtemos ya mas ſepultados en ignorancia que acarrea perdicion. Y pues Dios agora te declara lo que te conuiene ſaber y huyr, es mueſtra de que te quiere ſaluar, Por eſſo te es neceſſario leer con paciencia eſtas pocas planas, y conſiderar lo que en ellas te di-

FIG. 12.5. *The Antichrist Receives the Law from His Father the Devil*, Lucas Cranach the Elder, in *La imagen del Antechristo* (s.l. [Geneva, Jean Crespin], s.a. [1556], translated by Alonso de Peñafuerte, Madrid, BNE, U/11097(3)
From http://bdh-rd.bne.es/viewer.vm?id=0000095014&page=1

Though obviously derivative, they were no less dangerous for having an accompanying text in Spanish. In a similar way, images that appeared traditional, like Cranach's etching of the "Ascension" that depicted Christ's

CHRISTO.

A Chriſto el Señor con triumpho y gozo,
Le recibe el cielo para perpetuo repoſo.
S. Lucas XXIIII.

ANTECHRISTO.

En el infierno de fuego es por pena deuida,
El Antechriſto lançado para perpetua manida.
Apocalypſi XIX.

FIG. 12.6. *Christ's Ascension and The Antichrist surrounded by demons and His Father Satan as a Pope*, Lucas Cranach the Elder, in *La imagen del Antechristo* (s.l. [Geneva, Jean Crespin], s.a. [1556], translated by Alonso de Peñafuerte, Madrid, BNE, U/11097(3)
From http://bdh-rd.bne.es/viewer.vm?id=0000095014&page=1

body without a head or a face (Fig. 12.6), could have been seen by the sharp eyes of a Giraldo of Holland or an Esteban Jamete, who wanted to "see and consider." Not every image had to be as obvious as the one of the Antichrist surrounded by demons and his father Satan, portrayed as a pope with features very like those of Leo X (1513–21) (Figs. 12.5–6).

And we are still very far from understanding the meaning of a supposed (perhaps spurious) letter of 1571 from the Supreme Council of the Inquisition that forbade paintings of the Crucifixion, the Trinity, and twelve etchings of the Passion,[54] a clear although unknown and rare example of condemned painted images.

[54] Sánchez Castro,"La censura de la figuración artística en España," 39, apparently citing Madrid, Archivo Histórico Nacional, Inquisición, Libro 326.

13

Epic Temptation: Lope de Vega's Battle of Lepanto

Elizabeth R. Wright

It was a critical dismissal that echoed, if not round the world, at least in the seminar rooms, library carrels, and scholarly colloquia where Hispanists at work in the twentieth century set the canon for "Golden Age" theater studies. Marcelino Menéndez Pelayo thus judged Lope de Vega's only dedicated account of the Battle of Lepanto, the *comedia* titled *La Santa Liga (The Holy League)*:

> No hay fábula dramática de ninguna especie, sino una serie de escenas inconexas. Los dos primeros actos valen muy poco [...] Pero la tercera jornada se levanta en gran manera sobre las anteriores, y está animada de *espíritu verdaderamente épico*. (There is no dramatic plot at all, but a series of disparate episodes. The first two acts have little value [...] Yet, the third act surpasses the previous ones to a large extent, and is powered with a *true epic spirit*). [Emphasis added.][1]

The source of this surprise and vexation is not hard to pinpoint. By the time Lope composed *La Santa Liga*, the naval battle of October 7, 1571 had already been canonized as imperial Spain's naval battle par excellence – the *gran batalla naval*. As is the case with the majority of his plays, surviving documents do not allow us to pinpoint the exact dates for either the *La Santa Liga*'s composition or public performances. Nonetheless, a few clues suggest the drama took shape during a decisive transition point, both in Lope de Vega's literary career and in Spanish dynastic history.

[1] Marcelino Menéndez Pelayo, "Observaciones preliminares (*La Santa Liga*)" (1901), reprinted in *Obras de Lope de Vega*, vol. XXIV (*Crónicas y leyendas dramáticas de España*), ed. Marcelino Menéndez Pelayo (Madrid: Atlas, 1965), 101–24; citation 102. This article expands on a line of inquiry developed in Elizabeth R. Wright, "Enredos historiográficos: Lope ante Lepanto," *Anuario Lope de Vega* 18 (2012): 146–74 (www.revistes.uab.cat/anuariolopedevega).

Though the one extant edition from the seventeenth century did not appear in print until it was anthologized in the 1621 *Decimoquinta parte de las comedias de Lope de Vega Carpio* (hereafter *Parte 15*), best evidence suggests Lope wrote the play some two decades earlier, *c.* 1596–1603.[2] As such, the Lepanto drama took shape in the context of the transition period from the reign of Philip II to Philip III, when Lope branched out from his economic mainstay writing plays to seek recognition as the author of ambitious epic poems concerned with Spanish history. Epic poetry, in turn, would become the primary motor for his long-lived if ultimately unsuccessful effort to gain the position of salaried royal chronicler.[3] In spite of Lope's own abiding interest in adapting epic narrative modes to monumentalize notable events from Spanish history, Lope's *La Santa Liga* recounts Lepanto as the culmination of a chain of novelistic love intrigues set in an island-hopping plot. Action thus begins in the Ottoman *seraglio* and moves to the Turkish encampments of Cyprus during the siege of 1570–1, with the Muslim leaders and their trappings drawn to highlight alluring eastern exotica. In each locale, war councils intertwine with lovers' entreaties. Only Act 3 features the epic motifs and plot lines already associated with the 1571 naval battle by the time of the play's composition. Here, the playwright does proffer a lofty chronicle of war councils on each side of the battle lines at Lepanto, following directly the account in Antonio de Fuenmayor's 1595 biography of Pius V.[4] Continuing, a verse chronicle of the battle dramatizes the intense morning of fighting that transpired on October 7, 1571. Only in these scenes from the third and final act did Menéndez Pelayo find "true epic spirit," a conception rooted in nineteenth-century Spanish nationalism, or *españolismo*.[5]

[2] The composition date range of 1596–1603 follows the chronology of S. Griswold Morley and Courtney Bruerton, *Cronología de las comedias de Lope de Vega*, trans. María Rosa Cartes (Madrid: Gredos, 1968), 50.

[3] On Lope's turn to epic poetry writing at the end of the 1590s, see Elizabeth R. Wright, *Pilgrimage to Patronage: Lope de Vega and the Court of Philip III, 1598–1621* (Lewisburg, PA: Bucknell University Press, 2001); and Antonio Sánchez Jiménez, "Introducción," in *Isidro* (Madrid: Cátedra, 2010), 11–98; see esp. 22–3. My original concern with Lope's literary career *c.* 1598 was directly inspired by Richard Kagan's graduate seminars of 1992–3 on Habsburg historiography and his 1993 seminar paper on "Clio and the Crown," the starting point for his splendid 2009 book of the same title.

[4] See Antonio de Fuenmayor, *Vida y hechos de Pío V*, ed. Lorenzo Riber (Madrid: Real Academia Española, 1953), 226–9; Menéndez Pelayo records Lope's direct borrowings in his "Observaciones preliminares (*La Santa Liga*)," CX.

[5] On this project as it shaped North American Hispanism, see Richard L. Kagan, *Marcelino Menéndez y Pelayo y el hispanismo norteamericano* (Santander: Real Sociedad Menéndez y Pelayo, 2013).

In the century that followed Menéndez Pelayo's negative assessment of *La Santa Liga*, the play attracted relatively little attention. Within the relatively limited bibliography, the primary point of interest hinged on the seraglio love plots that drive the first two acts.[6] Frederick de Armas, for his part, revealed the play's intricate dramatization of two iconographic motifs associated with Titian: the portrait of Suleiman the Magnificent's favorite – known as Rossa Sultana or Roxelana – holding a weasel; and the allegorical depiction of the Holy League victory at Lepanto as the triumph of religion.[7] Two recent studies have drawn renewed attention to the question of epic. Guillem Usandizaga has shown how the drama records Lope's engagement with Alonso de Ercilla's *La Araucana*, a point to which I return ahead; for his part, Juan Udaondo has noted *La Santa Liga*'s evocation of the legacy of Torquato Tasso's *Gerusalemme liberata* (1581), while emphasizing that the playwright's epic motifs intertwine with flourishes drawn from other literary innovations, such as emblem literature and the new wave of artistic ballads in Spain known as the *romancero nuevo*.[8]

Here it bears emphasis that our enduring critical emphasis on assessing the epic dimensions within *La Santa Liga* would hardly have caught the playwright or his contemporaries by surprise. At the turn of the seventeenth century, epic poetry retained a central place in grammar school education and elite identity formation. Schoolmasters drew on the genre to inculcate Latin skills and military valor alike, prompting students to extract, memorize, and translate passages from classical epic, the *Aeneid* foremost. These epic exempla, in turn, served to model normative virtues for military leaders, *letrados*, and rulers. Capturing the prevailing

[6] See, for example, Albert Mas, *Les turcs dans la littérature espagnole du Siècle d'Or*, vol. I (Paris, Centre de Recherches Hispaniques, 1967), 393–407; and Miguel Renuncio Roba, "El mundo islámico en *La Santa Liga* de Lope de Vega / The Islamic World in *The Holy League* by Lope de Vega," *Anaquel de Estudios Árabes* 16 (2005): 205–17.

[7] See Frederick A. de Armas' "Lope de Vega and Titian," *Comparative Literature* 30 (1978): 338–52, as well as de Armas' "The Allure of the Oriental Other: Titian's *Rossa Sultana* and Lope de Vega's *La Santa Liga*," in *Brave New Words: Studies in Spanish Golden Age Literature*, eds. Edward Friedman and Catherine Larson (New Orleans, LA: University Press of the South, 1996), 191–208.

[8] See Guillem Usandizaga, "*La Santa Liga* y el Dios de las batallas," in *La representación de la historia contemporánea en el teatro de Lope de Vega* (Madrid and Frankfurt: Iberoamericana Vervuert, 2014), 125–53; Juan Udaondo Alegre, "Venus y Marte en Lepanto. Amor vulgar frente a virtud cristiana en *La Santa Liga* de Lope de Vega," in *Pictavia Aurea: Actas del IX Congreso de la Asociación Internacional "Siglo de Oro,"* Poitiers, July 11–15, 2011, eds. Alain Bègue and Emma Herrán Alonso (Toulouse: Presses universitaires du Mirail, 2013), 1157–66.

embrace of epic for elite education, the poet Gregorio Hernández de Velasco addressed his landmark translation of the *Aeneid* to Philip II, urging the king to study the first six books to hone his political skills and the second six for military arts.[9]

In view of the unparalleled prestige of epic at the turn of the seventeenth century and Lope's own ambitions to reach the pinnacle of royal patronage with epic poetry as his calling card, the question that informed Menéndez Pelayo's study of *La Santa Liga* remains one worth pondering today. Why give the Battle of Lepanto – an epic event if ever imperial Spain engendered one – such varied, comic treatment? In the pages ahead, I unpack this line of inquiry, examining *La Santa Liga* in relation to the playwright's multi-faceted mobilization of the emerging Lepanto historiography. Along the way, I reconsider the drama's variegated plot and shifting moods – the *escenas inconexas* that frustrated Menéndez Pelayo – in light of the era's diverse narrative proposals for recounting the naval battle at a distance of some three decades. For this task, I will take a decidedly literary path, though the ongoing scholarly discussions about Lepanto's long-term geopolitical significance inform my analysis.[10] Like the play itself, my examination comes in three parts. I first review how the first chronicles from Lepanto touched an epic nerve among poets and literary theorists. Then, I ponder Lope's comic take on the event, considering other narrative models on which he drew. Finally, I consider

[9] The advice to Philip II appears in the first complete Castilian translation of Vergil's *Aeneid, La Eneida de Virgilio*, translated by Gregorio Hernández de Velasco (revised ed., Zaragoza: Lorenzo y Diego Robles, 1586), prologue, n.n. Other key sources on epic and elite education include: Walter Ong, "Latin Language Study as a Renaissance Puberty Rite," *Studies in Philology* 66(2) (1959): 103–24; David Scott Wilson-Okamura, *Virgil in the Renaissance* (Cambridge: Cambridge University Press, 2010), 203–06; Andrew Wallace, *Virgil's Schoolboys: The Poetics of Pedagogy in Renaissance England* (Oxford: Oxford University Press, 2010), 178–227; and Craig Kallendorf, "Philology, the Reader and the *Nachleben* of Classical Texts," reprinted in *The Virgilian Tradition: Book History and the History of Reading in Early Modern Europe* (Aldershot: Ashgate, 2007), 137–56.

[10] The voluminous secondary source bibliography on Lepanto can be approached with John Francis Guilmartin, *Gunpowder and Galleys: Changing Technology and Mediterranean Warfare at Sea in the 16th Century* (1974; rev. ed. London: Conway Maritime Press, 2003); see esp. 235–68; and Niccolò Capponi's *Victory of the West: The Story of the Battle of Lepanto* (London: Macmillan, 2006); see esp. 253–86. On the long-term geopolitical significance, see the seminal article by Ottoman historian Andrew C. Hess, "The Battle of Lepanto and Its Place in Mediterranean History," *Past & Present*, 57 (November 1972): 53–73. Lepanto's resonance today is attested in the multiple printings of Alessandro Barbero's *Lepanto: La battaglia dei tre imperi* (Bari: Laterza, 2010).

points where Lope does confront the epic temptation built into Spain's *gran batalla naval.*

I. THE QUEST FOR MODERN EPIC

Across Europe, the initial news reports of the clash at Lepanto seemed an unmatched opportunity for epic narration. The first bulletins and letters that described the battle of October 7, 1571 recount a face-to-face confrontation between John of Austria, son of Charles V, and his Ottoman counterpart, Müezzinzade Ali Pasha, brother-in-law of the sultan Selim II. Their confrontation offered a chance to reimagine the clashes of great warriors of Antiquity, recalling Hector facing Achilles in the *Iliad*, or Aeneas against Turnus in the *Aeneid*. For instance, a letter by Nicolás Augusto de Benavides details how John of Austria, flanked by guards, corners Ali Pasha and demands he surrender. The proud Turkish commander haughtily refuses, proclaiming, "soy el Bajá y no me rendiré, sino que te mataré" ("I am the Pasha, and will not surrender but kill you instead"). As he raises his scimitar to smite the young Habsburg, a Spanish guard strikes him dead.[11] The scene suggests epic *furor*, to use the Vergilian term for Aeneas' warriorly rage. Most directly, it brings to mind the *Aeneid*'s final battle, where the Trojan pauses for one moment of hesitation, but then plunges his sword into a prostrate Turnus.

Beyond the appeal of the young John of Austria in combat as a Spanish Aeneas, Lepanto's geographic coordinates intensified the conviction that the naval victory merited epic treatment. Right away, mapmakers and chroniclers noted how the site where the clash between the massive Ottoman and Holy League fleets had transpired – in the waters of western Greece, at the Curzolaris islands – was just south of the site of the Battle of Actium of 31 BC. The largest galley battle of Antiquity, Actium was a touchstone for Renaissance poets, who scrutinized and emulated the vivid account that Vergil revealed as if etched on the shield of Aeneas, in *Aeneid* 8.675–731.[12] In light of the historical parallels between Actium

[11] See Nicolás Augusto Benavides, "Relación de la batalla de Lepanto ... dirigida a Lope de Acuña," October 10, 1571, Biblioteca Palacio Real (Madrid), II/2211, 56, http://avisos.realbiblioteca.es/?p=article&aviso=37&art=863 (accessed November 5, 2014), folios n.n.

[12] On the literary connection between Actium and Lepanto, see David Quint's paradigmatic *Epic and Empire: Politics and Generic Form from Virgil to Milton* (Princeton, NJ: Princeton University Press, 1993), esp. 21–49. Crucial considerations of Actium in Hispanic epic come from Elizabeth B. Davis, *Myth and Identity in the Epic of Imperial Spain* (Columbia, MO and London: University of Missouri Press, 2000), 61–97; and

and Lepanto, poets and literary theorists immediately grasped this historic opportunity to fulfill the deep-seated longing for a heroic poem suitable for modern times, to use the formulation French Hispanist Mercedes Blanco provides in her illuminating analysis of how this one event shaped literary practice for two decades.[13]

In fact, an astounding number and variety of poets weighed in with epic commemorations. In time for Lepanto's symbolically important first year anniversary, Juan Latino presented his *Austrias Carmen* (*The Song of John of Austria*), a two-book epic in Latin hexameters, for censors' approval. In parallel, Joan Pujol completed his *La singular y admirable victoria* in Catalan hendecasyllables. Within a half decade, Portuguese poet Jerónimo de Corte-Real completed his *Felicissima victoria*. Most famously, Alonso de Ercilla presented the second installment of his monumental epic of Spain's seemingly intractable campaign of empire-building in Chile, the *Araucana*, in 1578; its Canto 24 included a prophetic account of Lepanto. The epic impulse touched writers across the Atlantic as well. Thus, Francisco de Pedrosa, a native of Madrid long based in Guatemala, composed a six-book epic in Latin hexameters titled the *Austriaca* or *Naumachia*.[14] Italian poets also pondered Lepanto's epic dimensions. Giovanni Battista Arcucci, a jurist from Naples, prepared an epyllion (epic in miniature) titled the "Naupactean Victory," using the epic catalog to convey the multinational array of fighters who faced off at Lepanto. Likewise, Guglielmo Moizio, of Monferrato in the Piedmont region, crafted his "Song of the victory of the Christian fleet," building up to an arresting combat scene where John of Austria pursues, then slays, Ali Pasha.[15]

Notwithstanding this profusion of epic accounts, the cultural memory of Lepanto was also negotiated in lighter, even comic verse forms. In fact, one widely distributed collection of light verse inspired the

James Nicolopulos, *The Poetics of Empire in the Indies: Prophecy and Imitation in La Araucana and Os Lusíades* (University Park, PA: Pennsylvania State University Press, 2000), 176–215.

[13] See Mercedes Blanco, "La batalla de Lepanto y la cuestión del poema heroico," *Calíope* 19(1) (2014): 23–54.

[14] José López de Toro records the astounding number of poetic commemorations that appeared immediately after Lepanto in *Los poetas de Lepanto* (Madrid: Instituto Histórico de Marina, 1950), 24–5.

[15] For annotated English translations of the above-mentioned poems by Latino, Arcucci, and Moizio – plus poems by nineteen other neo-Latin poets who commemorated the battle – see *The Battle of Lepanto*, eds. Elizabeth R. Wright, Sarah Spence, and Andrew Lemons, I Tatti Renaissance Library, vol. LXI (Cambridge, MA: Harvard University Press, 2014).

above-mentioned Latinist of Guatemala, Francisco de Pedrosa, to compose his *Naumachia*. Writing to Philip II to request a subvention for this project, Pedrosa offered his Lepanto epic in Latin hexameters to rectify the slight to Spain's honor that had resulted when Italian poets marked the great victory with short, even comic poems.[16] The offending collection of light verse that reached Pedrosa across the Atlantic might be the *Raccolta* published in 1572 by the Venetian printer Sebastiano Ventura or the anthology titled *Il Trofeo della vittoria* of Luigi Groto, the "blind bard of Adria." Both collections of occasional verse appeared in print in very short order, before Lepanto's first anniversary.[17]

Taking a step back from the schoolmaster of Guatemala's wounded imperial zeal, we can now appreciate how the profusion of light, occasional poetry on Lepanto records the cacophony of celebratory voices that would be a crucial ingredient for the naval battle's deep and lasting cultural resonance. For instance, the compilation of verses from the Venetian print shop of Sebastiano Ventura includes a humorous poem attributed to one Gatian da le Codeghe ("Al mia Slin, ach'semia? ched Maumet?"), whose macaronic verses in Bolognese mimic the hissing consonantal clusters associated in European popular culture with Arabic and Turkish dialects spoken on Ottoman galleys.[18] This deliberately inarticulate staccato is far from the sublime heights of such epic touchstones as the galley clash at Actium in *Aeneid* 8 or the Trojan funeral games in *Aeneid* 5. But the celebratory songs of common people were an integral part of the victory celebrations. In fact, examining the emerging collective memory of Lepanto in Venice, musicologist Iain Fenlon notes how a first wave of officially organized liturgical and civic celebrations gave way to many ripples of popular commemorations that would give the naval battle its lasting cultural resonance.[19]

[16] Francisco de Pedrosa's *Austriaca sive Naumachia* survives in a single manuscript, BNE mss 3960. His memorandum to Philip II is transcribed in the *Colección de documentos inéditos para la historia de España* (hereafter *CODOIN*), eds. Martín Fernández de Navarrete et al., vol. III (Madrid: Viuda de Calero, 1843), 290–1. The passage referred above hails the story of Lepanto as "una materia la más alta y excelente que después del imperio romano se ha ofrecido," 291. (Orthography here and in other period sources has been modernized, except where a specific point of idiom would be lost by so doing.)

[17] *Raccolta di varii poemi Latini, Greci, e Volgari* (Venice: Sebastiano Ventura, 1572) and Luigi Groto (Il Cieco d'Adria), *Trofeo della Vittoria Sacra* (Venice: Sigismondo Bordogna and Francesco Patriani, 1572).

[18] See *Raccolta di varii poemi Latini, Greci, e Volgari*, fol. 45.

[19] Iain Fenlon, *The Ceremonial City: History, Memory and Myth in Renaissance Venice* (New Haven, CT: Yale University Press, 2007), 98–191.

From Venice we also find another crucial narrative paradigm for remembering Lepanto that will help make narrative sense of the island-hopping plot of *La Santa Liga*. As soon as news of a decisive naval victory over the Ottoman fleets reached the Serenissima, cartographers such as Giovanni Francesco Camoccio and Tommaso Porcacchi designed maps showing the exact location where the Holy League and Ottoman fleets clashed, as well as diagrams of their respective battle orders. As we know from the research of George Tolias and Jenny Jordon, the post-Lepanto *isolarii* (island atlases) plotted the eastern Mediterranean constellation of islands within a geopolitical chronicle of a frontier region where Ottoman encroachments threatened Venetian maritime power.[20] As the reader leafs through the post-Lepanto island atlases, the galley clash of October 1571 appears as the culmination of an island-to-island itinerary. In fact, the chronicle of the Holy League alliance and its victory at Lepanto shaped the publication history of the island atlases. For instance, the first edition of Tommaso Porcacchi's *L'isole piu famose del mondo*, from 1572, features a dedication to John of Austria dated January 1571, registering the expectation of a major clash between Christian and Muslim navies. The revised, post-Lepanto edition of 1576 incorporates a chronicle and some maps of the battle.[21] With this post-1571 reorientation of Mediterranean insular geography in mind, the *Santa Liga*'s seemingly disjointed plot takes on a previously unsuspected coherence.

II. AN *ISOLARIO* IN THREE ACTS

Action begins in Constantinople, transporting audiences in the *corral de comedias* to the same Ottoman seraglio depicted, for instance, in the diagram of Constantinople signed by cartographer Giovanni Francesco

[20] The post-Lepanto reorientation of the *isolarii* is discussed in Jenny Jordan, "Galley Warfare in Renaissance Intellectual Layering: Lepanto through Actium," *Viator* 35 (2004): 563–79; Jordan's "Imagined Lepanto: Turks, Mapbooks, Intrigue, and Spectacular in the Sixteenth Century Construction of 1571," unpublished dissertation (UCLA, 2004); George Tolias, "*Isolarii*, Fifteenth to Seventeenth Century," in *Cartography in the European Renaissance*, ed. David Woodward, vol. III, bk. 1, 263–84, of *The History of Cartography* (Chicago, IL: University of Chicago Press, 2007); and his "The Politics of the Isolario: Maritime Cosmography and Overseas Expansion During the Renaissance," *The Historical Review / La Revue Historique* 9 (2012): 27–52.

[21] Tommaso Porcacchi, *L'isole piu famose del mondo* (Venice: Simon Galignani & Girolamo Porro, 1576); for Lepanto, see 87–94. His first, 1572 edition – whose dedication to John of Austria is dated January 1571 – captures the air of expectation that preceded the naval clash: Tommaso Porcacchi, *L'isole più famose del mondo* (Venice: Simon Galignani and Girolamo Porro, 1572), prologue n.n.

Camoccio. In particular, the first intrigue hones focus on the harem plotted in Figure 13.1 (marked AA). In Lope's dramatic configuration of the seraglio, the sultan luxuriates in its sumptuous baths and relieves the burdens of his international empire with gallant hunting expeditions. It bears mention that this characterization of Selim II as a distracted ruler focused on hunting, baths, and his concubines tells us more about prevailing Spanish anxieties about kingship and European caricatures of "Selim the sot" than it does about this sultan's reign as recorded in Ottoman sources.[22] As the scene continues, the sultan strives to please his spirited concubine, Rosa Solimana. The name, of course, evokes Roxelana, the famed Slavic beauty who rose from slavery to become Suleiman the Magnificent's powerful favorite. Taking advantage of the abundant historical license allowed with stories set in distant lands, Lope adapts the famed seraglio love story of Suleiman's era for a story about his son and heir.[23]

Notwithstanding the harem interlude's heavy dose of Turkish exotica, the scene also takes on an informative dimension that correlates to the post-Lepanto island atlases. Selim's first monologue charts the Ottoman imperial aspirations with the same structuring element found within the *isolarii*. While protesting that he scorns worldly power without Rosa Solimana's love, the character charts the Levantine theater of Christian–Muslim rivalry:

No porque el Tigris pasé
y a Mesopotamia vi,
y el Tanais ensangrenté,
la gran Rodas destruí,
la firme Malta apreté. [...]
(Not because I crossed the Tigris and laid eyes on Mesopotamia and bloodied the Tanais, destroyed great Rhodes, and pressed unyielding Malta.)[24]

[22] For an analysis of Ottoman documents of the rule of Selim II, see Michel Lesure, *Lépante: la crise de l'empire ottoman* (Paris: Julliard, 1972), 55–7.

[23] Frederick A. de Armas, in "The Allure of the Oriental Other: Titian's *Rossa Sultana* and Lope de Vega's *La Santa Liga*," 203, underscores the purposeful nature of this historical "inaccuracy."

[24] Lope de Vega, *La Santa Liga*, in *Obras de Lope de Vega*, ed. Marcelino Menéndez Pelayo (1901), reprinted in *Obras de Lope de Vega*, vol. XXV (*Crónicas y leyendas dramáticas de España*), Biblioteca de Autores Españoles 224 (Madrid: Atlas, 1965), 231–78; citation 232 (*princeps* fol. 99r). Unless otherwise noted, all citations to the play refer to this edition, with orthography modernized except in cases where it would alter the meter. Parenthetical folio citations as *princeps* correspond to the play text published in the 1621 *Parte 15* (*Deimoquinta parte de las comedias de Lope de Vega* [Madrid: Alonso Pérez, 1621]), fol. 98v–121v; this is the only known seventeenth-century edition of the play. A complete digitalization of the textual witness preserved at the Biblioteca

FIG. 13.1. "Constantinople," Giovanni Francesco Camoccio, in *Isole che son da Venetia nella Dalmatia ... fino a Costantinopoli ...* Venice: Simon Pinargenti, 1573 (Biblioteca Histórica, "Marqués de Valdecilla," BH FLL 9665). The southernmost building is listed in the key (item AA) as the Grand Turk's seraglio, the setting for Act 1 of *La Santa Liga*

Mapping the geography of the Christian–Muslim conflicts that played out over the long sixteenth century, the catalog of Ottoman encroachment begins at the northeastern points of access to the Holy Land (Tigris, Mesopotamia, and the [Black Sea] Tanais), then moves westward to recall the 1522 capture of Rhodes and the 1565 Siege of Malta. Charting the realms where Ottoman initiative had checked European imperial ambitions, the play takes on an informative function akin to the *Isolarios*, even as it also serves up tantalizing morsels of Eastern exotica.

An Ottoman love intrigue likewise frames the Act 2 account of the 1570 invasion of Cyprus. Two commanders – Mustafa Pasha and Ali Pasha – fight among themselves over a beautiful Spanish captive. The

Histórica "Marqués de Valdecilla" of the Universidad Complutense de Madrid (call number FOA 245) is available on Google Books (search term "*Decimoquinta parte de las comedias*"). Juan Udaondo Alegre is preparing a much-needed scholarly edition for the edition of the *Parte 15* organized by the Prolope Group of the Universitat Autònoma de Barcelona.

aptly named Constancia resists the sexual advances of her captors and the pressure to convert to Islam. In parallel, the drama records the diplomatic negotiations that gave rise to the Holy League alliance. As in the scene above from Act 1, the melodramatic plot hewn to the popular appetite for colorful harem melodrama nonetheless provides factual information about Mediterranean geopolitics. Consider, for example, how Ali Pasha details the Ottoman naval campaign of 1570–1 that began with the attack on Cyprus:

... A Famagusta
vamos, si lo quiere el cielo:
vencida, habemos de dar
en Zante y Zefalonía,
Cherigo y Candia, en un día,
por tierra pienso allanar:
cobraré a Sopoto luego,
a Antibari y a Dulquino,
a Curcola y a Lesino,
y a Budoa pondré fuego.
En Cataro y en Corfú,
daré nunca visto espanto,
e iré desde allí à Lepanto.[25]
(... heaven willing we depart for Famagusta: this port conquered, we will alight on Zante, Cephalonia, Cerigo, Candia – pacifying within one day of landing. I will then claim my due with Sopoto, with Antibari, and with Dolicha, with Curzola, with Lesina, and then burn Budva. On Crete and on Corfu, I will sow fear as never before, and from there I will head to Lepanto.)

In poetic terms, the rapid-fire profusion of unfamiliar ports and islands trumpets Lope de Vega's artistic virtuosity. But the showmanship also serves the cause of history-making. After all, the space marked above spans the theater of warfare in the Aegean, Ionian, and Adriatic seas, where Ottoman fleets had retained the initiative from the 1522 capture of Rhodes through the 1570 siege of Cyprus. The spectator hearing these lines in the *corral de comedias* at the turn of the seventeenth century or reading the version published in the *Parte 15* anthology traces the island-to-island theater of conflict in a manner akin to a reader leafing from map to map through a Poracchi or Camoccio *isolario*.

[25] Lope de Vega, *La Santa Liga*, 264; *princeps*, in *Parte 15*, fol. 113r. My translation retains the period nomenclature, which often follows Venetian usage (Zante for Zakynthos, Sopoto for Aroania, Antibari for Bar, etc.). From the context, I take Cataro (the River Catarractus in Antiquity, now Anapodaris), as metonymically Crete (see Abraham Ortelius, *Synonymia Geographica* [Antwerp: Christopher Plantin, 1578], fol. 85r).

Finally, the love intrigues of Acts 1 and 2 fade completely as Act 3 narrates the naval battle, offering detailed recollections of the strategic deliberations of commanders on both sides of the battle lines. The account of the actual naval combat that transpired on the morning of October 7 comes into view in the drama on a canvas backdrop (*telón de fondo*). Allegorical personifications of the Holy League alliance members – Spain, Venice, and Rome – direct the public's attention to the fighting, in a manner that Frederick de Armas has shown follows Titian's iconography in the "Allegory of Religion."[26]

A rapid-fire profusion of assonant verses distills the long morning of fighting: the galleasses in the Holy League vanguard spew a blast of cannon fire; acrid smoke blackens the air; spreading flames consume men, masts, and prows alike; iron beaks shred and splinter wooden galleys; and soldiers dance to the deadly music of gunfire.[27] The battle culminates, as it did in many Spanish news bulletins, with the sight of the severed head of Ottoman admiral Ali Pasha displayed as a trophy on a Spanish pike. Had the play stopped here, the final note would most certainly have been epic. John of Austria triumphant beneath the severed head of Ali Pasha replays paradigmatic epic endings, most directly the final combat scene of the *Aeneid*, when Aeneas slays Turnus.

But the playwright steps away from this decidedly Vergilian setting to offer one final comic scene. A brief festive interlude serves as a coda, recounting John of Austria's triumphal entry into the Sicilian port city of Messina two weeks after the battle. The victorious fleet's return to its port of departure could itself have inspired scenes drawn from epic poetry. Planners of the event had devised a victory parade in the model of Roman *triumphi* of victorious generals, hailing Lepanto as a harbinger of John of Austria's "orientales victorias" (Eastern victories). Such language and imagery closely follow Vergil's prophecy of Julius Caesar's entry to Rome laden with Eastern plunder (*spoliis Orientis onustum, Aeneid* 1.289).[28] But the playwright here bypasses the epic potential of the triumphant Habsburg commander's return to Messina. Instead, his final verses are decidedly comic. Two buffoons stumble onto stage, calling out to the hero of Lepanto: "ascolta un poco, patrón / una bulleta en canzón, / e pillarete placer" ("listen a little, boss, to a ditty that'll pleasure you").[29] This closing

[26] De Armas, "Lope de Vega and Titian."

[27] Lope de Vega, *La Santa Liga*, 276 (*princeps*, fol. 120v).

[28] An idea of the elaborately scripted imperial choreography appears in the royal entry plans drafted by engraver Cesare Valentino, "Proyecto de puente conmemorativo en Mesina" (Rome, 1571, Palacio Real, IX/M/113, #106).

[29] Lope de Vega, *La Santa Liga*, 278 (*princeps*, fol. 121v).

tribute stands as far from the epic sublime as Messina is from Mataró, where drunken revelers might plausibly have yelled "ascolta" to John of Austria. Though hardly Vergilian, the comic scene rings true as it suggests how a victory parade would play out in unruly street celebrations.

In artistic terms, the revels depicted remind us that the playwright's foremost imperative was to offer entertainment for the notoriously restless urban theater-going public. Although we lack specific indices of the play's original production in a *corral de comedias* at the turn of the seventeenth century, an indication of the original performance context appears on the cover of the version published some two decades after the initial performance in the aforementioned *Parte 15* of 1621. The play's frontispiece recalls that the renowned actor and company director Baltasar de Pinedo memorably depicted the Sultan Selim ("representola Pinedo, y a Selín famosamente").[30] We are left to our own devices to imagine the blend of facial contortions, outsized gestures, and dazzling costumes through which this gifted mimic became a living figment of a mighty, but capricious sultan.

An idea of the flight of imagination elicited in the play's performances emerges in the era's popular costume books, whose woodcut images performed an analogous function to the actor in costume, inciting readers to conjure mental images of a distant and alluring Muslim adversary. Lope's imaginative depiction of the sultan and his favorite in the play's amorous scenes correlates to what we find, for example, in Cesare Vecellio's *Habiti antichi et moderni di tutto il mondo*, where a woodcut image of the Sultan Murat is paired with another of his favorite concubine. The descriptions that accompany the images urge the readers to stretch their imaginations to create a mental picture of the dazzling trappings of Ottoman grandeur of the Grand Turk. Vecellio's gloss notes that "one can neither describe nor imagine the sumptuousness of this great lord's dress"; of his favorite, the annotation stresses that "we must strive to imagine" the rich, bejeweled fabrics she dons in the seraglio.[31] On stage, the actors' gestures, sonorous monologues, and rich costumes would have assisted with this

[30] Lope de Vega, *La Santa Liga*, 231 (*princeps*, fol. 98v). On Pinedo's fame for "becoming" his characters, see the entry in Teresa Ferrer Valls, ed., *Diccionario biográfico de actores del teatro clásico español (DICAT)* (Kassel: Edition Reichenberger, 2008).

[31] Cesare Vecellio, *Habiti antichi et moderni di tutto il Mondo* (expanded ed., Venice: Sessa, 1598). See fol. 357v–358r for the Sultan Murat, and fol. 371v–372r for the favorite (Figures 13.2 and 13.3, respectively). For an annotated translation of both the 1590 and 1598 editions of Vecellio's costume book, see Margaret F. Rosenthal and Ann Rosalind Jones, eds., *The Clothing of the Renaissance World: Cesare Vecellio's Habiti Antichi et Moderni* (London: Thames and Hudson, 2008).

FIGS. 13.2 AND 13.3. "The Sultan Murat" and "The Turk's Utmost Favorite," respectively. In Cesare Vecellio, *Habiti antichi et moderni di tutto il Mondo* (expanded edition, Venice: Sessa, 1598; in Biblioteca Histórica, "Marqués de Valdecilla," BH FLL 26,141, fol. 357v and 371v, respectively). Each image in the costume book is followed by commentary that urges the reader to use the woodcut images to conjure images of luxury and adornment that defy easy description. Costumes and gestures of Baltasar Pinedo and the unknown actress who played Rosa Solimana would have added similar dimensions of marvel and exoticism for the audiences who saw *La Santa Liga* performed *c.* 1596–1603.

feat of imagination, guiding the public as they sought to envision the faraway ruler and his beguiling lover.

Shifting our perspective from ambiance and characterization to *La Santa Liga*'s plot development, some textual cues found in the version

FIGS. 13.2 AND 13.3 (*cont.*)

published in the *Parte 15* suggest how the soundscape conjured during performances in the *corral de comedias* brought coherence despite the shifts in location. For instance, in Act 2, a scene recalls how Ottoman commanders battered Nicosia's walls with heavy artillery during their 1570 siege and eventual capture of the Venetian fortress. A stage direction calls for the sound of clamoring small drums ("atabalillos").[32] Immediately following, another stage direction calls for actors to exit in "war mode," with fighting continuing off stage ("váyanse con guerra y fíngenla dentro").[33] Company members out of sight off stage would

[32] Lope de Vega, *La Santa Liga*, 339 (*princeps*, in *Parte 15*, fol. 111v).
[33] Lope de Vega, *La Santa Liga*, 260 (*princeps*, in *Parte 15*, fol. 112r).

thus be sounding percussive instruments and shouting. Immediately following, a third stage direction marks a scene change from Cyprus to Venice. It calls for the sound of music to signal the signing of the Holy League treaty ("suena música, y salen a publicar la Liga ...").[34] The off-stage echoes of percussion props conjuring menacing Ottoman drums on Cyprus would thus blend into the crescendo of sonorous brass and percussion that evoke a treaty-signing ceremony. Thus, sound effects provide continuity even as action jumps from Cyprus to Venice.

In positioning Lepanto as the culmination of a chain of events that began with the siege of Cyprus and encompassed the Holy League treaty negotiations, Lope worked within the historiographic paradigm set in the year after the battle by such chroniclers as Fernando de Herrera and Giovanni Pietro Contarini, in addition to the afore-mentioned post-Lepanto *isolarii*.[35] But within the *comedia* framework, the door opens to melodrama and parody. Readers familiar with the gruesome outcome of the Ottoman invasion, which culminated in the Venetian governor of Famagusta being flayed alive, might be taken aback by the lighter comic treatment given Mustafa Pasha in the drama. This general – portrayed in chronicles as shockingly bloodthirsty – is recast in the drama as a love-lorn suitor, distracted from his military goals by the unrequited love for a beautiful Spanish captive. These seeming lapses with the historical record incited Menéndez Pelayo's disdain. Yet a recent study by Aude Plagnard shows how such love intrigues had already become part of the Spanish epic framework. Specifically, poets of the previous generation, such as Juan Rufo and Jerónimo de Corte-Real, recounted the siege of Cyprus through a filter of the military and amorous rivalries among Turkish commanders. Plagnard notes how the insertion of love plots borrowed from other fictional genre – particularly the *romanzo* associated with Ariosto – mitigated the atrocious violence of the event. At the same time, such sexual escapades resonated with Eurocentric stereotypes of Turkish lasciviousness. She documents, moreover, how this same narrative pattern in Corte-Real influenced Cervantes' exemplary novel, *El amante liberal*.[36] Drawing from this same font of inspiration, Lope's *La Santa*

[34] See Lope de Vega, *La Santa Liga*, 260 (*princeps*, in *Parte 15*, fol. 112r).

[35] See Fernando de Herrera, *Relación de la guerra de Chipre y suceso de la Batalla Naval de Lepanto* (1572), in *CODOIN* 21 (1852): 243–382; and Giovanni Pietro Contarini, *Historia delle cose successe dal principio della guerra mossa da Selim ottomano a venetiani fino al dì della gran giornata vittoriosa contra turchi* (Venice: Francesco Rampazetto, 1572).

[36] Aude Plagnard, "Cautivas cristianas y enamorados turcos: el tratamiento épico de unos infortunios náuticos en la guerra de Chipre," *Criticon* 115 (2012): 125–45, esp. 129–30.

Liga portrays the famously bellicose Ottoman generals who devastated Cyprus as no match for love.

Yet even as *La Santa Liga* gives voice to stereotypes of Turkish leaders as both lascivious and cruel, the play's war councils offer counterbalancing glimpses of a Mediterranean *realpolitik* that transcended such demonizing impulses. In particular, the play attests to the grudging admiration among Spanish crown and military officials for Uluç Ali, the corsair-governor of Algiers. His wily escape from the devastation at Lepanto had intensified the already considerable fascination with this Calabrian-born convert to Islam. *La Santa Liga* accords him a respectful treatment as "rey de Argel," averting the negative labels of *pirate* or *renegade* one often finds in other sources.[37] In fact, the playwright's nuanced portrait of Uluç Ali pivots on a Vergilian allusion that attests to how Lope himself grappled with Lepanto's inescapable associations with epic motifs and settings.

III. EPIC TEMPTATION

In an Ottoman war council leading up to Lepanto, Uluç Ali voices his fears of the Catholic fleet which the crusading pope Pius V brought together. The playwright renders his prediction of Ottoman defeat by means of an intriguing hunting metaphor:

> El Papa es cazador, y con su mano
> nos pone como a pájaros la liga;
> los árboles y jarcias son las varas:
> caeremos, no dudéis.[38]
> (The pope is the hunter, and with his hand traps us as if spreading bird-lime, with masts and cables as the branches: we will fall, have no doubt.)

Ottoman weakness pivots on the purposeful ambiguity of the term *liga*. In the context of *La Santa Liga*, the term brings to mind the alliance of Spain, Venice, and the papacy. But within the hunting scene, *liga* also refers to the adhesive substance extracted from plants – such as

[37] Compare, for instance, the favorable view of Uluç Ali in *La Santa Liga*, 331–2 with the negative depiction of him as a "perfidious warlord" in Juan Latino's *Austrias Carmen*, ll. 1339–47, in Wright, Spence, and Lemons, *The Battle of Lepanto*. On the abiding fascination with Uluç Ali, see Emilio Sola Castaño, *Uchalí: El Calabrés Tiñoso, o el mito del corsario muladí en la frontera* (Barcelona: Edicions Bellaterra, 2010).

[38] Lope de Vega, *La Santa Liga*, 270; *princeps*, in *Parte 15*, fol. 117r. I depart from literal translation in order to clearly convey the bird-hunting metaphor, built on the double meaning of *liga*, as *league* and *bird-lime*.

the mistletoe berry or tree bark – to spread across branches in order to trap perching birds. The metaphor could have stood on its own for the many spectators or readers familiar with hunters' tricks. But for those who studied Vergil in grammar school, the hunting metaphor also calls to mind one of the most intensely studied epic passages. Specifically, the trapped bird about to fall prey to the hunter echoes the penultimate test of strength in the Trojan funeral games narrated in Book 5 of the *Aeneid*. There, Aeneas binds, then suspends a dove from the mast of a Trojan ship to serve as a target for an archery contest. The fluttering bird defies the first two archers, but then, with the third archer's masterstroke, falls lifeless. Lope adapts Vergil's scene-ending finite verb, *decidit*, with the *caeremos* that Uluç Ali intones; even more directly, Lope's line echoes the "cayó al instante muerta" that concludes this scene in the influential Castilian translation of the *Aeneid* by Gregorio Hernández de Velasco. An important signal for this emulation emerges from the verse structure, in that Lope positioned his verb "caeremos" in the same metric position at the beginning of the line as the verb *decidit* in the original Latin and the *cayó* in the Hernández de Velasco translation.[39]

This adaptation of the Trojan archery contest of *Aeneid* 5 epitomizes Lope's mastery of the Renaissance poetic discipline of *imitatio*. This term refers to the studied, even researched emulation of past masters that underpinned the most serious literary aspirations. Far from a mere copying of revered predecessors, the discipline demanded artful, even unexpected adaptations of the most prestigious bards.[40] Though there were many admired poets to emulate, Vergil's oeuvre remained the first port of call for epic poets.

But *imitatio* also required that writers come to terms with predecessors closer at hand. Quite fittingly, when firing begins at Lepanto, the playwright pays his respects to the most formidable Hispanic epic of the day, Alonso de Ercilla's monumental *Araucana*. Published in three installments (1569, 1578, and 1589), the sprawling poem recounts Spain's

[39] See Vergil, *Aeneid* 5.485–518 and fol. 107 of *La Eneida*, translated by Hernández de Velasco.

[40] On the discipline of Renaissance *imitatio*, see G. W. Pigman, III, "Versions of Imitation in the Renaissance," *Renaissance Quarterly* 33(1) (1980): 1–32, esp. 6–11; and Thomas M. Greene, *The Light in Troy: Imitation and Discovery in Renaissance Poetry* (New Haven, CT: Yale University Press, 1982). Analysis specific to poems of Lepanto appears in Davis, *Myth and Identity in the Epic of Imperial Spain*, 61–97. On Lope's overall engagement in *La Santa Liga* with diverse literary legacies, see Udaondo's "Venus y Marte en Lepanto. Amor vulgar frente a virtud cristiana en *La Santa Liga* de Lope de Vega."

seemingly intractable campaign to colonize Chile's fiercely independent Mapuche people. The second installment (1578) incorporated the chronicle of Lepanto into a prophetic vision that comes to life in a magician's magic globe. Two stanzas in particular shape *La Santa Liga*. The first transports the reader into the fog of war, at the moment where soldiers grab any loose item to use as a weapon:

Quién, faltándole tiros, luego afierra
del pedazo de remo o de la entena;
quién trabuca al forzado y lo deshierra
arrebantando el grillo o la cadena.
No hay cosa de metal, de leño y tierra
que allí para tirar no fuese buena,
rotos bancos, postizas, batayolas,
barriles, escotillas, portañolas.[41]

(One, running short on bullets arms himself with shards of oar or mast; another seizes the galley slave, unchaining him by breaking off the lock or chain. Not one item was untouched be it suited for firing, whether metal, wood, or clay, broken benches, upperworks, guardrails, barrels, hatches, portholes.)

Verses here bring to life the vicious hand-to-hand fighting that transpires as opposing galleys ram one another, giving rise to combat as intense as infantry war on land. Continuing, the poetic voice records protests that the violence defies description, but goes on to paint a vivid picture of battle:

No es posible contar la gran revuelta
y el confuso tumulto y son horrendo.
Vuela la estopa en vivo fuego envuelta,
alquitrán y resina y pez ardiendo,
la presta llama con la brea revuelta
por la seca madera discurriendo,
con fieros estallidos y centellas,
creciendo, amenazaba las estrellas.[42]

([...] It is impossible to recount the great disarray and the dizzying confusion and horrific clamor. The tow takes flight engulfed in raging fire – tar and resin and fish burning. The rapid flame intermixed with pitch rushes over the dry wood bringing fiery explosions and sparks. Rising higher they threaten the stars.)

[41] Alonso de Ercilla, *La Araucana*, ed. Isaías Lerner (Madrid: Cátedra, 1993), 677.

[42] Alonso de Ercilla, *La Araucana*, 677–8. Guillem Usandizaga demonstrates Lope's borrowings from Ercilla in "La contención del turco," in *La representación de la historia contemporánea en el teatro de Lope de Vega*.

Here, Ercilla captures the terrifying oxymoron of naval combat – fiery water. Flammable resin and galley wood have turned the Ionian seawaters into a massive fireball that threatens the heavens. Such poetic virtuosity garnered deep admiration among Spanish poets and positioned Ercilla as the gold standard for epic narrative in Castilian at the time of his 1594 death.[43]

Returning to Lope's *La Santa Liga* with this scene in mind, we can grasp how the *Araucana* informed the play's composition. Lope openly adapts the two stanzas I cite above. Italics record the direct lexical borrowings:

> Quién el pedazo del remo
> tira, o de entena quebrada,
> quién para *tirar el grillo*,
> los forzados desenclava;
> *batayolas, escotillas,*
> *barriles, bancos* y jarcias,
> *postizas y portanelas*
> rotas, sirven de arrojarlas;
> *alquitrán, pez y resina*
> *envuelta en fuego*, se clava
> entre la seca madera,
> y del agua brotan llamas.[44]
>
> (One throws a shard of oar or broken mast; another to fire a shackle unchains the galley slave; guardrails, hatches, barrels, benches and ropes, broken ballasts and portholes are rendered projectiles; tar, fish and resin engulfed in flame are engrained in dry wood and flames arise from seawater.)

The imagery and lexicon pay explicit homage to Ercilla's poetic legacy. First, soldiers improvise weapons from galley shards and chains, as seen in the first of two stanzas from *Araucana* 24 cited above. Then, the playwright renders the same watery inferno from the second stanza of the two stanzas alluded.

Yet just as the poet inclines to Ercilla to set the naval battle in motion, he swerves in another direction to conclude. Consider the contrast between Ercilla's lament for one of Spain's most high-profile casualties, Don Bernardino de Cárdenas, with the same moment rendered by Lope. Ercilla's marks the nobleman's death with consummate epic

[43] On Ercilla's prestige at the turn of the seventeenth century, see Mercedes Blanco, "Sur les frontières mouvantes de l'historiographie et l'epopée: l'*Araucana* d'Alonso de Ercilla (1569–1589)," in *La Renaissance des genres: Pratiques et théories des genres littéraires entre Italie et Espagne (XVe–XVIIe siècles)*, eds. Paloma Bravo et al. (Dijon: Editions Universitaires de Dijon, 2012), 241–65; and Isaías Lerner, "Lope de Vega y Ercilla: el caso de *La Dragontea*," *Criticón* 115 (2012): 147–57.

[44] Lope de Vega, *La Santa Liga*, 276; *princeps*, in *Parte 15*, fol. 121r.

gravitas: "fue con súbita furia en el camino / de un fuerte esmerilazo derribado, / cortándole con golpe riguroso / los pasos y designio valeroso" ("with the sudden fury of a gun blast he was struck down, cutting down with one cruel blow his deeds and lofty destiny"). Lope, in contrast, shifts from high-flown poetic lament to manifestly inarticulate laments: "¡Oh tragedia desdichada! / ¡Murió el gran don Bernardino, / Pasóle el pecho una bala!" ("Oh woeful tragedy! The great don Bernardino died his chest pierced by a bullet").[45] Coming from a poet of infinite linguistic dexterity, the avoidance of Ercilla's sublime "esmerilazo" in favor of the ruthlessly simple "pasóle el pecho una bala" is no accident. After all, Lope showed his ability to fit the *postizas*, *batayolas*, and *portanelas* of Ercilla's hendecasyllables into the shorter octosyllables that render the battle narration in *La Santa Liga*. Rather, the move from an obvious embrace of Ercilla's legacy to an equally conspicuous departure from it embodies a particularly sophisticated literary parody: Lope as poet-and-playwright gestures to Ercilla's famed poetic narration of Lepanto from *La Araucana* by pausing to note the story of one hero's death, a touchstone of epic narration. But then the playwright shifts back to the comic mode. This approach to and departure from an illustrious epic predecessor speaks to the blend of reverence, competition, and even wry humor that characterized sophisticated *imitatio*. Lope, in effect, records his debt to an illustrious predecessor, attests that he too is a poet skilled in epic narration, but then returns to the realm of comedy. That kingdom, after all, is not ruled by the epic style and artifice marked by Vergil and Ercilla, but by the prodigious comic talents of such actors as Baltasar Pinedo, the famed mimic who played Selim II in *La Santa Liga*.

To take full measure of the artistic attainment of *La Santa Liga*, we would do well to visualize an afternoon in a *corral de comedias* when the renowned actor became the Sultan Selim for one afternoon's entertainment. Yes, the staging of the great naval victory offered glimpses of the epic potential found within the chronicle of Spain's *gran batalla naval*. But the ongoing artistic quest for a great heroic poem for modern times belonged to another artistic domain. In the kingdom of comedy, power would come from the ability to engage and entertain a diverse public, conjuring an opulent seraglio, spinning a tortured love plot, and even recording the drunken ditties sung as victorious fleets returned to port.

[45] Citations from Alonso de Ercilla, *La Araucana*, 683; and Lope de Vega, *La Santa Liga*, 277 (*princeps*, in *Parte 15*, fol. 121r).

But in tandem with its function as mass entertainment, this *comedia* also attests to Lope de Vega's exploration of the medium's communication potential. By following the island-hopping plot to map out the eastern Mediterranean frontier between Catholic and Muslim maritime powers, spectators in the *corral de comedias* at the turn of the seventeenth century or readers of the version printed in the *Parte 15* of 1621 could glean specific information about contemporary geopolitics akin to that found in the post-Lepanto island atlases. The blend of entertainment and timely information about transcendent events of the day was an important impetus for the unprecedented popular success of the "new mode of writing *comedias*" (*comedia nueva*) that emerged in the late 1590s. In this sense, the likely composition date of *La Santa Liga c.* 1596–1603 is particularly significant. Written during a period of transition – both for Lope's literary career and for the configuration of royal power – the Lepanto play would prove a harbinger of Lope de Vega's conviction that the *comedia nueva* could serve to simultaneously inform the audience about important news events and entertain the notoriously restless urban playgoers. In time, even his own outsized ambitions for court patronage would become fodder for another *comedia* featuring *escenas inconexas*, *Los ramilletes de Madrid* of 1615. Like *La Santa Liga*, its seemingly disparate episodes – on closer inspection – adhere by means of a clever logic rooted in both poetics and politics.[46] I thank the dedicatee of this book for the inspiration and analytical tools needed to grasp and enjoy a kind of *comedia* that stymied even the great Menéndez Pelayo.

[46] See Lope de Vega, *Los ramilletes de Madrid*, ed. Elizabeth R. Wright, in *Parte XI* de las *Comedias de Lope de Vega*, 2 vols, coord. Laura Fernández (Madrid: Gredos, 2012), I, 469–619.

14

Staging Femininity in Early Modern Spain

Marta V. Vicente

It was a quarter past three at the Coliseo de la Cruz, one of Madrid's main playhouses in the late eighteenth century, and there were no signs that a performance was about to begin.[1] On that afternoon in early December 1764, the theater diva María Ladvenant was to perform *Andromaque*, by Jean Racine; translated into Spanish as *Andrómaca; o al amor de madre no hay afecto que lo iguale* ("Andromaque; or no other affection equals mother's love").[2] *Andrómaca* tells the story of a widow who agrees to sacrifice her virtue by marrying her captor in order to spare her son's life. After the wedding, Andromaque attempts suicide to remain faithful to her deceased husband.

[1] This case is housed in the Archivo Histórico Nacional (AHN) Consejos, Lib. 1354, 1764, fol. 1754 (Alcaldes de Casa y Corte). Although many private homes in Madrid were known to hold informal performances inside closed doors, there were only three official playhouses in Madrid: the Caños del Peral, the large Coliseo de la Cruz, which after its renovation in 1754 could house up to 1,500 spectators, and the Coliseo del Príncipe on the central street of calle Príncipe and also rebuilt in 1745; René Andioc, *Teatro y sociedad en el Madrid del siglo XVIII* (Madrid: Castalia, 1987); Phillip B. Thomason, *El Coliseo de la Cruz, 1736–1860. Estudio y Documentos* (Woodbridge: Tamesis, 2005).

[2] In June 1764, *Andrómaca* had played for thirteen consecutive days. In December, Ladvenant put this classic drama back on stage. The play had several translations during the second half of the eighteenth century and it is almost impossible to know which version was represented in a given time; *Andrómaca* was translated first by Margarita Hickey, then by José Clavijo y Fajardo, and later by Pedro de Silva; Ana Cristina Tolivar Alas, "Traducciones y adaptaciones españolas de Racine en el siglo XVIII," *Investigación Franco-Española: estudios* 1 (1988): 177–90. According to Juan Antonio Ríos, de Silva's translation turned the classic tragedy into "an absurd heroic comedy," that gained popularity on the Spanish scene; Juan Antonio Ríos, "La tragedia francesa," in *El teatro europeo en la España del siglo XVIII*, ed. Francisco Lafarga (Lleida: Universitat de Lleida, 1997), 63–85. Unless otherwise indicated, translations are my own.

The play was scheduled to start at 2:30 pm, as the laws prescribed.[3] Laws regulating theater performances also mandated a play to be divided into three acts and two in-between acts: one being the comic short piece known as *sainete*, and the other the popular *tonadilla*. The *tonadillas* were sensual, satirical, and risqué songs that were the rage in the Spanish theater of the late eighteenth century. The short *tonadilla* (from five to ten minutes) concluded with a dance and told a story, which was usually a satirical comment on social events. At the Coliseo de la Cruz, Ladvenant herself sang the eagerly awaited *tonadilla*. The second interlude featured Ramón de la Cruz's *sainete* "El Sarao" ("The Shindig").[4] Among the eager spectators at the Coliseo de la Cruz was city officer Don Benito Balado de Ponte, while the constable Don Juan Estevan de Salaberri attended the performance in the neighboring theater of Coliseo del Príncipe.[5]

Don Benito was growing impatient with the delay of the performance and, without anyone's assistance, he hurried backstage to the actors' dressing rooms.[6] There, he found María Ladvenant, who since her arrival from Valencia in 1759 had taken the Madrid theater scene by storm. In 1764, at twenty-four years of age, Ladvenant, also known as *La Divina* (The Divine), was the company's *autor*.[7] This meant she was in charge of her own theater company and received orders from the *Junta de Teatros* (Board of Theaters) regarding the specific requirements for each performance. La Divina directed rehearsals and actors' auditions, worked as the company treasurer, and, in general, approved everything that took place on stage.[8] Ladvenant was also an accomplished actress in comedy, drama, and even in the new popular genre of *tonadillas*.

[3] Municipal laws regulating theater entertainment in Madrid decreed the times and places where performances had to take place: 2:30 pm in winter and 4:00 pm in summertime; Emilio Cotarelo y Mori, *Iriarte y su época* (Madrid: Sucesores de Rivadeneyra, 1897), 68–9; Emilio Cotarelo y Mori, *María Ladvenant y Quirante: Primera dama de los teatros de la corte* (Madrid: Sucesores de Rivadeneyra, 1896), 171.

[4] Cotarelo y Mori, *María Ladvenant*, 166–71. According to Ramón de la Cruz those in-between acts were intended to work as a contrast to the play represented, thus easing the performance.

[5] Don Benito was officer of the *sala de alcaldes de la casa y corte*.

[6] Actresses regularly complained dressing rooms at the Coliseo de la Cruz were too small, and with little privacy; Thomason, *Coliseo de la Cruz*, 108.

[7] The *autor de comedias* was virtually the director of a theater company. She obtained the permits for the performance, hired actors, distributed their roles, arranged their wardrobe, and in general orchestrated the entire production of the play, from beginning to end; John Allen, "Gaspar de Torres, autor de comedias," in *Comedias y comediantes: Estudios sobre el teatro clásico español*, eds. Manuel Diago and Teresa Ferrer (Valencia: Universitat de València, 1991), 337–48.

[8] Cotarelo y Mori, *María Ladvenant*, 63.

FIG. 14.1. "María Ladvenant, first lady of the Spanish theater, born in Valencia, died in Madrid on April 1st, 1767 at the age of twenty-five." Museo de Historia de Madrid

María Ladvenant was undoubtedly the star of her company, leading other popular actresses of Madrid's theater scene, such as Paula Martínez Huerta, second actress in the play under contract in Ladvenant's company during the theater season of 1764–5.[9] Proof of these actresses' popularity was the fact that every night soldiers had to escort Paula as well as Ladvenant's sister, Francisca, also a member of the company, as they left the theater so their fans (or *apasionados*) would not harm them. The three

[9] *Ibid.*

women benefited from a celebrity culture that granted them unusual freedom, and which sometimes allowed them to make demands that authorities considered unacceptable. In fact, the uppity attitude of the actresses was well known among authorities and the public. Only a year earlier, in 1763, the great actress, *tonadillera* and professional rival to Ladvenant, Mariana Alcázar, had defied municipal authorities by demanding to take a leave from performing and still retain her place as "first lady" in her company. She also demanded a role for her sister, Antonia. Being denied both of these favors, Alcázar refused to perform, thus breaking her contract, for which she was given house arrest.[10]

With this in mind, knowing the reputation of these "first ladies" and that he was in the dressing room of the most revered artist of the time, nevertheless Don Benito asked Ladvenant to start the comedy right away "because it was time." The officer reported La Divina responded "with great contempt" and told him "to shut up and that another actress, named Paula, had not arrived yet."[11] But when Paula arrived fifteen minutes later, still the play did not start. Don Benito went to the dressing rooms a second time. There, he found Ladvenant, Paula Martínez Huerta, and a third actress, Casimira Blanco, known as *La Portuguesa* (The Portuguese), famous *tonadillera* and muse to the well-known writer José Cadalso.[12] The women told Don Benito to "deal with it" and started singing a *tonadilla*, mocking the officer. La Portuguesa pointed out: "El Señor must have quarreled at home and now he wants us to pay for it." Finally, the play started at 3:54 pm, almost an hour and a half after its scheduled time.

Don Benito reported to the constable, and a few days later, La Divina received notice to "from now on obey and keep exactly what is mandated by his Majesty."[13] The notice probably included a copy of the edict published a year earlier, on April 12, 1763, that listed all "the precautions ordered to be observed" by actors and *autores*. In addition to the

[10] *Ibid.*, 105–09.

[11] Born in Valencia, Paula Martínez Huerta arrived in 1761 at Madrid from Cadiz to work at the company of María Hidalgo. After the death of Ladvenant, Paula became the first lady of the company of Nicolás de la Calle. She also had her sisters, Pepita and Francisca, in Madrid working as actresses as well as her brother, José Martínez, and her father, José Martínez Huerta; Emilio Cotarelo y Mori, *Don Ramón de la Cruz y sus obras: ensayo biográfico y bibliográfico* (Madrid: J. Perales y Martínez, 1899).

[12] Cotarelo y Mori, *Iriarte y su época*, 85.

[13] AHN Consejos, Lib. 1354, 1764, fol. 1754 (Alcaldes de Casa y Corte), fol. 1759 "Prevengase a Maria Lavenan que en adelante cumpla y guarde exactamente lo mandado por SM 3-XII-64."

scheduled time of the performance, these precautions emphasized a few other rules that the three actresses had certainly ignored many times. First, in each representation, and particularly in the *sainetes* and *tonadillas*, the company author had to prevent the actors from performing indecent and provocative dances and songs that "could cause the smallest scandal." Second, actresses had to show decency in their wardrobe, being allowed to dress as men only from the waist up. Third, actresses and actors had to refrain from performing "encores of dances and *tonadillas*," and from yielding to the request "that a particular actor or actress would perform them." Moreover, the edict explained, the constable and his officer were entitled to use whatever measures were necessary to restrain those theater aficionados.[14]

Don Benito's encounter with the three actresses and the edict both suggest the difficulties authorities faced when trying to shape theater performances into a spectacle that could serve as an educational tool for the public. The theater was "after the pulpit" the best place from which to influence people.[15] From this lay pulpit, the writers and intellectuals of the Spanish Enlightenment, known as *ilustrados*, were hoping to transmit to the masses the "Enlightenment project": a way of thinking and seeing the world "without any other light besides one's own reasoning."[16] Scholars have argued whether and when such an Enlightenment project became part of the intellectual legacy of Spain.[17] Yet, the Spanish Enlightenment as a "way of thinking," more than a particular philosophical, ideological, or even political movement, shared striking similarities with its European counterparts.[18] Spanish *ilustrados*, much as the French *philosophes*, had as one of their main goals to liberate individuals from the yoke of their uncontrolled emotions by using their reason. The use of reason also had a public dimension: all festivities and distractions had to educate people on how to control their own emotions and urges by presenting a moral message intended for social utility. In

[14] The 1763 edict was merely repeating and expanding its previous 1753 version; AHN Consejos, Lib. 1340, 1753, fol. 189 (Alcaldes de Casa y Corte).

[15] Nicholás Fernández de Maoratín, quoted in Jesús Rubio Jiménez, "Censura y teatro en el siglo XVIII o la verdad de la mentira," *Cuadernos de Ilustración y Romanticismo* 19 (2013): 53–84.

[16] Benito Jerónimo Feijóo, *Teatro crítico universal* (Madrid, 1726–40), v.7, prologue to the reader.

[17] Anthony Pagden, "The Reception of the 'New Philosophy' in Eighteenth-Century Spain," *Journal of the Warburg and Courtauld Institutes* 51 (1988): 126–40.

[18] Joaquín Álvarez Barrientos, *Ilustración y Neoclasicismo en las letras españolas* (Madrid: Síntesis, 2005), 15.

contrast to the Baroque theater, with its exuberant imagination, the new theater of the eighteenth century had to display a rightful use of imagination towards creating a piece that would educate the audience in good moral principles.[19] Nothing was superfluous. As Jean-Jacques Rousseau put it in his *Letter to d'Alembert*, amusements "are only permissible insofar as they are necessary." To him, "every useless amusement is an evil for a Being whose life is so short and whose time is so precious."[20] Spanish *ilustrados* shared that very same concept; all theater had to be morally useful, or else it bore the evils of idleness and excess.

The theater performance that the *ilustrados* envisioned had to construct a credible fantasy, with characters behaving according to their social status, age, and sex, which guaranteed the appearance of reality. In 1694, José Barces had already pointed at the pedagogical aim of the theater play and the role of imagination that must accompany a perception of reality.[21] To achieve this, the narrative of the play had to be presented so the performance appeared as natural and logical. If imagination was uncontrolled and unleashed, it bore the "false apprehension, judgment or discourse of something that has not reality or fundament." This could jeopardize the crucial "good taste," a paramount concept in the thought of the Enlightenment. Good taste allowed audiences to choose between what was truthful and morally correct from what was false and superficial. As the eighteenth-century Spanish author and theorist Ignacio de Luzán pointed out, it all came in how closely the artistic expression was able to imitate nature while controlling reason.[22] This was an ultimate aim of the theater and its educational aspect: to show its audiences that the social interactions narrated on stage were only part of the natural world.

A credible fantasy on stage had to follow what scholars have labeled neo-classic lines, whereby the narrative of the play displayed the three classic units of action, place, and time. A play had to have a single narrative, occurring in one geographical place and within a limited time. In fact, to eighteenth-century critic and member of the Junta de Teatros, Santos Díez González, the essence of good theater was found in imitating one sole action that offered a credible example for spectators to follow

[19] Begoña Lolo, *Paisajes sonoros en el Madrid del siglo XVIII: La tonadilla escénica* (Madrid: Ayuntamiento, 2003); Jesús Pérez Magallón, *El teatro Neoclásico* (Madrid: Laberinto, 2001), 16–17.

[20] Jean-Jacques Rousseau, *Letter to d'Alembert and Writings for the Theater*, trans. Allan Bloom, Charles Butterworth, and Christopher Kelly (Hanover, NH: University Press of New England, 2004), 262.

[21] Peréz Magallón, *Teatro neoclásico*, 25.

[22] *Ibid.*, 229.

in their private lives.[23] Artificiality had to be avoided, as well as breaking the action into two scenes happening at different times, which could only distract the spectator from the content of the play, diluting its emotional force. The illusion of reality also demanded all action to occur within one short period of time. The content of a play was also meant to promote decency or *decoro*, verisimilitude, and imagination linked to reason, all of which fostered good taste.[24]

In this way, the theater became the expression of the core ideals of the Enlightenment in trying to find the perfect relationship between nature and society. Nature in its nakedness was not something the *ilustrados* took as a model; instead the nature they had in mind was a domesticated one. When the theater imitated nature, it represented all its beauty as well as usefulness.[25] Equally what was represented expressed social relations and models that were natural, such as how women and men had to behave in society, a direct translation of how males and females of the different species interrelated and acted in nature. Consequently, the ideal character of a woman in comedy and drama was in essence displaying the ideal female virtues: loyalty, honor, honesty, obedience, and resignation. No matter how complex the story was, it had to display the ideal social relations men and women were expected to follow.

Ilustrados faced the challenge of marrying the goals of the Enlightenment project with the reality of the theatergoers who in viewing the performance were ultimately seeking a pleasurable experience. In the second half of the eighteenth century, the theater had become a mass entertainment. After decades of limited performances throughout Spain as direct consequence of the religious campaign that considered theater to be a pernicious form of entertainment, in the 1760s theater performances were back on public stages.[26] This recovery accompanied a series of proposals for reforms to guide performances into the educational tool that the theater was meant to be.[27] The criticism focused first

23 Josep María Sala Valldaura, *Los sainetes de González del Castillo en el Cádiz de finales del siglo XVIII* (Cadiz: Fundación Municipal de Cultura, 1996), 34.

24 Peréz Magallón, *Teatro neoclásico*, 43.

25 *Ibid.*, 32.

26 The popularity of the theater, and in particular of the spectacle of *sainetes* and *tonadillas*, made minister Pedro Rodríguez de Campomanes issue a law in 1767, authorizing theatrical representation in all Spain; Antonina Rodrigo, *María Antonia, la Caramba: El genio de la tonadilla en el Madrid goyesco* (Madrid: Prensa Española, 1972), 39.

27 José Mariano Nifo, *Idea política y Cristiana para reformar el actual teatro de España*, (Madrid, 1769); see René Andioc, "De la comedia nueva a la reforma del teatro," *Cuadernos de Ilustración y Romanticismo* 19 (2013): 7–25.

on the performance of *autos sacramentales* (Eucharistic plays), in which, according to critics, actors and actresses of dubious personal reputation were representing biblical characters. These religious plays were prohibited in 1765. In 1788 another kind of play, claimed to display uncontrolled imagination, the comedies of magic and saints (*comedias de magias y santos*), was also banned. All these performances, where the observer demanded a complete spectacle able to "charm the eyes, and to an extension their ears," revealed a view of the theater incompatible with the ideal of the *ilustrados*.[28]

The popularity of the theater in the late eighteenth century had also turned it into a visual spectacle with a form and content difficult to control. Many actors and actresses became the equivalent of modern celebrities, whose whereabouts audiences eagerly followed, and whose personas created curiosity, praise, or condemnation, but never indifference. This was not a novelty; popular actors in the Spanish theater of the seventeenth century, such as the well-known Juan Rana, had been praised and criticized by writers of their time.[29] Long before the eighteenth century, critics had regarded the theater as an arena, like the pulpit, where "the people" could be instructed and educated. There were, however, important differences between the eighteenth-century theater scene and previous ones. First, eighteenth-century theater productions placed greater emphasis on the regulation of the theater as part of the educational program of the Enlightenment. Second, it was during the late eighteenth century that small theater pieces such as the *sainetes* and *tonadillas* found unprecedented popularity and threatened to jeopardize the didactic efforts of the *ilustrados*. The writer and personal friend of María Ladvenant, José Clavijo y Fajardo, believed that for an uneducated public, a good theater performance had the power to "reform a country and keep it reformed while the one that offers bad examples is able to pervert it or keep it perverted."[30] The short *sainetes* and *tonadillas* generally fell into the category of "bad examples" – first, because they were filled with elements that interfered with the scripted story, and, second, because the great popularity of these short pieces made the play itself and its contents irrelevant. In the *tonadilla*, *La Maja Limonera* (The Maja Lemon Seller) (*c.* 1765), when the maja, that idealized character that represented the essence of the Spanish character, tells her audience that a majo, the

[28] René Andioc, *Teatro y sociedad*, 49.

[29] Peter Thompson, *The Triumphant Juan Rana: A Gay Actor of the Spanish Golden Age* (Toronto: University of Toronto Press, 2006).

[30] José Clavijo Fajardo, *El Pensador* (Madrid, 1762–7), vol. IX.

maja's male counterpart, has been chasing her, she does it in a sensual and cheeky way: "A certain majo/ Has been chasing me/ The more I watch him, the more I see him / In case he comes closer/ I'll sit here/ See my delicious lemons/ Come on, boys/ They are big/ Like horses/ Who will take them? / Come on/ They are running out/ And they are very good."[31] At the end of the *tonadilla* the audience, likely still smirking at the thought of the *tonadillera*'s lemons being "as big as horses," felt quite unready to go back to the more serious and educative play at hand.

The *tonadillas* became part of what Theodor Adorno has labeled the "culture industry" of the eighteenth century and the Enlightenment project.[32] This popular culture began in the seventeenth century, with the growing availability of gazettes and newspapers that fostered the exchange of opinions. By the eighteenth century, these publications had established a life of their own.[33] In his *Fábulas gacetales* (*Journalistic Fables*) (1759) Benito Jerónimo Feijóo warned the reader that gazettes bore little fidelity to reality and that newspapers could spread lies.[34] Gazette writers did not check their sources and allowed themselves to be dragged into sensationalism. To Feijóo, writers of gazettes chose news "because of what they have that is surprising and flashy, because it satisfies the appetite for what is new and foreign, taking advantage of the delight that the outrageous news produce." By falling for these tricks, the public became the *vulgo* (the populace), uneducated masses unable to exercise the crucial good taste. The public could be educated, but the *vulgo*, drawn by low passions and instincts, could only be worthy of "deception."[35]

The Enlightenment project was aiming at turning "the populace," into the educated public. However, paradoxically, such an attempt threatened to destroy the original value of the theater performance.[36] It became a fruitless effort, since most *ilustrados* agreed that "the populace" did not have "good taste" and preferred *tonadillas* and *sainetes* to plays that would educate them. To Juan Sempere y Guarinos, "the vulgo only likes light roles that distract them with the novelty and do not burden

[31] "La Maja Limonera" (*c.* 1765), *tonadilla* transcribed in José Subirá, *La tonadilla escénica* (Madrid: Tipografía de Archivos, 1930), 28–9.

[32] Theodor Adorno, *The Culture Industry: Selected Essays on Mass Culture* (London: Routledge, 2001).

[33] Francisco Sánchez Blanco, *La mentalidad Ilustrada* (Madrid: Taurus, 1999), 280.

[34] Feijoo, *Teatro Crítico Universal*, VIII, 5.

[35] Sánchez Blanco, *Mentalidad Ilustrada*, 291.

[36] Álvarez Barrientos, *Ilustración y Neoclasicismo*, 144.

them with reasoning."[37] What writers thought entertained the masses – the excess, expressed in the visual spectacle of the sensual body, and the laughter of the *tonadillera* – only told a story that did not imitate what was natural and realistic. Instead, it constructed a reality that confused the imaginary with the ordinary. Scenes did not follow a principle of causality, but worked in juxtaposition in a story plot where the coherence of argument was not important. Many *tonadillas* displayed a narrative plot that combined different stories at once, some of which were imaginary and humoristic. In "El Canapé" ("The Loveseat") (1769), the *tonadillera* sang: "I have purchased a loveseat/ This morning/ And it has told to me every little thing/ That goes on/ About its first owner/ Who was the mistress/ Of a certain suitor/[and she was] much courted." The *tonadillera* exclaimed not only maids knew about all the gossip in town, but "now even seats tell all."[38] Next, she was ready to tell her audience all the indiscretions her newly acquired loveseat had revealed to her.

Tonadillas like "El Canapé" displayed a perspective of the theater that disregarded a rational view of life to favor a scenography that was entertaining and promoted what was surprising and unexpected. They disregarded the norms of the plot narrative, telling stories that were disconnected and happened in different places and different times. Moreover, the *tonadilla* fulfilled an "awaited desire," as the Latin root of the word "spectator" indicates.[39] To please the audience, the *autor de comedias* favored performances of *tonadillas* that contained popular music and comic scenes to provoke "excessive laughter," a subversive aspect of the entertainment piece that seemed to go against the good taste, so dear to the *ilustrados*.[40] The performance was crucial, and a skillful *tonadillera* could change just with her intonation the meaning of the lyrics she sang. Thus, the *tonadillera* María Antonia Fernández, better known as "La Caramba," became famous for her provocative and sensual singing of the word "caramba" (meaning "good heavens," but also an euphemism for "dammit"), with which she always made sure to conclude her *tonadillas*.[41] Accordingly, the Junta de Teatros prohibited the inclusion of the word "caramba" in a *tonadilla*, when "the interpreter made a pause

[37] Juan Sempere y Guarinos, *Ensayo de una biblioteca española de los mejores escritores del reinado de Carlos III* (Madrid: Imprenta Real 1785), 38–39.

[38] "Ya hasta los asientos todo lo parlan."

[39] Sala Valldaura, *Los sainetes de González del Castillo.*

[40] *Ibid.*, 36.

[41] Antonina Rodrigo, *María Antonia La Caramba el genio de la tonadilla en el Madrid goyesco* (Motril: Ayuntamiento, 2011).

between the first and second syllable," a pause that probably brought up the excitement and laughter of the *tonadillera*'s audience.[42]

A theater performance had to show "clarity" and "verisimilitude" and follow "rules," the most important of which was the imitation of nature to make sure a work of art was "reasonable."[43] Instead, the *tonadillas* went against all these precepts. The *tonadillera* and her performance promoted an imagination with no "reality or fundament," bringing on all the evils of a "useless amusement." Frivolous *tonadillas* were performed along with plays such as *Andrómaca*, whose plot was to transmit Greek values that the *ilustrados* championed: female virtue and maternal dedication. *Ilustrados* believed when the *tonadillas* were performed along with a Greek tragedy, they could confuse audiences who would become unable to discern which one was the model to follow and which to disregard. This was particularly true when the actress who performed the role of a virtuous mother and wife was the same who sang, sometimes cross-dressed, about love and lust.[44] Moreover, the well-known reputation of actresses such as Ladvenant as promiscuous lovers and less-than-devoted mothers could only make matters worse. And yet, it was this part of the performance that people came to see and enjoy. As Leandro Fernández de Moratín, son of Nicolás Fernández de Moratín, argued, the spectator slept through the performance of the truthful art form, only to wake up "to see some voluptuous dance, light, French in other words." According to Moratín, both the audience and the actors believed the play was unimportant and instead promoted all the artificiality of the theater scene, from the illumination and decoration to the songs and dances.

The writer and journalist Francisco Mariano Nipho agreed with Leandro Fernández de Moratín that theatergoers were only interested in seeing and enjoying and not learning. They went to the comedy just to see the "tonadillas that arouse gently and dangerously the sensual fire ... they excite the inner tickling of our sickening and fleshy satisfaction; many go, and to tell the truth, almost everyone goes."[45] For Tomás Sebastián y Latre, *sainetes* and *tonadillas* spoiled the illusion of the theater play,

[42] Biblioteca Nacional (BN) Mss. 14076 "Censura de la palabra caramba."

[43] David Gies, *A History of Theater in Spain* (Cambridge: Cambridge University Press, 2012).

[44] Ladvenant performed the *tonadilla* written by Luis Misón, *Los Jardineros* (The Gardeners) (1761). She sang it with Mariana Alcázar; one of them was crossed as male gardener while the other performed as his female lover; *Spanish Theater Songs: Baroque and Classical Eras*, ed. Carol Mikkelsen (Van Nuys, CA: Alfred Music, 1998), 18–21.

[45] Andioc, *Teatro y sociedad*, 50.

"and the poet loses then all the work that it had taken him to arouse (*imflamar*) the spectator in each act." He sought a solution to this problem by proposing that the *sainetes* and *tonadillas* and all short pieces be performed at the end of a play as a form of finale.[46] His proposal was never put into practice.

When first performed on stage, in 1757, the *tonadilla* was a simple song and demanded a simple staging with little room for complicated scenes and changes to the neo-classic demands of action, place, and time. However, as the *tonadilla* gained autonomy it became a short theater piece in itself with a more complex story line. Actors and actresses felt free to improvise to deliver what the public liked the most: a mixing of the story with real events of daily life. The *tonadilla*, therefore, was able to construct a fictive reality by putting together parts whose sum totaled a fantasy. It rejected the construction of a linear narrative where virtue is rewarded and a separation of the audience from the action. Instead, in the *tonadilla* the plot moved back and forth. In the *tonadilla* one could find "a storm, and then a court martial, followed by a dance, and later a funeral."[47] Moreover, the narrative of the *tonadilla* also challenged the separation of the private life of the *tonadilleras* from their onstage character.

During her performances, the *tonadillera* blurred the line between her professional and private life. She inserted references to real incidents in her performance, going back and forth between the pretended and the real self. By creating a story that mixed up their personal life with current news and their audiences' expectation of what was exciting and gossipy, the *tonadilleras*' spectacle questioned the role of imagination in creating a credible fantasy about real life. It was the construction of the fantasy of a hypersensual woman that the public entertained and approved. The dangers of the *tonadilla* were not only in the fantasy of the story told but also in the fact that such fantasy became embodied in the sensual body of the *tonadillera*. The *tonadillera* engaged verbally and sometimes even physically with her audience, taunting them and asking them to respond. Her movement, clothing, and laughter all evoked the fantasy of the sexualized female; a fantasy that questioned the boundaries between what was real and imaginary.

[46] Tomás Sebastián y Latre, *Ensayo sobre el teatro español* (Zaragoza, 1772), 36.
[47] Leandro Fernández de Moratín, *La Comedia Nueva* (Barcelona: Crítica, 1994 [1792]).

Clergymen had long seen in actresses confirmation of their longstanding denunciation of the immorality of the theater, which had led to temporary prohibitions of theatrical representations, while also questioning women's most valued virtue: modesty. The *tonadillera* went further to upset the values of modesty. With her artifice she questioned the premise that society and its artistic expression must imitate the harmony in nature. With their provocative attire, flashy jewelry, and sensual dances, actresses and *tonadilleras* like Ladvenant and La Portuguesa portrayed an exaggerated and sexually dangerous type of femininity that also contradicted the ideal of womanhood prescribed by the *ilustrados*. *Tonadilleras* expressed through their bodily performance – with daring vocabulary, luxurious and risky attire, sensual dances and uncontrolled laughter – a femininity that was sexualized and non-maternal. It was a far different notion of beauty from moral rightfulness and usefulness. The beauty and appeal of the actress was linked to uncontrolled emotion and excess expressed in the laughter that the spectator identified with the *tonadillera*.

The most threatening aspect of the *tonadilla* was the way it used imagination as a tool, guided not by reason but by irrational emotion and sensation. This excessive aspect of imagination was best expressed in the use of laughter, in particular of the *carcajada*, the outburst of laughter. The carcajada represented everything the Enlightenment project was not: the relaxed lack of order that could upset the hierarchy at any time by making irrelevant all what was put in front of the audience. The *tonadilleras*' misuse of laughter, transformed into *carcajada*, expressed the danger of uncontrolled emotions. The *ilustrados* approved of laughter, or *risa*, as a natural expression of an emotion. However, they condemned the *carcajada* as an expression of excess and uncontrolled laughter, thus unnatural and artificial. In his *Desengaños al teatro español* (*Disenchantments to the Spanish Theater*) (1763), Nicolás Fernández de Moratín criticized the indiscriminate use of laughter as a form of excess, defending instead the use of what is "natural" in all forms of art. The Spanish Royal Academy of Language (*Diccionario de Autoridades*) echoed the *ilustrado*'s point by modifying the definition of the term "carcajada." The 1729 edition of the *Diccionario* had a merely physical and almost anatomical definition: the *carcajada* was "the sound that the sudden and impetuous laughter makes when it expels the air from the lungs." Decades later, in 1780, the new definition contained a moral connotation absent in 1729. The *carcajada* became "the strong and excessive laughter." The adjective "excessive," connoting uncontrollable or poorly subdued emotion, was an important addition.

The excessive component made the laughter of the *tonadillera* dangerous. As the French feminist Hélène Cixous suggests, with her laughter a woman is shaking "the old grounds of logic, overcoming the heaps of reading, blowing the law to pieces, and making rubble of man's property."[48] Like the laughter of the Medusa, the *carcajada* of the *tonadillera* is excessive and foments desire. To Cixous, with their laughter women return to their bodies. And, I would add, by performing with their bodies in such a hyperfeminine way (which includes the adornment and the laughter) the *tonadilleras* questioned the place of women in the Enlightenment project. The excess of the *tonadillera* was not only unnatural, but also against nature.

The performance of the *tonadillera*, from the uncontrolled laughter to the sensual dances, seemed to endanger the Enlightenment project. However, the criticism against the *tonadillas* and the actresses who performed them was not unanimous, and many *ilustrados* became part of the paradox that resulted from praising the ideal values of motherhood and virtue while raging about the *tonadillera* of the day. Many *ilustrados* fell under the spell of the *tonadillera* and her magnetic charm, that acted not so much as a Medusa but as sirens luring those unhappy *ilustrados*. Cases like the famous minister and writer Gaspar Melchor de Jovellanos were not uncommon. Of him it was said that "without missing his first obligation of conferences and literary exercises, a sound, clear and tuned voice was calling him to imitate the arias and dances of María Ladvenant, while accompanying himself with a guitar."[49] This attraction, in part, corresponded to the fact that, for some, the *tonadillera* represented the true Spanish character, which was far from the aristocratic affectations of the French. Ladvenant was one of those *tonadilleras*, although in Juan de la Cruz's famous engraving of La Divina, dressed in "theater costume in the old Spanish style," far from the humble ideal of the maja, the diva from Valencia appears as richly dressed in a fashion that still brings to mind the French style.

Ilustrados like Tomás Iriarte or Félix María de Samaniego praised composers of *tonadillas* like the well-known Luis Misión as creators of a "national music."[50] And if composers could create a music that was national

[48] Hélène Cixous, "The Laugh of the Medusa," *Signs* 1(4) (Summer 1976): 875–93; originally published in French as "Le rire de la Medusa" translated by Keith Cohen and Paula Cohen.

[49] Vicente Cueva Díaz, "Jovellanos y la música de su tiempo,"*Boletín Jovellanista* 5 (2004): 38.

[50] Subira, *La tonadilla escénica*, I, 17.

FIG. 14.2. *Theater costume in the old Spanish style* (portrait of María Ladvenant), by Juan de la Cruz Cano y Olmedilla, Museo de Historia de Madrid

and patriotic, the *tonadilleras*' depiction of the majas and majos seemed reason enough to allow and encourage the performance of *tonadillas* in Spanish theaters. As the *tonadilla* "De fuera vendrán" ("They will come from abroad") sang: "The foreign ways/ Corrupt Spain so much/That only majos and bullfights/ Show we are Spaniards."[51] It was the maja in the *tonadillera* that allowed *ilustrados* to tolerate their excesses. On one hand, the *tonadilleras* symbolized a form of counter-enlightenment; the spectacle of their body, voice, and movement represented everything the *ilustrados* criticized and feared would jeopardize their educational program.[52] On the other, the *tonadillera* could also symbolize this *españolidad*; because of their popularity, the *tonadillera* could represent – even promote – the true view of the Spanish woman for the *ilustrados*.[53]

In 1790 the minister Gaspar Melchor de Jovellanos summarized all these points. To him the "true theater" (*la verdadera comedia*) had not only to elevate the Spanish above all that was foreign, but also "to improve the education of the citizens" and promote "respect to all hierarchies, to laws and authorities, to conjugal fidelity, parental love, and the children's tender obedience to their parents."[54] To do so he recommended eliminating from all plays anything that would not follow this agenda. In particular, he recommended removing "all the entremeses, sainetes and tonadillas, as well as the puppets, clowns, and magic lanterns, all which are examples contrary to virtues that place as heroes rogues and explore the laughter of the 'low comedy' of vulgar character."[55] To critics like Jovellanos, the *tonadilleras* could corrupt the viewers' hearts by promoting and encouraging excessive and dangerous emotions.

Besides the content and form of the *tonadilla*, its critics focused in particular on the relationship between actress and public. The performer was nothing without her audience. Clavijo y Fajardo expressed this in 1762 when he wrote in *El Pensador* (*The Thinker*): "her self-esteem, her

[51] This *tonadilla* was composed in 1797 by Pablo del Moral; Lolo, *Paisajes sonoros*.

[52] On the Spanish Counter-Enlightenment, see Paul Ilie, *The Age of Minerva (vol. I): Counter-Rational Reason in the Eighteenth Century: Goya and the Paradigm of Unreason in Western Europe* (Philadelphia, PA: University of Pennsylvania Press, 1995).

[53] Jesús Torrecilla discusses how travelers to Spain who left a written record of their perceptions reinforced this aim of the *ilustrados* to promote such "alegria desfrenada de las clases bajas ... el revuelo, la chispa, el colorido y los bailes apasionados," as elements that characterized such *españolidad*; *España exótica: La formación de la imagen española moderna* (Boulder, CO: Society of Spanish and Spanish-American Studies, 2004), 7.

[54] Elisabeth Le Guin, "'Tonadillas' and 'diplomacia' in Enlightenment Madrid," *Early Music* 40(3) (August 2012): 421–40.

[55] Gaspar Melchor de Jovellanos, *Memoria para el arreglo de la política de los espectáculos y sobre su origen en España* (Madrid, 1790).

beauty and adornment is not a small thing: with all, if it does not have an end to which it is directed, if there isn't anyone to please with the adornment and beauty, one and the other would be dull, and it would be just the same if she were one-eyed, cross-eyed or disheveled."[56] A key to the success of this relationship was the construction of the *tonadillera*'s persona on and off stage as an accessible celebrity. The public and private life of *tonadilleras* like Ladvenant merged, and her performances played into this duality. Theatergoers expected La Divina to engage in dialog with the public and the public to interact with the *tonadillera* by offering approval of the performance.

She would sing:

> "Ay, dear gentlemen/ Beloved of my soul pay attention/ Di, di, di, di, di, dirindínsi, si/ To entertain you I sing it with pleasure dear gentlemen of my heart/ Di, di, di, di, di, dirindín (si, si)/ Forgive me if you don't like it, sir/ You don't need to get angry/ I have already lingered for too long/I must go it is late."

She would finish her short act with a *seguidilla* (Spanish folk dance), always interacting with the public:

> "Here I finish my tonadilla/ Dear gentlemen, clap your hands, since I earn my living with this/ To feed my poor family."[57]

Everything about the *tonadillera*'s appearance and everything she taught and transmitted – the uppity attitude, her movements, even her laughter – were visual elements that constructed her femininity. Such was the power of her performance that the word *tonadillera* became in the late eighteenth century synonymous with "woman." It was a particular type of woman, marked by physical traits from the wardrobe to bodily movements. In 1722 cardinal Luis Belluga stated that the problem was not simply that women showed or even insinuated the flesh in between their layers of clothing. The problem was also in the clothing itself: "with the various artifices with which [women] adorn themselves ... with the multiplicity of *sainetes*, laces and ribbons with which they adorn their sleeves, wearing them naked, showing the lows [of their dresses], showing not only their shoes, but much more. I would like to ask weren't they more modest dress[ed] as men, with their capes and hats the way many do nowadays in Spain."[58] The frills of feminine adornment made Belluga long for cross-dressing.

[56] José Clavijo Fajardo, *El pensador* (Madrid, 1762), 110.
[57] Biblioteca Histórica de Madrid. Mus. 100–06.
[58] Luis Belluga y Moncada, *Contra los trajes y adornos profanos* (Murcia, 1722), 787–8.

In an undated edict of the first half of the eighteenth century, Inquisitor-General Francisco Pérez de Prado also attacked entertainments that acted as "incentives to sensuality." To the inquisitor the spectator became excited by the "lascivious movements and shakes unworthy of the Christian century, and with the scandalous touching between contact of men to women, either on the hands, of the embracing of the arms, chest to chest, on the back, or on the knees … and other frivolous inventions; with which the virginal modesty of the maids and the honor of the married women are corrupted."[59] Their vocabulary made open sexual references, their appearance and attire upset proper social belonging. This formula of the sensual singer born to please the audience became so successful that in 1767 writer Bernardo de Iriarte stated: "We can consider the representation of our plays a mere pretext for the *sainetes* and *tonadillas*."[60] In fact, authors intentionally divided the play into more acts than laws allowed so they could have more *sainetes* and *tonadillas*.[61] By 1767, the *tonadilla* had already taken over the theater world of the main Spanish cities. The theater performance per se had taken on a life of its own that was designed to please rather than to educate, and by doing so it defied the Enlightenment project's vision of the theater as an educational tool.

Yet the threatening femininity displayed by the *tonadilleras* was also a construction of the *ilustrados*. The celebrity aspect of the *tonadillera*, which certainly could be subversive and transmitted to younger generations, was only achieved by a few. It was an important group of perhaps a dozen or twenty women altogether who made their mark in the theater world of late eighteenth-century Spain, and in particular of Madrid. It is also important to note that for the working *tonadillera* – the overworked, underpaid actress – the fancy wardrobe was a burden. For instance, in 1768 María de Guzmán, known as "La Guzmana," complained to the Junta de Teatros that her two years of non-stop performance of *tonadillas* and *zarzuelas*, for which she had to purchase all her costumes, had drained her finances and she asked to be exempted from singing. But the Junta ordered her to keep on singing and in compensation raised her

59 Francisco Pérez de Prado, *Defensa canónica de la potestad decretoria y executiva que por el derecho de Jesucristo y de su Iglesia tienen los obispos sobre sus súbditos* (1734).

60 BN Mss 9327, *Informe de D. Bernardo de Iriarte al Conde de Aranda sobre las Comedias* [en informe dirigido al Conde de Aranda quien le había encargado revisar el repertorio de los teatros].

61 Andioc, *Teatro y sociedad*, 49; Nicolás Fernández de Moratín, *Desengaño al teatro español* (Madrid, 1763).

salary to two reales daily.[62] The documentation echoes La Guzmana's complaints. Even for *tonadilleras* such as La Divina or La Portuguesa, singing and acting was a job which, like many others in the eighteenth century, required the help and collaboration of family members. It was common for actors and actresses to rely on family members not only to access the theaters of the main cities, but also to keep their jobs and advance their careers. María Ladvenant herself, the most famous of all *tonadilleras*, was at the center of a large family of actors and actresses who also made their name in the theater world.[63]

Ilustrados and ordinary individuals, who were usually men, considered even the job of the "working *tonadillera*" could endanger the family structure and alter a wife's expected role. Testimony to this exists in many documents from impatient husbands, eager to bring back their wives from the theater to the family household outside Madrid. This is the case of Mariano de la Rosa from Ciudad del Puerto de Santa María, in Cadiz, where famously the best *tonadilleras* came from. His wife, Josefa de la Figuera, had left for Madrid to work for the "Compañía de Comicas" (The Actresses Company). In March 1773, Mariano de la Rosa complained his wife's absence not only made her "not comply with the conjugal union and marital cohabitation to which she is obliged," but brought her "the freedom that marital separation brings." In a memorial directly addressed to the monarch, rather than to the Junta de Teatros, the hurt husband asked to have his wife back from Madrid to her hometown so she could comply with her marital obligations. The Junta, and not the monarch, answered the memorial denying Mariano de la Rosa's wish.[64] The irony was that the *tonadillera* was promoted and sponsored by the same *ilustrados* who wrote about the value of women in the family and the need for a woman to remain at home to raise their children. Jovellanos' muse was in fact a *tonadillera*, and he himself perhaps secretly dreamt of becoming one. Part of the *tonadillera*'s charm, even among men like Jovellanos, was this excessive display of femininity that challenged some of the basic tenets of femininity based on a body sexually confined within the safety of the married life.

62 BN Archivo de Francisco Barberi n. 79 22 March 1768 (Junta de Corregimiento).

63 Her parents, Juan Ladvenant and María Quirante, were actors as well as her brother and sister, Rafael and Francisca, and some of her cousins; Andrés Peláez Martín, "En el mundo de las tonadilleras," in *Manuel García: de la tonadilla escénica a la ópera española (1775–1832)* (Cadiz: Junta de Andalucía, Universidad de Cádiz, 2006), 66.

64 BN Mss. 14016 n. 15.

Besides its importance to the history of the Spanish Enlightenment, the study of the *tonadilleras* also helps us rethink the place of the eighteenth century in the cultural formation of the Spanish empire. Oftentimes, the eighteenth century is cast aside as an unwelcome reminder of the end of the empire and its limited achievements. Yet, it is in this period that some of the long-debated themes of previous centuries resurge again seeking for their resolution. One of them was the need for the Spanish monarchs, with the geographical and human disparity of the empire, to find a cohesive identity for their nation. This quest for an identity sometimes had to resort to an imaginary ideal of harmony in the formation of the empire. This is an element that the historian Joan Scott explored in 2001 in her "Fantasy Echo: History and the Construction of Identity."[65] As she saw it, the fantasy of a common identity told as a cohesive historical narrative holds groups and their political relevance together by extracting "coherence from confusion," and by reducing "multiplicity to singularity." Fantasy is then the shared image (or images) that unites. The Spanish empire and the identity it promoted were made up by these series of images that provided a sense of unity, continuity, and even fusion as a way of fixing inconsistencies and contradictions.

The fantasy of unity in the Spanish empire worked at different levels, some of which have already been explored in the articles of this volume. One was the element of religion, and how an Iberian religiosity was in fact a construct that fed the monarchy's particular needs of representation. We can find other fantasies in the legal arena, and its expected unifying body of law on both sides of the Atlantic. A third component appears in the cultural world since fantasy was, in fact, a term much used in relation to the *tonadillera* as the previous pages have illustrated. Partly as a response to the efforts of the *ilustrados* and their quest to construct a Spanish identity distinctive and separate from the French influences, the *tonadillera* embodied the expectations of the authentic *españolidad*: simple, unaffected, and genuine.[66]

The fantasy of an identity, however, also had the potential of becoming subversive, threatening to break up the intended unity. In fact, with her elusive personality and excessive femininity, the *tonadillera* was able to question a unified Spanish identity. On stage, the *tonadillera*, although

[65] Joan W. Scott, "Fantasy Echo: History and the Construction of Identity," *Critical Inquiry* 27(2) (Winter 2001): 284–304.

[66] Rebecca Haidt, "Los Majos, el 'Españolísimo gremio' del teatro popular dieciochesco: sobre casticismo, inestabilidad y abyección," *Cuadernos de Historia Moderna* 10 (2001): 155–73.

perhaps not consciously, explored the fantasy of *españolidad* by overplaying the role. Instead of a symbol of cultural unity, with her performance she invited her audience to rearticulate the bases of the *españolidad* that she represented. In her imaginary world she replaced identities that were meant to be permanent, clear, and well defined with images that appeared unstable and fluid. With her performance, the *tonadillera* symbolized the end of the efforts of a united empire that demanded a homogenous, clear, and unquestioned *españolidad*. It is not coincidental that the high of the *tonadillas* in the Spanish theater happened during the 1770s and 1780s, just a few decades before the American wars of independence. The "imaginary domain" that the *tonadillera* inhabited, that place where the self, in its entire spiritual, emotional, and creative dimension is formed, had the potential to recast the bases of the old empire as it was coming to an end.[67]

[67] Drucilla Cornell, *The Imaginary Domain: Abortion, Pornography and Sexual Harassment* (New York: Routledge, 1995).

Conclusion: The History of Early Modern Spain in Retrospect*

John Elliott

The past fifty years have seen a transformation in the ways in which historians have come to approach and understand the history of early modern Spain, a period running roughly from the marriage of Ferdinand and Isabella in 1469 to Napoleon's invasion of the Iberian Peninsula in 1808. Much of this transformation reflects the wider transformation of the historical enterprise itself since the ending of the Second World War, but it also reflects the enormous changes that have overtaken Spain since the death of General Franco in 1975. Spanish and non-Spanish historians alike have contributed to this major rethinking of the Hispanic past. Their interaction and collaboration has done much to enrich and enliven an area of historical study that, outside the realms of art history and the history of literature, had been relatively neglected by international scholarship in the preceding half century and, not surprisingly, was badly affected by the country's political and economic problems, and the intellectual isolation of its historians, in the aftermath of the Spanish Civil War.

The Civil War drove many Spanish intellectuals and historians into temporary or permanent exile, impoverishing Spanish intellectual life for a generation, while simultaneously enriching that of the countries which gave them refuge, notably the United States and the states of Latin America. Some of these exiles turned their attention to the history of colonial Hispanic America, while others continued to reflect on the history of their native land, now viewed from a distance. Two of the most

* I am grateful to Professor Geoffrey Parker for his comments on an earlier draft of this chapter.

famous of these, Américo Castro (1885–1972) and Claudio Sánchez Albornoz (1893–1982), became locked in a controversy over the character and course of the Spanish past that generated much discussion, both inside and outside Spain, and gained Castro in particular a wide following among North American hispanists.[1] For all the differences between them, however, they were engaged in a similar pursuit, and one that had occupied Spanish intellectuals to an almost obsessive extent ever since 1898, when Spain lost the last relics of its American empire. Their concern was with the singularity of Spain, the differences that had set it apart over the centuries from its European neighbors, and the development of national characteristics and forms of life that might help to account for those differences.

In reality theirs was a form of history that, in its search for the essence of Spain, harked back to the preoccupations and assumptions of nineteenth-century historians, for whom national character, which they were inclined to see as unchanging, constituted an important element of historical causation. Yet the publications of these two major figures in the world of Spanish émigrés had the great merit of raising questions and offering hypotheses about aspects of the Spanish past that resonated with historians of a new and very different age. In particular, Américo Castro's emphasis on the consequences of the long period of coexistence, or *convivencia*, between Jews, Muslims, and Christians in medieval Spain, set off a train of enquiry and discussion in the 1950s and 1960s about the relationship between ethnicity, religion, and authority that has stimulated an enormous amount of fruitful research, given rise to numerous books and articles, and is well represented in this volume.

International historical scholarship in the aftermath of the Second World War, however, was moving in a different direction. Outside Spain, Marxism and the social sciences were exercising a growing influence over historians, who increasingly turned their attention to social and economic history as prime instruments of historical explanation. The *Annales* school in France was in the ascendant, and in 1949 Fernand Braudel published his great book on the Mediterranean in the age of Philip II, which was to be enormously influential among historians of early modern Spain, above

[1] Américo Castro, *La realidad histórica de España* (Mexico City: Porrúa, 1954); English trans. by Edmund King, *The Structure of Spanish History* (Princeton, NJ: Princeton University Press, 1954). Claudio Sánchez-Albornoz, *España: un enigma histórico* (Buenos Aires: Editorial Sudamericana, 1956). See also Richard L. Kagan (ed.), *Spain in America. The Origins of Hispanism in the United States* (Urbana, IL: University of Illinois Press, 2002), 14.

all in encouraging them to think about the geographical, economic, and social forces that shaped the reign of that most enigmatic of monarchs.[2] The impact of these forces, however, did not necessarily demand a macro-historical approach in the manner of Braudel. One of the most enduring and valuable consequences of the new historiography was to be found in the new direction it gave to regional, local, and municipal studies, as exemplified by Pierre Vilar's massive history of the agrarian and industrial development of early modern Catalonia, or Bartolomé Bennassar's reconstruction of Valladolid in the age of Philip II, with its rich account of the city's economic life, social structure, and patterns of behavior.[3]

Although in the long run such works were to prove highly influential in Spain itself, native Spanish historians were, at the time of their publication, in many ways cut off from this new movement in the field of historical writing. Travel abroad was difficult, and for some impossible; censorship made it hard to acquire foreign books, especially if they betrayed any Marxist influence; the regime relentlessly propagated a vision of the Spanish past that insisted on Spain's unique role as the upholder of transcendental values against the determined assaults of Marxism, atheism, and secularization; and there was close surveillance of, and interference in, appointments to university chairs. At the same time severe financial constraints restricted the opportunities for archival research. New historical influences did, however, break through into the academic world, and Braudel's *Mediterranean*, in particular, had a massive impact, even though it did not appear in a Spanish translation until 1976.[4]

To some extent this Braudelian impact has tended to overshadow the important contributions being made by Spanish historians in spite of all the difficulties they faced. In Seville, for instance, Ramón Carande (1887–1986) was at work on his major study of the finances of Charles V,[5] while in Barcelona Jaume Vicens Vives (1910–60), who would do much to introduce a new generation to the *Annaliste* approach and methodology, was embarking on his fundamental revision of the history both of

[2] Fernand Braudel, *La Méditerranée et le monde méditerranéen à l'époque de Philippe II* (Paris: Colin, 1949); Spanish translation, *El Méditerraneo y el mundo méditerraneo en la época de Felipe II*, 2 vols. (Mexico City: Fondo de Cultura Económica, 1976).

[3] Pierre Vilar, *La Catalogne dans l'Espagne moderne. Recherches sur les fondements économiques des structures nationales*, 3 vols. (Paris: S.E.V.P.E.N., 1962); Bartolomé Bennassar, *Valladolid au siècle d'or* (Paris: La Haye, Mouton et Cie, 1967).

[4] Manuel Moreno Alonso, *El mundo de un historiador. Antonio Domínguez Ortiz* (Seville: Fundación José Manuel Lara, 2009), 253.

[5] Ramón Carande, *Carlos V y sus banqueros*, 3 vols. (Madrid: Revista de Occidente, Sociedad de estudios y publicaciones, 1943–67).

Spain and his native Catalonia.[6] Other major historians, too, were emerging alongside Vicens Vives – notably the economic historian Felipe Ruiz Martín (1915–2004) and the cultural historian José Antonio Maravall (1911–86), both of whom spent some time in Braudel's Paris, while Antonio Domínguez Ortiz (1909–2003) was quietly working away in the archives, and was to become the dominant figure in rewriting the social history of early modern Spain during the second half of the twentieth century.

The historical scene in Franco's Spain was therefore not as bleak as it is sometimes depicted, but the prevailing conditions in the country, especially during the first half of the regime, left the field wide open to foreign scholars, many of whom would find themselves working almost alone in archives whose vast and often unsuspected riches still remained largely unexploited. A number of French, British, and North American historians, finding themselves to be members of an increasingly crowded profession in search of new fields to conquer, began to wake up to the possibilities offered by the history of Spain, and not least the Spain of the early modern period. But in moving into Spanish history they tended to come up against a wall of prejudice. The *leyenda negra* – the *Black Legend* of a cruel, fanatical, and backward Spain – had etched itself into the European consciousness, and was particularly strong in the Anglo-American world, where religion sharpened and perpetuated deep-seated anti-Spanish sentiments that reached back to the age of Spanish predominance in the sixteenth century.[7] These sentiments were compounded by the nationalist victory in the Civil War, which left some historians, and particularly those who had participated directly or emotionally in the conflict, with a firm determination never to set foot in Spain as long as General Franco remained in power.

In consequence, non-Spanish historians who decided to work on the history of early modern Spain found themselves faced with a double challenge. They were writing for a public which was largely ignorant of Spain, and as a rule had little or no interest in finding out more about its history and culture. At the same time they had to dispel longstanding prejudices that impeded the acceptance of an unprejudiced reading of the Spanish past. These obstacles in the way of a better public understanding of Spain

[6] See Josep M. Muñoz i Lloret, *Jaume Vicens i Vives. Una biografía intel.ctual* (Barcelona: Edicions 62, 1997), ch. 2.

[7] For attitudes to Spain among United States scholars in the nineteenth and twentieth centuries, see Richard L. Kagan, "Prescott's Paradigm: American Historical Scholarship and the Decline of Spain," *The American Historical Review* 101 (1996): 423–46.

and its history could not, and cannot, be lightly dismissed. Not only was it difficult to make the history of an alien and unfamiliar country accessible to foreign readers, but it was also necessary to avoid the temptation to replace a black legend with a white. For all the work of two or three generations of foreign historians, prejudices still linger, and the effort to write an impartial history of early modern Spain without turning it into an apology has undoubtedly affected the way in which they have treated the subject, and remains a continuing challenge.

In spite of such problems, foreign scholars contributed impressively in the second half of the twentieth century to the expansion of early modern Spanish history into areas of interest beyond the narrowly political, institutional, and diplomatic which had dominated so much of its earlier treatment. Social and cultural history, especially in the shape of what came to be known as the history of *mentalités*, were the special beneficiaries of international scholarly attention, which brought with it not only fresh concerns but also the new insights that inevitably come when outsiders look at the history of a nation other than their own. Perhaps the most important consequence of this internationalizing process, however, was to subvert and erode the sense of Spanish exceptionalism that had so profoundly affected the way in which the Spanish past was traditionally addressed. Spain was "different," or so at least Franco's tourist ministry took pleasure in proclaiming. But foreign historians well versed in the history of their own countries could not fail to notice the similarities between aspects of the Spanish past and that of their homelands. Idleness, for instance, which was so often presented as a peculiarly Spanish vice, was common to all the societies of an early modern Europe in which employment opportunities were limited and were heavily influenced by the agrarian cycle.[8]

Differences, then, began to be diluted as soon as the history of early modern Spain was set into a comparative and European context. This process of dilution came at a time when Spain itself was undergoing major changes. The modernization of its backward economy was gathering pace, and economic transformation was to be followed by a profound political transformation. Franco's death in 1975, the restoration of the monarchy, the transition to democracy, and the subsequent entry of Spain into the European Union changed Spain's status in the eyes of the

[8] The mythology that led to the creation and perpetuation of the traditional theme of Spanish "idleness" has been admirably explored and exposed by Ruth MacKay, *"Lazy, Improvident People". Myth and Reality in the Writing of Spanish History* (Ithaca, NY and London: Cornell University Press, 2006).

world. It also started to change the way in which Spaniards saw themselves. A new generation of historians enjoyed the freedom to travel that its predecessor lacked, and took advantage of the range of opportunities now becoming available to engage and interact with foreign colleagues in ways of which historians who had lived through the Civil War and its aftermath could only dream.

This inevitably began to affect the way in which Spain's historians began to envision and write about their national past, and, in due course, led to a more discriminating view of the publications of foreign scholars. At a time when the Spanish academic world was emerging from its prolonged isolation, there was an instinctive tendency to view uncritically the fruits of foreign scholarship, on the assumption that foreign products were inherently superior to the home-produced variety. In reality, as was only to be expected, the contributions made by non-Spanish historians varied enormously in quality and importance, as a new and more self-confident generation of Spanish scholars would come to realize.

Foreign influences, however, did much at a critical moment in the emergence of contemporary Spain to change perceptions about the Spanish past. With the history of early modern Spain ceasing to be framed in terms of the country's alleged exceptionalism, standard topics began to be seen in a rather different light. The age-old question of the causes of Spain's seventeenth-century "decline," for instance, which still loomed large in the historical agenda of the 1950s and 1960s, had become much less of a preoccupation by the turn of the twentieth and twenty-first centuries. There were various reasons for this.[9] As many historians began to view the seventeenth century as an age of general European, and possibly even global, crisis,[10] Spain's economic and social troubles came to appear as part of a much wider phenomenon, rather than one that was confined to Spain and was explicable in terms of purely Spanish failings and errors of policy.

Closer study, too, of policy-making and the policy-makers in Madrid during the first half of the century, and of the writings of the *arbitristas*, the economic projectors and moralists who tried to attract their attention, made it clear that, instead of being a concept retrospectively imposed by historians, "decline" was one of the concepts employed by contemporaries as they struggled to understand and address the domestic and

[9] For the theme of decline as both a Spanish and a universal phenomenon, see J. H. Elliott, *History in the Making* (New Haven, CT: Yale University Press, 2012), ch. 4.

[10] See Geoffrey Parker, *Global Crisis: War, Climate Change and Catastrophe in the Seventeenth Century* (New Haven, CT: Yale University Press, 2013).

international problems of their country. Meanwhile, the apparent paradox that Spain's century of "decline" was also its *siglo de oro*, the golden age of its arts, raised some fascinating questions about the intellectual and artistic impact of court culture and aristocratic and clerical patronage – topics that became lively issues in international historical scholarship in the later decades of the twentieth century, and that had been largely ignored by Spanish art historians reared in the positivist tradition.[11]

The reassessment of "the decline of Spain" was also encouraged by the country's transformation in the closing decades of the twentieth century. With its increasing approximation to the political culture of other West European nations and the advent of an unprecedented degree of economic prosperity, the pessimism generated by the sense of "decline" as a constant of the Spanish experience gave way to the kind of optimism about the country's prospects and achievements that had for so long been in short supply. Viewed from this new, more confident, standpoint, perhaps the most striking characteristic of the *monarquía hispánica* was not, after all, its many undoubted failings but its capacity for long-term survival, and its resilience in the face of adversity.

The cultural vitality of the new Spain may also have helped encourage a reassessment of traditional assumptions about the country's intellectual stagnation and its scientific and technological backwardness in earlier periods. Closer examination of the evidence suggested that early modern Spaniards, faced with the immense challenge of controlling, understanding, and interpreting the New World of America, developed preoccupations, concerns, and interests that were not necessarily those of their European contemporaries. Their contributions to the scientific enterprise had therefore tended to be ignored or underplayed by historians of science and philosophy primarily interested in identifying those features of early modern European civilization that served as more obvious signposts on the road to "modernity."[12]

A further consequence of the new political order established by the 1978 Constitution was a revival of interest in earlier constitutional arrangements, and more specifically those that prevailed during the two centuries of Habsburg rule. Post-1978 Spain was a Spain of autonomous communities, and this required a change in the relationship between them and the central government in Madrid. This new-style relationship evoked

[11] See Jonathan Brown, *In the Shadow of Velázquez* (New Haven, CT: Yale University Press, 2014), ch. 2.

[12] See María M. Portuondo, "On Early Modern Science in Spain," in this volume (Chapter 8).

parallels with the Spain that existed before the imposition of Bourbon rule in 1700, a Spain with a substantial degree of self-government at the king's command. In the Spain of the House of Austria the king of all was famously the king of each, respecting the distinctive identity, customs, laws, and traditions of each individual territory. The structure and problems of the *monarquía* in the sixteenth and seventeenth centuries thus acquired a new actuality in the post-Franco era.

In this new era the constitutional arrangements introduced in 1978 gave a fresh immediacy to a question that had always been present, if at times latent, in the writing of Spain's history – that of the persistent tension between aspirations for unity and hankerings after autonomy. How far should historians treat early modern Spain as a single entity, and how far as a plurality, a collection of *patrias*, each with its own distinctive history? Spain had a long tradition of regional, local, and civic historiography, with chroniclers proudly proclaiming the unique virtues and achievements of their own particular *patria*.[13] The Romantic Movement and nineteenth-century nationalism gave fresh impetus to histories of "the Spains," with Catalan historians, for example, developing a historiography that was increasingly inclined to cast Catalonia as the victim of a Castilian-dominated centralizing state. In the 1940s and 1950s Vicens Vives and others would challenge this narrowly nationalist view of the Catalan past,[14] but the question of how to handle the respective claims of national, regional, and local histories remained, and remains, unresolved. The effect of the acceptance in the 1978 Constitution of the rights of the autonomous communities, in conjunction with the creation of new universities and university history departments in their territories, was to reinforce a pluralist vision of the Spanish past, and all too often to limit historical horizons to national, linguistic, or provincial boundary lines, as if the history of a community could be understood in isolation from the larger political entity of which it formed a part.

A shrinking of horizons in some areas has fortunately been accompanied by their expansion in others. The renewed interest in the workings of the *monarquía* has led to a growing appreciation that the history of Habsburg and early Bourbon Spain cannot properly be understood if it is confined simply to Spain itself. As a pluralist, composite state the *monarquía* included all, or part of, the Netherlands, Spain's Italian

[13] See Richard L. Kagan, *Urban Images of the Hispanic World, 1493–1793* (New Haven, CT: Yale University Press, 2000).

[14] Muñoz i Lloret, *Jaume Vicens*, ch. 1.

possessions, and, for sixty years, Portugal. All of these fell within the orbit of the royal government in Madrid. The various kingdoms and territories were connected by numerous official and unofficial ties, and events in one part of this extended *monarquía* could all too easily have repercussions in the others. While some historians have long appreciated the need to study their own special areas of interest within the context of the Spanish monarchy as a whole,[15] early modern Belgian, Dutch, Italian, Portuguese, and Spanish historians, together with those from Latin America, have become much more aware in recent years of the importance of this wider contextualization. This in turn has led to significant collaborative enterprises between historians of different nationalities, and to a number of anthologies that explore systems of government, images of power, or viceregal courts across the whole range of the *monarquía*.[16]

These connected histories have not been limited to Spain's European possessions. In line with the turn taken by international historical scholarship towards Atlantic history,[17] historians of Spain and its American empire have similarly been moving away from the traditional compartmentalization between histories of Spain and the Indies towards a history that integrates the two sides of the Spanish Atlantic. Again, this was far from being a new development. Transatlantic trade always required a genuinely Atlantic treatment, as the massive study of Seville and the Atlantic by Pierre and Huguette Chaunu reminds us,[18] but the history of the Spanish Atlantic has acquired a new momentum in recent years as part of the wider movement in the world of international scholarship towards the writing of connected and transnational histories. The migration of people across the Atlantic, for instance, has attracted much attention.[19]

[15] For example, Rosario Villari, *Un Sogno di libertà: Napoli nel decline di un impero, 1585–1648* (Milan: Mondadori, 2012), revising and extending his *La rivolta antispagnola a Napoli: le origini, 1585–1647* (Bari: Laterza, 1967).

[16] Examples include Massimo Ganci and Ruggiero Romano, eds., *Governare il mondo. L'impero spagnolo dal XV al XIX Secolo* (Palermo: Società Siciliana per la Storia Patria, Istituto di Storia Moderna, 1991); Antonio Álvarez-Ossorio Alvariño and Bernardo J. García García, eds., *La monarquía de las naciones. Patria, nación y naturaleza en la Monarquía de España* (Madrid: Fundación Carlos de Amberes, 2004); and Pedro Cardim and Joan-Lluís Palos, eds., *El mundo de los virreyes en las monarquías de España y Portugal* (Madrid and Frankfurt: Iberoamericana, Vervuert, 2012).

[17] See Bernard Bailyn, *Atlantic History. Concept and Contours* (Cambridge, MA: Harvard University Press, 2005).

[18] Pierre and Huguette Chaunu, *Séville et l'Atlantique*, 8 vols. (Paris: Colin, 1955–9).

[19] Ida Altman, *Emigrants and Society. Extremadura and Spanish America in the Sixteenth Century* (Berkeley, CA: University of California Press, 1989); Auke P. Jacobs, *Los movimientos migratorios entre Castilla e Hispano-América durante el reinado de Felipe III, 1598–1621* (Amsterdam: Rodopi, 1995).

So, too, has the transatlantic movement of books;[20] and the continuing ties between settlers and their relatives who remained at home have been extensively documented in valuable collections of their correspondence.[21]

The Spanish monarchy, as a transnational and Atlantic empire, lends itself especially well to this kind of integrated treatment. As the first genuinely global empire, however, its reach extended far beyond the Atlantic to embrace the world. This makes its study ideally suited to what has become an increasingly popular form of integrated history, global history. If global history is regarded both as the history of globalization and as the study of intercontinental and transnational connections, the *monarquía* scores highly on both counts.[22] Spaniards and Portuguese as navigators and explorers were the prime movers in early modern Europe's transoceanic expansion – an expansion that made possible the movement of commodities around the world, with precious metals among the first and most important. Since Earl J. Hamilton published his seminal work in 1934 on the impact of American silver on Spanish and European prices, there has been a growing appreciation of the global reach and impact of the silver extracted from the mines of Mexico and Peru.[23] But globalization was not confined to material commodities. Together with people and goods it also embraced religion and culture. The *monarquía hispánica* was the *monarquía católica*. As such, its divinely appointed mission was to spread the Christian gospel across the world. The consequences, both intended and unintended, of this massive intercontinental

[20] See Carlos Alberto González Sánchez, *Los mundos del libro. Medios de difusión de la cultura occidental en las Indias de los siglos XVI y XVII* (Seville: Diputación de Sevilla, Universidad de Sevilla, 1999), and Pedro J. Rueda Ramírez, *Negocio e intercambio de libros con América en la carrera de Indias (siglo XVII)* (Seville: Diputación de Sevilla, Universidad de Sevilla, 2005).

[21] Enrique Otte, *Cartas privadas de emigrantes a Indias, 1540–1616* (Seville: Consejería de Cultura, Junta de Andalucía, 1988). Some of these letters were published in English by James Lockhart and Enrique Otte, eds., *Letters and People of the Spanish Indies. The Sixteenth Century* (Cambridge: Cambridge University Press, 1976). A further collection in Rocío Sánchez Rubio and Isabel Testón Núñez, eds., *El hilo que une. Las relaciones epistolares en el viejo y el nuevo mundo, siglos XVI–XVIII* (Mérida: Junta de Extremadura, Editora Regional, 1999).

[22] For a very recent example, involving Spanish-Portuguese collaboration, see Carlos Martínez Shaw and José Antonio Martine Torres, eds., *España y Portugal en el mundo, 1581–1668* (Madrid: Polifemo, 2014).

[23] Earl J. Hamilton, *American Treasure and the Price Revolution in Spain, 1501–1650* (Cambridge, MA: Harvard University Press, 1934); Dennis Flynn and Arturo Giraldez, "China and the Spanish Empire," *Revista de Historia Económica* 14 (1996): 309–38; Kenneth Pomeranz, *The Great Divergence. China, Europe, and the Making of he Modern World Economy* (Princeton, NJ: Princeton University Press, 2000), 269–74.

transfer of culture and beliefs offer great opportunities for future research. Globalization brought in its train hybridity and cultural as well as human *mestizaje*, stimulating in the process new forms of creativity.[24]

Times have changed, and never more so than in the last half century. It is sufficient to glance at Roger B. Merriman's *The Rise of the Spanish Empire in the Old World and the New*, first published between 1918 and 1934 and reprinted as late as 1962,[25] to realize the extent and nature of the historiographical transformation that this has brought about. Merriman's was essentially a political, diplomatic, military, and administrative narrative, heavily based on published contemporary sources and secondary works. Narrative no longer holds the dominant position it enjoyed in Merriman's time – a change that reflects general trends in professional historical writing since the mid-twentieth century – and there has been a massive increase in the number of primary sources consulted or brought to light by three generations of historians. Moreover, as the preceding pages have indicated, the range of topics now regarded as integral to the subject, the amount of archival research they have inspired, and the quantity of publications, many of them of high quality, which they have generated, suggest the degree to which historians of early modern Spain are operating in a very different world from the one known to Merriman.

Studies of the Inquisition, for instance, have become an industry in themselves, and have made possible a knowledge of the mental and social worlds of Jews, Muslims, and *conversos* far exceeding that which Américo Castro could command.[26] The social history of early modern Spain has flourished, even if economic history, which enjoyed its heyday in the 1940s, 1950s, and 1960s, is not the force it once was, although impressive work has been done, and continues to be done, not least in the study of royal finances and their impact, and in that of demographic history.[27] Cultural history, into which the study of religion has been

[24] See Serge Gruzinski, *Les Quatre Parties du Monde. Histoire d'une mondialisation* (Paris: La Martinière, 2004).

[25] Roger Bigelow Merriman, *The Rise of the Spanish Empire in the Old World and the New*, 4 vols. (New York: Cooper Square Publishers, 1918–34; repr. 1962).

[26] See the valuable survey and bibliographical essay in James S. Amelang, *Historias paralelas. Judeoconversos y moriscos en la España moderna* (Madrid: Akal, 2011). The bibliographical essay is not included in the English version of the book, *Parallel Histories. Muslims and Jews in Inquisitorial Spain* (Baton Rouge, LA: Louisiana State University Press, 2013), but the "Select Bibliography" is still impressively long.

[27] See, for example, Juan E. Gelabert, *La bolsa del rey. Rey, reino y fisco en Castilla, 1598–1648* (Barcelona: Crítica, 1997) and Bartolomé Yun, *Marte contra Minerva. El precio del imperio español, c. 1450–1600* (Barcelona: Crítica, 2004). For population studies, in addition to the standard works by Jordi Nadal, *La población española (siglos XVI a*

excessively subsumed, and the history of representation – focused on royal entries, festivals, and images of power[28] – are currently the most fashionable approaches to the Hispanic past, very much in line with recent trends in international scholarship. The cultural turn is not yet played out. It is, however, the rich diversity of approaches that most impresses.

Some sense of this diversity can be gained from this volume of essays, ranging as they do from aspects of urban life to the realities of living in an ethnically mixed society, and from scientific and historical enquiry to the world of images and books. Three of the contributors are Spanish but the remainder are citizens of the United States, and they naturally tend to reflect the approaches and interests of contemporary North American scholarship. These are not necessarily identical with those of their Spanish and European colleagues, some of whose priorities are different, although there is often a considerable transatlantic overlap. It is striking, for instance, that the volume contains no political history, and does not engage with any of the issues about Spain as a composite monarchy, or the global reach of Spanish power, which have figured so prominently in the recent work of Spanish and European historians.

This does not in any sense detract from its value as offering lively examples of some of the best North American scholarship on early modern Spain, but it does suggest something of its limitations as a representation of current work in the field. History has always been a house of many mansions, but the time is coming when some large-scale reconstruction will be needed. Innumerable extensions have been added, sometimes at oblique angles to the original structure, and much attractive and ingenious redecoration has taken place, but at the price of some loss of architectural coherence. The challenge that now faces historians of early modern Spain is to integrate the best of the new with the best of the old.

This will not be an easy undertaking. So much has been published, and from such a variety of standpoints, that the way forward is far from clear. Yet recent developments suggest something of what is required. The study of art, thought, and culture need to be set not only within a broad

XX) (Edición corregida y revisada; Barcelona: Ariel, 1984) and Vicente Pérez Moreda and David-Sven Reher, eds., *Demografía histórica en España* (Madrid: El Arquero, 1988), see the *Revista de Demografía Histórica* (formerly the *Boletín de la Asociación de Demografía Histórica*) which publishes the results of current research.

[28] For an engaging Spanish example of the genre, see Teofilo F. Ruiz, *A King Travels. Festive Traditions in Late Medieval and Early Modern Spain* (Princeton, NJ: Princeton University Press, 2002).

political framework but also in a more specific temporal context if their full implications are to be grasped. That context is political, as well as economic, social, religious, and intellectual.

The history of the early modern state, the subject of a brilliant paper that Vicens Vives wrote for the Eleventh International Historical Conference in 1960 but did not live to deliver,[29] has tended since then to be overshadowed by other topics which seemed more attractive to later generations. Yet the history of the state, and, in its wider sense, the history of power, remains fundamental to our understanding of the past. Old-style political history, however, as Vicens was already pointing out in 1960, is no longer adequate. If it is to be persuasive and acceptable to a new age, the insights that have come through new research into economic, social, and cultural history, must inform it. Political and institutional history need to embrace the attitudes and assumptions of rulers and ruled, in other words "political culture" in the broadest sense of that term, together with the formation and behavior of elites, and the activities and impact of individuals as well as those of social groups.

The biographical approach to the Spanish past has not traditionally found great favor among professional historians in Spain itself, with the exception of Manuel Fernández Álvarez (1921–2010), whose biographies of Charles V and Philip II have enjoyed great popularity in his home country;[30] but in general, where political biography is concerned, much of the running has been made by non-Spanish historians.[31] There are signs, however, that attitudes are changing and that a biographical approach is now gaining respectability among the Spanish historical community.[32] In particular, the Real Academia de la Historia's fifty-volume *Diccionario biográfico español*, despite its difficult gestation and the controversy attending its birth, has opened for today's historians of early modern Spain a range of possibilities for the deeper exploration of

[29] Jaime Vicens Vives, "Estructura administrativa estatal en los siglos XVI y XVII," *XIe Congrès des sciences historiques. Rapports*, IV, Stockholm-Uppsala (1960), 1–24. Reprinted in M. Batllori and E. Giralt, eds., J. Vicens Vives, *Obra dispersa*, 2 vols. (Barcelona: Vicens Vives, 1967), II, 359–77.

[30] Manuel Fernández Álvarez, *Carlos V, el César y el hombre* (Madrid: Espasa, Fundación Academia Europea de Yuste, 1999); *Felipe II y su tiempo* (Madrid: Espasa-Calpe, 1998).

[31] See, for example, the publications of Geoffrey Parker on Philip II, including the most recent, *Imprudent King. A New Life of Philip II* (New Haven, CT: Yale University Press, 2014).

[32] Straws in the wind are the series *Los hombres del rey* published jointly by the Centro de Estudios Europa Hispánica and Marcial Pons Historia, and the creation of the Unidad de Estudios Biográficos in Barcelona.

politics, institutions, and society that were simply not available to preceding generations.

The diversity and vitality of the essays in this volume indicate something of the enormous potential for creative work that the study of early modern Spain offers both now and in the future. The new tools available for research, and in particular the digitization of Spanish archives, have created a totally new working environment, although one that carries dangers, such as inadequate contextualization and the risk of information overload. But, above all, in spite of the massive changes that have occurred since the Second World War, the subject itself remains not only endlessly rewarding but also of high historical importance. This, after all, was the country and society that ushered in a new age of European and world history by venturing overseas, conquering, colonizing, and exploiting non-European peoples, and establishing an empire that survived for three centuries and more; that provides an early and sobering warning of the kind of challenges that face the multi-ethnic and multicultural societies of today; that was a pioneer among Western societies, in developing mechanisms for bureaucratic government over large areas of territory; that gave birth to philosophers, writers, and thinkers who led the way in reflecting on the consequences of globalization, and in devising new concepts of human rights and a new system of international law; and that produced a civilization enduringly rich in works of literature and art. Much has been achieved over the past half century in exploring the complexities of this endlessly fascinating society, but more, much more, still remains to be done.

Index